Technology, Humans, and Society

Toward a Sustainable World

A volume in the Academic Press
SUSTAINABLE WORLD Series

Technology, Humans, and Society

Toward a Sustainable World

Richard C. Dorf

University of California, Davis

ACADEMIC PRESS

A Harcourt Science and Technology Company

San Diego   San Francisco   New York   Boston   London   Sydney   Tokyo

This book is printed on acid-free paper. ∞

ACADEMIC PRESS
A Harcourt Science and Technology Company
525 B Street, Suite 1900, San Diego, California 92101-4495, USA
http://www.academicpress.com

Academic Press
Harcourt Place, 32 Jamestown Road, London NW1 7BY, UK
http://www.academicpress.com

Library of Congress Catalog Card Number: 00-111080

International Standard Book Number: 0-12-221090-5

PRINTED IN THE UNITED STATES OF AMERICA
01 02 03 04 05 06 HP 9 8 7 6 5 4 3 2 1

CONTENTS

CONTRIBUTORS

Numbers in parentheses indicate the pages on which the authors' contributions begin.

Braden Allenby (133, 476) AT&T Engineering Research Center, Basking Ridge, New Jersey 07920

Mark Atlas (126) Carnegie Mellon University, Pittsburgh, Pennsylvania 15213

Robert U. Ayres (84) Center for the Management of Environmental Resources INSEAD, F-91490 Milly-la-Forest, France

John J. Berger (302) Renewable Energy and Natural Resources Consultant, El Cerrito, California 94801

Karen Blades (133) December Technology, 491 Crescent Street, #102, Oakland, California 94610

David Bodansky (271) Department of Physics, University of Washington, Seattle, Washington 98195

Noellette Conway-Schempf (120) Automatika, Inc., Pittsburgh, Pennsylvania 15238

Robert Costanza (17, 190) Institute for Ecological Economics, University of Maryland, Solomons, Maryland 20688-0038

Richard Counts (405) Institute of Transportation Studies, University of California, Davis, California 95616

Michael De Alessi (348) Center for Private Conservation, San Francisco, California 94114

Charles DeLisio (451) Makato Architecture & Design, Pittsburgh, Pennsylvania 15203-1721

Anthony Eggert (405) Institute of Transportation Studies, University of California, Davis, California 95616

Joseph Fiksel (111) Battelle Memorial Institute, Columbus, Ohio 43201

Richard Florida (126) Carnegie Mellon University, Pittsburgh, Pennsylvania 15213

Indur M. Goklany (465) Independent Consultant, Vienna, Virginia 22182

Eban S. Goodstein (52) Department of Economics, Lewis and Clark College, Portland, Oregon 97219-7899

Michael R. Hagerty (435) Graduate School of Management, University of California, Davis, California 95616

Garrett Hardin (30) Department of Biological Sciences, University of California, Santa Barbara, California 93106

Barrett Hazeltine (101) Engineering Division, Brown University, Providence, Rhode Island 02912

Paul Heyne (59) Department of Economics, University of Washington, Seattle, Washington 98195

Peter Huber (80) The Manhattan Institute, New York, New York 10017

Thomas Jacobson (460) Department of Environmental Studies and Planning, Sonoma State University, Rohnert Park, California 94928

Daniel M. Kammen (310) Energy and Resources Group, University of California, Berkeley, California 94720-3050

Eric F. Karlin (23) School of Theoretical and Applied Science, Ramapo College, Mahwah, New Jersey 07430-1680

David C. Korten (197) PCD Forum, Bainbridge Island, Washington 98110-1972

Thomas G. Kreutz (417) Center for Energy and Environmental Studies, Princeton University, Princeton, New Jersey 08544

Lester B. Lave (43, 120) Graduate School of Industrial Administration, Carnegie Mellon University, Pittsburgh, Pennsylvania 15213

Joel Makower (457) The Green Business Letter, Oakland, California 94610

Juan Martinez Alier (17) Department of Economics and Economic History, University of Barcelona, E-08035 Barcelona, Spain

Ronald Mascitelli (206) Technology Perspectives, Northridge, California 91326

H. Scott Matthews (43) School of Industrial Administration, Carnegie Mellon University, Pittsburgh, Pennsylvania 15213

Christopher M. McDermott (174) Lally School of Management, Rensselaer Polytechnic Institute, Troy, New York 12180

Roger Messenger (297) Department of Electrical Engineering, Florida Atlantic University, Boca Raton, Florida 33431

Norman Myers (475) Green College, Oxford University, Oxford, England

Joan M. Ogden (417) Center for Energy and Environmental Studies, Princeton University, Princeton, New Jersey 08544

Martin J. Pasqualetti (290) Department of Geography, Arizona State University, Tempe, Arizona 85287–0104

Mukund R. Patel (324) Faculty of Engineering, U.S. Merchant Marine Academy, Kings Point, New York 11024

John Peet (235) Department of Chemical and Process Engineering, University of Canterbury, Christchurch, New Zealand

Alan Pilkington (389) School of Management, Royal Holloway, University of London, Egham, Surrey, England

David Pimentel (356) College of Agriculture and Life Sciences, Cornell University, Ithaca, New York 14853

Marcia Pimentel (356) Division of Nutritional Sciences, Cornell University, Ithaca, New York 14853

Walter V. Reid (352) Millennium Assessment Project, Seattle, Washington 98103

William Samuelson (214) School of Management, Boston University, Boston, Massachusetts 02215

Olman Segura (17) International Center for Political Economy, National University of Costa Rica, Costa Rica

James R. Sheats (146) World E-Services, Hewlett Packard, Palo Alto, California 94304

Margaret M. Steinbugler (417) Center for Energy and Environmental Studies, Princeton University, Princeton, New Jersey 08544

Derby A. Swanson (165) Applied Marketing Science, Waltham, Massachusetts 02451

ABOUT THE AUTHOR

Richard C. Dorf is professor of management and professor of electrical and computer engineering at the University of California, Davis. Professor Dorf is the author of *Modern Control Systems, Electric Circuits,* the *Manufacturing and Automation Handbook,* the *Electrical Engineering Handbook,* the *Engineering Handbook,* and the *Technology Management Handbook.*

Professor Dorf has extensive experience with education, industry, and government. He has served as advisor to numerous firms, financial institutions, and governmental agencies. He has also been a visiting professor at the University of Edinburgh, Scotland, the Massachusetts Institute of Technology, Stanford University, and the University of California, Berkeley. He is a Fellow of the IEEE.

PREFACE

The sustainability of the world depends on the interaction between economic activity and the natural environment. Economic activity is organized by government and business to provide humankind with food, shelter, and other human necessities such as health, security, and mobility. The concept of a sustainable world must rest squarely on a global framework for economic vitality, environmental quality, and social justice. The means of building, organizing, and operating within such a global framework depend on sound decisions, responsible actions, appropriate technologies, and thoughtful governments. The purpose of this book is to address the issues that illuminate a pathway to a sustainable world.

This book is written for the reader who wishes to address the issues of sustainability with consideration of the environmental, social, and economic issues. Thus, the book is directed toward the scientist, businessperson, engineer, manager, governmental regulator, lawmaker, environmentalist, or citizen who is concerned with the future of our world. With the assistance of highly qualified contributors, we strive in this volume to address a broad array of matters and provide a framework that could lead to a sustainable world.

We recognize the various filters that business, technology, science, and society at large bring to the issues. Nevertheless, we attempt to describe the foundations of science, technology, and economies that might, if appropriately organized, lead to a sustainable future.

After suggesting a framework for organizing a sustainable future, we then turn to a description of the critical matters of natural resources and the related technologies to process and transmit energy, materials, water, and wastes. Then we turn to the critical matter of creating a sustainable agricultural system that can adequately feed the world while nurturing the earth.

The design of a sustainable transportation system that enables mobility with limited impact on the world's air and land is a challenge addressed in several dimensions. Next, we consider the role of performance indicators for tracking quality of life and measures of progress and prosperity. Social entrepreneurship is one means of providing leverage arising from human creativity applied to social needs. We then address the matter of architecture and buildings and present several visions for the future. Finally, we summarize the potential for a sustainable future and reinforce our hope that, with sound methods and understanding, we can successfully travel the path toward a sustainable world.

I thank the contributors to this volume, whose thoughtful insights provide the reader with a solid foundation for thinking about the need to incorporate sustainability into all aspects of daily life. I also wish to acknowledge the exemplary assistance of my editor, Joel Claypool, and the skilled production team of Julio Esperas and Lorretta Palagi.

Davis, California
December 2000

One unexpected benefit of the Apollo missions to the moon was the photographs that the astronauts took of the earth from space. These images of the blue-green planet, the only place in the universe that humankind can call home, became a symbol of environmental consciousness around the globe. PHOTOGRAPH COURTESY OF CORBIS.

CHAPTER 1

Sustainability, Economics, and the Environment

1.1 Sustainability

Perhaps the most discussed question of the year 2000 is whether the human species will survive the new millennium. The pace of global population growth, rampant air and pollution in the developing world, and the limited resolve of the United States to lead the environmental revolution it fostered may spell significant trouble for the quality of human existence. Social forces have been set in motion such as the desire for electricity, telephones, automobiles, and clean water. Once humans experience these things, they want more of them, regardless of the ecological cost.

The list of possible solutions to environmental degradation includes improving technology, using resources more efficiently, minimizing waste, and even shifting taxes to penalize pollution. But these solutions are bogged down in disputes.

Sustainability of the world depends on the interaction between economic activity and the natural environment. Economic activity affects the natural environment and, vice versa, the state of the natural environment affects economic activity. In addition, the level of human impact on the natural environment is now such that its capacity to support future economic activity at the level required by the expected human population and its aspirations is in question. The resulting issues are characterized by ignorance and uncertainty. The human impact on the environment is now such that global responses are necessary. However, the sustainability problem is quite complex.

The emerging challenge is to develop a **sustainable global economy**: an economy that the earth is capable of supporting indefinitely. The depletion of natural resources and increasing pollution may overcome the planet's capabilities. In meeting our current needs, we may be destroying the ability of future generations to meet their needs.

Many of the proposed approaches to sustainability seem utopian or ideological, feasible only on a small scale. Many fail to address the underlying changes in the economy that are at the root of our most complex social and environmental problems.

Several definitions of a sustainable economy are provided in Table 1.1. The key idea of sustainability is the power to maintain and support future generations without compromising their quality of life.

Perhaps the most complete and useful definition is provided by Daly, who states that sustainable economy is attained when (Daly, 1996):

> Rates of use of renewable resources do not exceed regeneration rates; rates of use of nonrenewable resources do not exceed rates of development of renewable substitutes; rates of pollution emission do not exceed assimilative capacities of the environment.

The concept of a sustainable economy provides a framework for discussing problems and solutions. The concept is rooted in the hypothesis that economic viability, environmental quality, and social justice are interrelated in complex ways. It suggests that achieving any one of these outcomes requires attention in some way to all three. Controversy revolves around exactly how these areas of concern can be joined — and, for that matter, whether they even can be. Some avoid the inherent tensions between the three broad issues by retreating to the primary concerns of their particular discipline and limiting consideration of the other factors. Some environmentalists, for example, talk of sustainability as an ecological concept with reduced concern for economic demands or social needs. Some entrepreneurs see tremendous business opportunities in consumer demand for "green" products without fully acknowledging the potential environmental or social impacts of expanded consumption. Furthermore, some social service advocates still see environmental concerns as a luxury that must be deferred until the more desperate needs of the poor are provided for.

The important sources of environmental thought are religious tradition, Malthusian economics, and environmental planning. Many religious traditions identify humankind within nature, whereas Judeo-Christian tradition is often seen as promoting exploitation or alternatively stewardship of nature.

Malthusian economics portrays the spiraling population growth as one that will eventually outstrip the earth's resource base. Famine and ecological ruin are the specters that haunt this view of human development. Although Malthus's most dire forecasts have not been realized, the idea underlying his theory permeates much thinking about population limits and resource conservation.

Table 1.1 Selected Definitions of a Sustainable Economy

- An economy that equitably provides opportunities for satisfying livelihoods and a safe, healthy, high quality of life for current and future generations
- A mixture of proactive visioning and sound economic reasoning
- Meeting the needs of the present without compromising the ability of future generations to meet their needs
- Maximizing the benefits of economic development, subject to maintaining the services and quality of natural resources over time
- A fair balance among economic, environmental, and social goals while efficiently using renewable and nonrenewable resources

Environmental planning is concerned with identifying the carrying capacities or growth limits for specific parcels of land use and population thresholds beyond which development cannot be sustained by the natural environment.

For many, sustainability means the ability to meet current and future needs while maintaining a good quality of life. One vision describes a sustainable world as providing goods and services (outputs) using a set of inputs to provide a good quality of life while sustaining the ecological basis for this activity (Daly, 1996). Of course, this description requires a consensus on the quality of life, a reasoned statement of needs or outputs, and a definition of ecological sustainability. For many, the concept of a good life lies at the heart of environmental conflict.

The matter of ecological sustainability leads to the issue of carrying capacity. The earth's **carrying capacity** is the earth's ability to provide inputs for a given level of human use over time. Carrying capacity often refers to the population that a given ecology can support. Carrying capacity is often described as the earth's ability to assimilate waste outputs. In fact, carrying capacity may be best described as the earth's ability to provide required inputs and outputs as well as assimilate the waste outputs for a given population.

A model of the economy, as shown in Figure 1.1, will help to cement these relationships. The inputs to the economy are the natural capital, intellectual capital, financial capital, and technology. The outputs are the desired benefits and the undesired waste. Both types of outputs impact the quality of life.

The **intellectual capital** of an organization is the talents of its people, the efficacy of its management systems, the effectiveness of its customer relations, and the technological knowledge employed and shared among its people and processes. Intellectual capital is knowledge that has been formalized, captured, and used to produce a process that provides a significant value-added product or service. Intellectual capital is useful knowledge that has been recorded, explained, and disseminated and is accessible within the firm (Stewart, 1997). The intellectual capital of an economy is the sum of all intellectual capital generated by its member organizations. Examples of

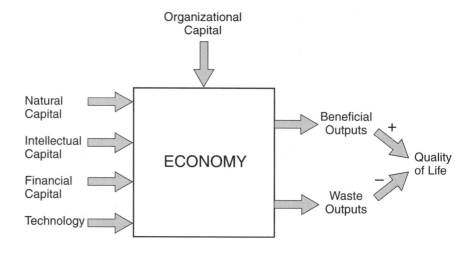

Figure 1.1 **A model of the economy.**

companies' intellectual capital are patents, databases, design methods, software codes, brands, hardware designs, manufacturing methods, machines, and tools.

The source of intellectual capital is twofold: human capital and organizational capital. Human capital is the combined knowledge, skill, and ability of the company's employees. Organizational capital is the hardware, software, databases, methods, patents, and management methods of the organization that supports the human capital. Therefore,

$$\text{Intellectual capital} = \text{Human capital} + \text{Organizational capital.}$$

Organizational capital is the hardware, software, databases, methods, laws, systems, management structures, and architectural arrangements for organizations and people in an economic system. The organizational capital of the economy will influence the efficiency of production of the outputs. Improved technology and better organizational capital can reduce the need for national inputs and the resulting waste. There may be limits to input use and waste control, but improving technology may postpone the appearance of these limits beyond the foreseeable future.

Natural capital refers to those features of nature, such as minerals, fuels, energy, biological yield, or pollution absorption capacity, that are directly or indirectly utilized or are potentially utilizable in human social and economic systems. Because of the nature of ecologies, natural capital may be subject to irreversible change at certain thresholds of use or impact (Clayton, 1996).

The pro-growth response to limits is to expect technology to overcome them. As the resource inputs are increasingly demanded, market prices will rise and thus drive technological change for substitutes. For example, as petroleum reserves are depleted, the market price of oil will rise and provide a drive for replacement sources such as renewables like solar energy.

There appears to be reasonable acceptance of environmental limits, but less agreement on what the limits are and how to ensure that these limits are not exceeded. Nevertheless, there is solid agreement that economic growth (EG), environmental preservation and protection (EP), and social equity (SE) can be mutually reinforcing goals and policies to achieve these goals should be integrated. A simple summary of these triple goals is to state that

$$QoL = EG + EP + SE, \tag{1.1}$$

where QoL is quality of life. Economic growth is not a singular goal but rather one of three balanced goals.

It is the role of national policies and organizations to structure and manage the organizational capital and to obtain and develop the best human capital in order to secure outstanding intellectual capital, which leads to wealth creation and good quality of life.

How technology is handled and organized is often more important than the specific technological efficiency. This is obvious, for instance, for transportation. It is more important to organize a transport system so that trains and trucks can carry cargo in both directions rather than make the engine of the train or truck more efficient.

Referring to Figure 1.1, the most optimistic economic plan is for a nation, or the world as a whole, to use human capital, financial capital, organizational capital, and technology to minimize the required natural capital inputs and the waste outputs while achieving the maximum beneficial outputs subject to the ecological constraints of nature and the goal of sustainability. Ultimately, people use their intellectual capital and skills to maintain their society. **Living capital** is the sum of the whole of mankind's accumulated usable knowledge, their capacity for choice, and the store of embodied energy, as well as nature's active potential to create and sustain itself in yet more complex and able forms.

1.2 Environmental Impacts

As we noted in the preceding section, the necessity for sustainability requires that we substitute intellectual capital (IC) technology for demand on natural capital (NC). A significant cutback in material and energy inputs to economic processes is the one sure way of cutting back on wastes and pollutants that are beginning to overwhelm the assimilative capacity of our environment.

The developed nations with about 20% of the world's population (P) consume about 80% of the current natural capital flow. If all peoples were to consume at the level of affluence of the developed nations, a fivefold increase in environmental impact might occur. Over the same period of development the population might double from the current 6 billion to 12 billion persons. We can propose a simple model of total environmental impact (EI) as

$$EI = P \times A \times M,$$

where P is total population, A is per-capita affluence as measured by consumption in turn multiplied by a measure of the damage done by the means (M) employed in supplying each unit of that consumption. Each of these factors reflects an incremental charge. For example, $P = (1 + \Delta P/P_0)$ where ΔP is the change in population and P_0 is the initial value.

Let us consider the case where population doubles during the next 50 years and the affluence increases by fivefold. If the means employed in supplying each unit of that consumption increase remain unchanged ($M = 1$), then the increase in environmental impact is

$$EI = P \times A \times M = 2 \times 5 \times 1 = 10.$$

Thus, a tenfold increase in EI would occur, which would greatly stress the environment.

If, on the other hand, we could change the means of production by reducing the environmental impact resulting from the means employed in supplying each unit of consumption by a factor of 1/10, then EI remains unchanged since

$$EI = P \times A \times M = 2 \times 5 \times (1/10) = 1.$$

Assume that the impact caused by the means of production (M) is reduced (improved) by a factor M, where

$$M = 1/(T \times IC),$$

where T is an incremental improvement in technological efficacy and IC is an incremental improvement in intellectual capital. If $T = 4$ and $IC = 2.5$, then

$$M = 1/(4 \times 2.5) = 1/10.$$

Then, with this improvement in M we have

$$EI = P \times A \times M = 2 \times 5 \times 0.1 = 1$$

and the environmental impact remains constant in spite of the increases in population and worldwide affluence. This calculation is not meant to be accurate but simply illustrative of our goal to mitigate the effect of population increase and affluence improvement with significantly improved technology and intellectual capital. Population growth and equitable affluence for developing countries might require a tenfold increase in natural capital throughout. The resulting level of environmental impact may overwhelm the earth's waste assimilation capacity as well as draw down the earth's natural resources. However, in principle, a tenfold improvement in mitigation measures and technology is feasible given the opportunities for energy and materials conservation and new ways of organizational processes and methods as well as better means of accessing and improving human capital.

1.3 Population

Population growth is an important factor in worldwide sustainability. Clearly, population cannot grow indefinitely until it exceeds the world's carrying capacity. A majority of all people of earth live in the six most populous nations: China, India, the United States, Indonesia, Brazil, and Russia. The advancement of medical and health sciences and technologies has resulted in reduced death rates and extended life expectancies. Unfortunately, death rates declined without immediately affecting birth rates. The consequence is an explosive increase in world population reaching 6 billion in 1999.

As the world's population grows, it is becoming increasingly urban. In 1950, just 30% of the world's population lived in cities. By 2000, this number reached 50%. This change is in response to opportunity. Those who live in rural areas, in general, have lower incomes and fewer chances to improve themselves economically. Cities, for all their faults, are hubs of industry, creating wealth and jobs.

Over one or two generations, the newly urbanized workers go to school, learn new skills, and adopt middle class habits and outlooks. Large families are not economically beneficial to middle class urban dwellers, since children must be housed, clothed, and fed. Children are an expensive luxury, rather than an economic necessity, for city dwellers. Thus, as societies urbanize and industrialize, and as education levels rise — especially among women — net birth rates tend to fall (Bloom, 2000).

Thus, as developing societies shift to urban societies we can expect a fall in birth rates as illustrated in Figure 1.2. Eventually, the net rate of population increase may stabilize at a replacement rate and the world population may stabilize at 10 billion persons towards the end of this century (Potts, 2000). The contributing factors of stabilization of population growth rate are urbanization, the improved status of women in some lands, the widened availability and enhanced safety of contraceptive techniques, and the speedy flow of ideas and information (Morrison and Tsipis, 1998).

Malthus believed that populations could increase geometrically while agricultural productivity could only increase arithmetically. He postulated that population growth would lead to rising food costs due to increasing demand. This, he stated, would lead to widespread malnutrition and rising mortality. He did not envision the way that technology coupled with organization capital would transform the productivity of the economy. Thus far, the rise in the world's population has been more than compensated for by the rise in technology and organizational capital.

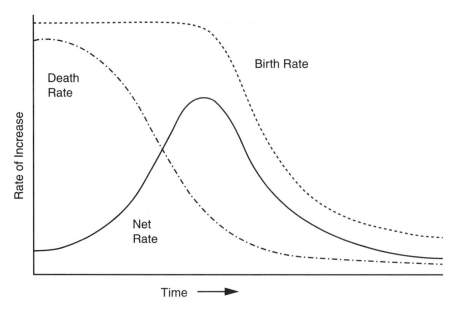

Figure 1.2 **Population growth rate.**

Nevertheless, perhaps one-fifth of the existing world's population is malnourished as illustrated in Table 1.2. This problem is a challenge to needs for structural reorganization of the distribution system. It also is difficult to foresee how population growth can be reconciled with widening aspirations toward high levels of consumption and resource-intensive lifestyles worldwide.

In his recent book, Amartya Sen, recipient of the Nobel Prize for economics, argues that economic development translates into better lives for people only if social programs lead to certain liberties: freedom from hunger, illiteracy, premature death from lack of health care, and the tyranny of undemocratic governments, which tend to ignore the poorest (Sen, 1999a).

Table 1.2 **Malnourishment**

Nation or region	Percent of population undernourished
Central Africa	48
Southern Africa	44
East Africa	42
Caribbean	30
South Asia (w/o India)	25
India	22
Central America	16
China	12
Middle East	11
South America	8

SOURCE: Data from *The Economist*, Oct. 16, 1999.

The population growth and resource use in various regions is illustrated in Table 1.3. Note that the primary growth in population is expected in the lesser developed regions. Resources are seen to be more available to the United States and Canada and that region is also responsible for heavy per-capita emissions of carbon dioxide. A pathway to worldwide affluence and sustainability will suggest lower carbon dioxide emissions and energy consumption per capita in the developed world and reduced population increases in the less developed world. Reduced population growth is attainable. As an example, the number of births per woman in India has declined from six in 1950 to three in 2000. A set of illustrative goals for these worldwide factors is provided in Table 1.4. These goals can only be achieved with great effort and coordination. These goals call for improved economic and resource levels for developing regions with reduced per-capita energy consumption and emissions levels for the developed regions. The projected population size of eight large nations is given in Table 1.5. These projections illustrate the world's population challenge. Managing the pathway to achieving these goals will be challenging indeed.

Table 1.3 **Population and Resources**

Region	1999–2005 Expected annual population increase (millions annually)	1998 Income per capita ($US)	Energy consumption (million BTUs per person annually)	Freshwater availability (millions of gallons per person annually)	Cropland (acres per person)	Carbon dioxide emissions (tons per person annually)
Africa	20	$650	15	2.2	0.7	1.2
Asia	51	$2,490	33	1.4	0.4	2.5
Europe	0.5	$13,710	144	2.5	1.1	9.4
Latin America	7.5	$3,710	49	9.6	0.9	3
Oceania	0.5	$15,430	181	19	4.5	12
United States and Canada	2.5	$27,100	357	5.6	2	22

SOURCE: Data from National Geographic Society, 1998.

Table 1.4 **Possible Regional Goals for Population and Resources Achieved by 2030**

Maximum percentage population increase annually	Minimum income per capita annually ($US)	Maximum energy consumption per capita (million BTUs annually)	Minimum freshwater per capita (millions of gallons annually)	Cropland (acres per person)	Maximum carbon dioxide emissions (tons per person annually)
1%	$4,000	150	2.0	1.0	5

Table 1.5 Population Size in Millions

| | 1998 | | 2050 | |
Rank	Country	Population	Country	Population
1	China	1,255	India	1,533
2	India	976	China	1,517
3	United States	274	Pakistan	357
4	Indonesia	207	United States	348
5	Brazil	165	Nigeria	339
6	Pakistan	148	Indonesia	318
7	Russia	147	Brazil	243
8	Japan	126	Bangladesh	218

SOURCE: Data from United Nations, *World Population Prospects,* 1996.

1.4 Sustaining the Environment

An **environment** is a locale, surroundings, or region. Thus, we talk of our local environment such as Napa Valley or a region such as the Mexico City metropolitan area. The concept of environmental unity implies that environments are linked since what we do in one region may affect other regions and ultimately the total planet. An **ecosystem** is a community of organisms and its environment. A sustainable ecosystem is one from which we harvest a resource that is still able to maintain its essential functions and properties (Botkin, 1998).

If we are to sustain the quality of the earth's environment, the biochemical cycles need to operate within healthy ranges of transfer and storage. For example, it is generally agreed that the United States needs to control the level of sulfur released into the atmosphere by electric power generating plants using coal as a fuel.

The environment of the earth consists of the soil and solids, the oceans, the atmosphere, and energy reserved from the sun and used and redistributed by living things. The complexity of human activities such as agriculture, forestry, fishing, materials processing industries, and others causes diverse interactions and effects. The goal of sustaining the earth's ecosystem seeks to maintain the viability of life on our earth. To achieve this overall goal, we restate some key principles in Table 1.6. The first principle states the requirement that the conditions necessary for the operation of the processes underlying an ecosystem must be maintained. An example of a violation of this principle is the construction of Egypt's Aswan High Dam in the Nile River valley. The alteration of this ecosystem expanded the habitat for snails that host the flatworm *Schistosoma mansoni*. As a result the proportion of people in the Nile River valley with *Schistosomiasis* increased from 5% in 1968 to 77% in 1993.

Another example of the violation of this principle is the use of DDT (dichloro-diphenyl-trichloro-ethane) as an agricultural chemical. In the late

1950s Rachel Carson brought to the public attention the fact that DDT spraying was resulting in the death of songbirds, bringing about a "silent spring" that led to the banning of DDT in 1972.

Table 1.6 **Principles of Ecosystem Sustainability**

1. The necessary conditions for processes in an ecosystem must be maintained.
2. Operation must be within the carrying capacity of the ecosystem.
3. Harvesting rates should not exceed the regeneration rate.
4. Waste emissions should not exceed the assimilative capacity.
5. The rate of exploitation of nonrenewable resources should be equal to or less than the rate of development of renewable substitutes.

Most of us experience, from time to time, a violation of principle 2 in Table 1.6 when the number of automobiles on a major road exceeds the carrying capacity of that road and gridlock occurs.

Perhaps the clearest example of principle 3 is visible when the harvest of trees in an ecosystem exceeds the regeneration rate. This principle implies a **sustainable yield threshold**. Ecosystems such as fisheries or forests have a harvest threshold above which the regeneration rate cannot sustain the harvest. If the harvest from a fishery exceeds that threshold for an extended period, stocks will decline and the fishery may abruptly collapse.

RACHEL CARSON

Rachel Carson was a scientist who worked at what is now the U.S. Fish and Wildlife Service. With the success of her second book, The Sea Around Us, *in 1951, she began to write full time.*

With a bestseller to her credit, she used that freedom, and her reputation as a best-selling author, to write Silent Spring, *her powerful examination of modern agriculture and its use of the insecticides that had come out of American chemical weapons research during World War II. Following its publication in 1962, Carson was denounced as a hysteric, not just by the chemical companies but also by government officials.*

Knowing what she was up against, however, she had laid out the evidence meticulously. DDT, then sprayed routinely on beans, peanuts, tomatoes, and other crops, had lethal consequences for fish as well as birds, she demonstrated. Used to ward off Dutch elm disease, it killed the earthworms that ate fallen elm leaves, and the robins that ate the earthworms. In falcons and other species, it produced thin-shelled eggs that hatched before their time.

As a result of the pressure that originated with Carson's book, the Nixon administration ordered a domestic ban on DDT in 1972. The DDT ban proved to be a godsend to falcons, eagles, and pelicans, among other species.

An example of the violation of the fourth principle is the waste disposal of the plastic six-pack rings that break down and disintegrate very slowly. Surveys have shown that the rings can strangle birds and fish. Furthermore, the worst culprit for fish may be discarded plastic fishing line.

The fifth principle calls for development of renewable substitutes for nonrenewable resources. This principle can be illustrated by the goal of substituting solar, hydro, and nuclear energy sources for fossil fuel sources. The rate of development of renewable substitutes needs to exceed the incremental exploitation rate of nonrenewables over the next several decades so that we can ultimately expect the use of renewable sources to exceed that of nonrenewable sources.

Beyond the five principles of ecosystem sustainability, it is important to preserve the stability of an ecosystem. Rich **biodiversity** is an important factor to preserve the stabilizing functions of an ecosystem. The concept of biodiversity encompasses three types of biological variation: (1) genetic diversity, (2) habitat diversity, and (3) species diversity. Factors that tend to increase biodiversity include a physically diverse habitat and moderate amounts of disturbance. Factors that tend to decrease biodiversity include environmental stress and a severe limitation or exhaustion of an essential resource.

Damaged ecosystems can often be restored, in part, to their former state. A positive example is the restoration of Lake Erie, which was pronounced "dead" in the 1960s. Scientists concluded that the cause of the destruction of the lake's ecosystem was excessive phosphorus emissions into the lake from municipal wastes. As a result of U.S. and Canadian efforts, by 1985, the annual release of phosphorus had been reduced by 84% (Botkin, 1998). Populations of fish rebounded and the lake ecosystem recovered much of its diversity. However, the blue pike may become extinct.

An advocate of the necessity for biodiversity in plant and animal populations was Aldo Leopold. His essays were published in 1948 as *A Sand County Almanac*. He advocated a comprehensive view of ecology and valued biological diversity. He wrote:

> We abuse land because we regard it as a commodity belonging to us. When we see land as a community to which we belong, we may begin to use it with love and respect. There is no other way for land to survive the impact of mechanized man, nor for us to reap from it the esthetic harvest it is capable, under science, of contributing to culture.

There are limits to the benefits of technology and financial capital to substitute for essential services provided by natural capital. Examples are the hydrological cycle, biological waste assimilation, and detoxification as well as natural scenery and biodiversity. Thus, unlimited population growth is unsustainable over the long term. The total land available for habitation and cultivation is finite and clearly can support only a limited (although large) number of persons.

1.5. Global Warming and the Greenhouse Effect

Scientists have been warning of the buildup of potentially climate-destabilizing "greenhouse gases" in the atmosphere. These gases include carbon dioxide, methane, and nitrous oxide. It is alledged that industrial and automobile emissions are causing a rise in the atmosphere temperature, a process called *global warming*. A succession of major weather events during the 1990s such as floods and heat waves have been tied by some to the buildup of carbon dioxide in the earth's atmosphere. Carbon dioxide is the heat trapping gas of greatest concern and about six billion tons are discharged into the air throughout the world every year. Carbon dioxide traps solar radiation in the same way a greenhouse traps the sun's warmth; thus the term **greenhouse effect**. The United States is responsible for about 20% of these emissions.

Many scientists estimate that a doubling of carbon dioxide emissions due to the increase in worldwide economic growth could lead to an increase in temperatures at the earth's surface. The impacts of this rise in temperature could be significant to the world's ecosystem, thus the advocacy for the control of carbon dioxide emissions as documented in the U.N. Framework Convention on Climate Change, a treaty signed in 1992.

Because every act of combustion produces carbon dioxide, the developing nations are reluctant to limit their use of energy as they create industrialized economies. But the industrialized nations note that most of the increase in carbon dioxide emissions over the next half-century is likely to be generated by energy production in those rapidly modernizing economies.

Scientists are unsure what will work best to limit or reduce the concentration of carbon dioxide in the atmosphere. For example, there is disagreement on the importance of carbon sinks such as the trees of North America.

The abatement required to stabilize or reduce carbon dioxide concentrations is immense and may require a fundamental transformation of global energy infrastructure. No clear path to such a future is evident. A serious effort to limit climate change will require multiple technological paths to be pursued.

Masters (1997) provides a useful discussion of the science of global atmospheric change. The computer climate models used to estimate climate change depend on causal links between human activity and climate effects. Life has existed on the earth for millions of years. In that time climate has fluctuated with ice ages and eras of heat. Global climate depends on many factors interacting in subtle and complex ways. Nevertheless, many scientists have concluded tentatively that there is a discernible human influence on global climate.

Various models indicate different levels of carbon dioxide in world air, but one author estimates growth from a carbon content level of 770 gigatons in 1995 increasing to 890 gigatons in 2025 (Morrison and Tsipis, 1998).

Global thermostatic processes are probably robust, but will not necessarily maintain the current conditions. The atmosphere and climate have been at

different equilibrium states in the earth's past. If greenhouse gases double in
the next 50 to 100 years, then, the average temperature might rise by 2°C.
However, it is possible the temperature rise may be distributed unevenly over
the earth. Changes may result in melting of polar ice and a rise in ocean levels. The generally agreed on approach to the reduction of greenhouse gas
emissions is to reduce the number of emission sources, to change the source
technologies, and to increase the number of emission assimilators. Thus, a
simple, but difficult to implement, plan might call for fewer automobiles
with internal combustion engines, fewer power plants using fossil fuels, and
better technologies for these automobiles and power plants.

1.6. Economics

The economy is a subsystem of the environment and depends on the environment both as a source of raw material inputs and a "sink" for waste outputs. Economic growth and environmental protection are mutually
reinforcing goals. **Economics** is the study of mankind in the ordinary business of life (Mankiw, 1997). Economics can be defined as the study of how
society manages its scarce resources. Clearly, society is interested in effectively managing its material and environmental resources. The five principles of economics described in Table 1.7 are the key principles that apply to
economic analysis and the environment. Making decisions that involve the
environment requires trading off one goal against another as noted in the
first principle. One trade-off society faces is between efficiency and equity.
Efficiency is the property of reaping the most benefit from scarce
resources, such as natural gas supplies. **Equity** is the property of distributing
economic prosperity fairly among the members of society.

A **market economy** allocates resources through the decentralized decisions of individuals and firms as they interact in markets. Although markets
are a good method of allocating resources, sometimes markets fail to do so.
One possible cause of market failure is an **externality** which is the impact
of one or many person's actions on the well-being of a bystander. Smog
from commuters' autos may impact the health of children in a school
located next to the main road. Governments can sometimes improve these
conditions — in this example, they could relocate the school.

Table 1.7 Five Principles of Economics and the Environment

1. Decisions involving economics and the environment involve trade-offs.
2. Markets are usually an efficient way to organize economic activity.
3. Governments can sometimes improve market outcomes.
4. People respond to incentives.
5. A nation's standard of living depends on its ability to efficiently produce goods
   and services.

SOURCE: Adapted from Mankiw, 1997.

Almost all variation in living standards among countries is explained by **productivity**, which is the quantity of goods and services produced from the sum of all inputs such as hours worked and fuels used.

Rational people make decisions by comparing costs and benefits. Incentives can be used to change costs or benefits and often can be used to modify economic behavior. Public policies that involve incentives can help to protect the environment.

A **negative externality** such as smog from autos can be reduced if government provides regulations or positive incentives. Regulations include emission standards and incentives may include gasoline taxes that increase the cost of auto travel, thus providing an incentive to reduce auto travel.

A tax used to reduce the effect of a negative externality may be more efficient than a regulation. Consider the goal of reducing the smog level in the Los Angeles basin. One approach is to enact a regulation that requires each industrial plant to reduce its emissions by 10% for each of the next 5 years. An alternative would be to levy a tax on the amount of pollution above the desired standard. Another alternative is to use tradable **pollution permits**. Every industrial plant would be issued a tradable permit to generate a set level of pollution for each of the ensuing 5 years. Then, a market would emerge for these permits and they would be bought and sold. The firms that can reduce pollution only at a higher cost will buy permits from the firm that can more effectively reduce the pollution and thus is able to sell the permit. The final allocation of these pollution permits will be efficient.

Consider a group of industrial firms generating emissions. Government can limit emissions by regulation, taxes, or tradable permits. Regulations often proscribe methods and limits and may miss providing incentives for innovative solutions. Taxes on pollution emissions can reduce the level of pollution, but governments have difficulty estimating the tax level to attain a specified emission limit. Tradable permits issued at the desired level may encourage firms to innovate new solutions and sell some of their permits.

Free **public goods** such as parks, scenic vistas, and the air in a valley are not subject to a price and thus the market economy does not allocate them well. It is the role of government to allocate scarce public resources such as parks, ocean beaches, roads, and clean air and water. Government uses **cost–benefit analysis** to estimate society's costs and benefits of a proposed regulation or project. This analysis can be very difficult to accomplish. How can you calculate the costs and benefits of controlling emissions of carbon dioxide by means of proposed regulations or incentives? The benefits and costs would be very difficult to calculate.

Common resources are available free of charge to anyone who wants to use them and one person's use of them reduces other people's employment of them. Consider a Town Commons in the center of a town, available for sheep grazing. This common land belongs to all and everyone is free to graze their sheep on the commons. As economic activity grows, more and more sheep are placed on the commons and eventually the commons become barren. This example illustrates the **tragedy of the commons**, which explains why common resources can be used more than is desirable. Examples of common resources include clean air and water and roads.

Any realistic set of governmental policies needs to integrate environmental and economic considerations. However, it must be recognized that government will not always arive at an optimum or even workable solution. Governments engender their own bureaucratic self-interests and inadequacies. In many cases, government has instituted environmental policies that result in costs that far outweigh the benefits. To use cost–benefit analysis, one can use a discount of future benefits costs. However, this method may lead to postponement of policies for large cost and large benefit projects that have the costs today and the benefits at a much later time. Take as an example the benefit of a project to avoid the impact of global warming, which provides a big benefit (B) 50 years hence, but costs (C) this year. If we discount the benefit by a discount rate (R), in order to proceed we require that

$$C < \frac{B}{\left(1+R\right)^{50}}.$$

If B = \$100 billion, C = \$2 billion, and R = 0.10, we have failed the requirement that

$$C < \frac{B}{\left(1+R\right)^{50}} \quad \text{since } (1+R)^{50} = 1.1^{50} = 117.4.$$

We also need to consider the idea of substitution. Goods are **substitutes** when the price increase of one good leads to an increase in the demand of another good. An example of substitutes is agriculturally produced methanol as a substitute for gasoline as an automobile fuel. Substitution can occur at seven levels as shown in Table 1.8. An example of beneficial substitution is the use of solar energy as a substitute for fossil fuel energy in electric power plants.

We seek to avoid a bifurcated view between the environment and the economy. Some support and encourage the use of natural resources with less attention to the unwanted effects of that process. On the other hand, others deplore the exhaustion of resources and pollution with little credit to the important process of creating wealth for all. A summary of the underlying reasons for environmental degradation is given in Table 1.9.

Table 1.8 **Levels of Substitution**

Level	Example
1. Raw material	The same fuel may be obtained from different sources such as gas from oil wells and coal.
2. Material	Aluminum may substitute for copper.
3. Component	Semiconductors substitute for vacuum tubes.
4. Subsystem	Electric motors replace internal combustion engines.
5. System	Shuttle services replace private cars.
6. Strategic	Different strategies can lead to the same goal.
7. Values	Values lead to new strategy.

Table 1.9 **Selected Underlying Reason for Environmental Degradation**

- Increased demand for water, biological resources, and energy services as a result of economic and population growth
- Choice of inappropriate technologies
- Inadequate government policies for the management of the use of water, biological resources, and energy for sustainable development
- Inability to internalize environmental externalities into market prices, that is, appropriately reflect the costs of environmental degradation in market prices
- Inability to determine the long-term consequences of development activities
- Inadequacy of markets and governments in their national income accounting to recognize the true value of natural resources

1.7 Envisioning and Implementation of a Sustainable Society[1]

By Robert Costanza,[*] Olman Segura,[†]
and Juan Martinez Alier[+]

[*]*Director, Institute for Ecological Economics*
University of Maryland

[†]*International Center for Political Economy*
National University of Costa Rica

[+]*Department of Economics and Economic History*
University of Barcelona

1.7.1 OVERVIEW

Ecological economics represents a new, transdisciplinary way of looking at the world that is essential if we are to achieve sustainability. Getting "down to earth" and making sustainability operational requires the integration of three elements: (1) a practical, shared *vision* of both the way the world works and of the sustainable society we wish to achieve; (2) methods of *analysis* and modeling that are relevant to the new questions and problems this vision embodies; and (3) new institutions and instruments that can effectively use the analyses to adequately *implement* the vision.

The importance of the *integration* of these three components cannot be overstated. Too often when discussing practical applications we focus only on the implementation element, forgetting that an adequate vision of the world and our goals is often the most practical device to achieving the vision, and that without appropriate methods of analysis even the best vision can be blinded. The importance of *communication* and *education* concerning all three elements also cannot be overstated.

---

[1]This section is based on the introductory chapter of *Getting Down to Earth,* Washington, DC: Island Press, 1996.

1.7.2 INTRODUCTION

This section is a synthesis and overview of the major ideas contained in the book *Getting Down to Earth,* which is divided into three major sections dealing with vision, analysis, and implementation.

Within the diversity of points of view, there are some basic points of consensus, including (1) the vision of the earth as a thermodynamically closed and nonmaterially growing system, with the economy as a subsystem of the global ecosystem. This implies that there are limits to biophysical throughput through the economic subsystem; (2) the future vision of a sustainable planet with a high quality of life and fair distribution of resources for all its inhabitants (both humans and other species) within the material constraints imposed by 1 (above); (3) the recognition that in the analysis of complex systems like the earth at all space and time scales, fundamental uncertainty is large and irreducible and certain processes are irreversible, requiring a fundamentally precautionary stance; (4) that institutions and management should be proactive rather than reactive and should result in simple, adaptive, and implementable policies based on a sophisticated understanding of the underlying systems which fully acknowledge the underlying uncertainties. This forms the basis for policy implementation which is itself sustainable.

1.7.3 VISION

There are several elements we have combined under the heading of "vision," some of which are "positive," having to do with theories and understanding about how the world works, and some of which are "normative," having to do with how we would like the world to be. The relationship between positive and normative, like the relationship between basic and applied science or between mind and body, or logic and emotion, is best viewed as a complex interaction across a continuum, rather than a simple dichotomy. To some extent, we can change the way the world is by changing our vision of what we would like it to be. Likewise, the strict dichotomy between basic and applied science has often proven to be more a hindrance than a help in developing useful understandings of complex systems, as has the mind–body dichotomy.

Visionaries and theorists have often been characterized as mere impractical dreamers. People become impatient and desire action, movement, measurable change, and "practical applications." But we must recognize that action and change without an appropriate vision of the goal and analyses of the best methods to achieve it can be worse than counterproductive. In this sense a compelling and appropriate vision can be the most practical of all applications.

One of the major differences between ecological economics and conventional academic disciplines is that it does *not* try to differentiate itself from other disciplines in terms of its content or tools. It is an explicit attempt at pluralistic integration rather than territorial differentiation. This is admittedly confusing for those bent on finding sharp boundaries between academic disciplines, but we believe it is an essential revisioning of the problem.

Ecological economics does not aim at analyzing or expressing ecological, social, and economic relationships in terms of the concepts and principles of any one discipline. It is thus not merely ecology applied to economics, nor is it merely economics applied to ecology. It is a transdisciplinary approach to the problem that addresses the relationships between ecosystems and economic systems in the broadest possible sense, in order to develop a deep understanding of the entire system of humans and nature as a basis for effective policies for sustainability. It takes a holistic systems approach that goes beyond the normal narrow boundaries of academic disciplines. This does not imply that disciplinary approaches are rejected, or that the purpose is to create a new discipline. Quite the contrary. What ecological economics rejects is the "intellectual turf" model that unfortunately still holds sway in much of academia.

Ecological economics focuses more directly on the problems facing humanity and the life-supporting ecosystems on which we depend. These problems involve (1) assessing and insuring that the scale of human activities is ecologically sustainable; (2) distributing resources and property rights fairly, both within the current generation of humans, between this and future generations, and between humans and other species; and (3) efficiently allocating resources as constrained and defined by 1 and 2 above, and including both marketed and non-marketed resources.

1.7.4 HOW THE WORLD WORKS: BIOPHYSICAL CONSTRAINTS

In *Getting Down to Earth,* John Holmberg, Karl-Henrik Robèrt, and Karl-Erik Eriksson use a holistic systems perspective, the laws of thermodynamics, and the basic shared values of society to develop four basic principles that are necessary "system conditions" for a sustainable society. These are (paraphrased slightly):

1. Mined substances must not systematically accumulate in the ecosphere.
2. Anthropogenic substances must not systematically accumulate in the ecosphere.
3. Natural capital must be preserved and conserved.
4. Resource use must be efficient and fair with respect to meeting human needs.

System conditions 1, 2, and 3 relate to maintaining an ecologically sustainable scale of the economy within the ecosphere, while condition 4 relates to the requirements for a fair distribution of wealth and resources and an efficient allocation as described above.

Herman E. Daly extends this line of reasoning in Chapter 3 of *Getting Down to Earth* to address the issue of the optimal size or scale of the human economy within the ecosphere. He points out that "consumption" is the dissipation of ordered matter, and that natural systems and natural capital represent a primary source of the value added to products. At some point the transformation of natural into human-made capital costs us more in terms of natural capital services lost than it gains us in terms of human-made capital services gained. Material growth in the economy beyond this point is "anti-economic." Sustainability requires that at this point we shift from natural-capital-consuming material growth to natural-capital-preserving qualitative development. As the economy matures, like an ecosystem changing from early, weedy succession to mature forest, the emphasis shifts to producing as much value per unit of consumption as possible, rather than producing and consuming as much as possible.

Another perspective on this problem is provided by Joseph A. Tainter in *Getting Down to Earth*. He takes an historical perspective and shows that civilizations have tended to collapse in the past because their increasing complexity led to decreasing marginal returns to complexity and ultimately to negative returns to complexity. We have created an aversion to this history by focusing only on the positive benefits of economic growth (and increasing complexity) and ignoring the negative effects like natural capital depletion, increasing crime, congestion, etc. Hopefully, the ecological economics vision will help us to change this cavalier attitude toward history, to learn where we are in this process of evolving complexity, and to develop in such a way as to remain sustainable rather than collapsing as so many of our predecessor civilizations have done.

While in the North the vision of the environmental problem has focused on global environmental change, the relationship between population and resources, intergenerational equity, and the internalization of environmental costs, in the South the problems have been seen as more focused on poverty, massive ecological degradation, and intragenerational equity. These differences are relevant, but we emphasize the general importance of natural processes in adding value to the economy and the common need to minimize throughput while maximizing ecological productivity if the goal is sustainability. Different social structures will be required to meet these goals in the North and in the South. The benefits of drawing on indigenous/local ecological knowledge systems and the experience of grassroots organizations are that they conserve and use nature respectfully as a cultural contribution to other species and themselves. Their vision of how the world works will be essential in generating the technological and institutional innovations necessary for sustainability.

1.7.5 SOCIAL GOALS: HOW WE WOULD LIKE THE WORLD TO BE

A broad, overlapping consensus is forming around the goal of sustainability. Movement toward this goal is being impeded not so much by lack of knowledge, or even lack of "political will," but rather by a lack of a coherent, relatively detailed, shared vision of what a sustainable society would look like. Developing this shared vision is an essential prerequisite to generating any movement toward it. The default vision of continued, unlimited increases in material consumption is inherently unsustainable, but we cannot break away from this vision until a credible alternative is available. The process of collaboratively developing this shared vision can also help to mediate many short-term conflicts that will otherwise remain irresolvable.

Donella Meadows in *Getting Down to Earth* lays out (1) why the processes of envisioning and goal setting are so important (at all levels of problem solving); (2) why envisioning and goal setting are so underdeveloped in our society; and (3) how we can begin to train people in the skill of envisioning and begin to construct shared visions of a sustainable society. Several general principles emerged, which include:

1. In order to effectively envision, it is necessary to focus on what one really wants, not what one will settle for. For example, the lists below show the kinds of things people really want, compared to the kinds of things they often settle for.

Really want	Settle for
Self-esteem	Fancy car
Serenity	Drugs
Health	Medicine
Human happiness	GNP
Permanent prosperity	Unsustainable growth

2. A vision should be judged by the clarity of its values, not the clarity of its implementation path. Holding to the vision and being flexible about the path is often the only way to find the path.
3. Responsible vision must acknowledge, but not get crushed by, the physical constraints of the real world.
4. It is critical for visions to be shared because only shared visions can be responsible.
5. Vision has to be flexible and evolving.

Probably the most challenging task facing humanity today is the creation of a shared vision of a sustainable and desirable society, one that can provide permanent prosperity within the biophysical constraints of the real world in a way that is fair and equitable to all of humanity, to other species, and to future generations. This vision does not now exist, although the seeds are there. We all have our own private visions of the world we really want and we need to overcome our fears and skepticism and begin to share these visions and build on them until we have built a vision of the world we want. We have sketched out the general characteristics of this world — it is ecologically sustainable, fair, efficient, and secure — but we need to fill in the details in order to make it tangible enough to motivate people across the spectrum to work toward achieving it. The time to start is now.

1.7.6 ANALYSIS

Even the fragmented and vague vision of a sustainable and desirable world which we can currently articulate begins to change the kinds of analysis that are most appropriate for operationalizing it. In particular, we need models that:

1. Acknowledge the biophysical constraints on economic systems;
2. Analyze the linkages between ecological and economic systems at several interacting scales;
3. Acknowledge the fundamental uncertainties in our ability to predict the behavior of complex adaptive systems.

Change from the current trajectory of development is not an option — it is a necessity. The problem cannot be solved by small incremental corrections, as many economists and politicians recommend, but requires fundamental changes in the goals of society and the rules of the game. Consider the distribution issue from the perspective of the distinction between material economic growth and real improvements in the quality of life. In the North, quality of life may actually be decreasing with increased material consumption while in the South, poverty requires material growth to at least some minimum standards. We consider some strategies for achieving sustainable improvement of quality of life in this broader context, which is dependent on changing the distribution of resources.

1.7.7 IMPLEMENTATION

Finally, we are concerned with the implementation of the ecological economics vision, including various institutional changes necessary for sustainability. These cover a broad spectrum including property rights regimes, NGO activities, community-based management, and implementing UNCED.

Failures can occur in the current system. A typology of three basic types of failures includes (1) transaction failures; (2) empowerment failures; and (3) government failures. Each of these has several sub-types and reasons for the failure. For example, transaction failures are due to either (1) market system failures caused by missing markets or market performance failures; (2) negotiation failures due to missing parties or asymmetries in bargaining power; or (3) preference failures due to missing knowledge or incomplete preferences or time preference bias. Of these, only the first sub-type has been given much attention by economists, and empowerment and government failures are hardly recognized at all. Institutional reform is necessary to address all of these failures in an integrated way in order to achieve sustainability.

1.7.8 CONCLUSIONS

Integrating a shared vision of a sustainable and desirable world with adequate analysis and innovative implementation is the "full package" necessary to achieve sustainability. All three aspects of this task need much improvement, but their integration is lagging furthest behind. Ecological economics is helping to foster the transdisciplinary dialogue necessary to pull the package together and move forward toward the newly articulated goals.

1.8 Dynamic Aspects of Sustainability

BY ERIC F. KARLIN
Professor of Plant Ecology and Dean
School of Theoretical and Applied Science
Ramapo College

The concept of sustainability is strongly associated with permanence — that is, human populations able to live permanently in an area, cultivating the soil in such a way that that it remains arable for future generations, establishing permanent nature preserves, etc. Indeed, the master models of sustainability, natural ecosystems, have sustained life on earth for more than 3.5 billion years. Talk about permanence! But what exactly has been sustained over this time? Just what has been permanently maintained? Interestingly, life, and the ecosystems that support life, are not static; they have changed significantly over time. Forests did not exist 500 million years ago; and flowering plants, which dominate most present-day terrestrial ecosystems, had not yet evolved when dinosaurs first walked the planet's surface some 200 million years ago. In fact, the ability to change, to evolve, is a major factor in life's persistence. Particular species and ecosystem types are not permanent; extinction is their ultimate fate.

It is life itself that is permanent, maintained by natural ecosystems. What kind of life, and how much of it exists, varies. The dimension of change must be addressed in the planning of sustainable human-based ecosystems. Static and inflexible ecosystems will not be sustainable in an environmental context that continually fluctuates.

1.8.1 THE DYNAMIC NATURE OF ENVIRONMENTS

One of ecology's basic principles is that the environment is continually modified in response to variations in abiotic variables (weather, earthquakes, fire) and by the actions of living beings. For instance, a tree canopy results in a shaded ground surface. Likewise, cattle (*Bos taurus*) in a field not only eat many of the plants growing there; they also trample the ground and their cow pies dot the landscape. In some cases the environmental changes are minimal, while in others the changes are significant. Sometimes the change is brief and transient, but sometimes it is persistent and long term. In every case, the environment undergoes constant change. Some 18,000 years ago much of Canada and the northeastern United States was covered by extensive glaciers, often exceeding 1500 meters in depth. That glaciated landscape differed dramatically from the boreal conifer forests and temperate deciduous forests present there today.

When environmental changes are abrupt or large scale, the ecosystem is unlikely to be sustained. In such cases, existing species are replaced by those which can better tolerate the new environmental setting; life remains, but the ecosystem is significantly altered. When environmental changes are minimal, the ecosystem may sustain the present assemblage of species for long periods, changing only slightly over time. But even small or gradual changes, if continued long enough, can eventually result in significant modifications.

As each of the many components of the complex web of variables comprising an ecosystem's environment changes — so will the ecosystem dependent on it. Ecosystems are thus continually changing systems. In some cases the change is rapid and readily apparent (e.g., forest regeneration after a fire). In other cases, the change is slow-paced and not readily apparent in the short term. Douglas fir (*Pseudotsuga menziesii*) dominates many old-growth forests of the Pacific Northwest. But if such a forest were to stay intact long enough (about a millenium), Douglas fir, which is a successional species, would eventually be replaced by western hemlock (*Tsuga heterophylla*), which is a "climax" species.

1.8.2 RESILIENCY

The sensitivity of ecosystems to perturbation (ecosystem resiliency) is also important. How capable is the system of surviving and regenerating after a minor or major disruption? A truly sustainable system has a fair degree of resiliency built in; it can withstand most of the environmental variations it may be confronted with. Some ecosystems, however, may have minimal resiliency; they are so fragile that a minor perturbation could cause their demise.

Resiliency is especially critical in the preservation of rare ecosystem types. A rare wetland type in Ohio may be the only known location for three critically endangered species and for many additional rare species. If subjected only to minimal perturbations, the wetland might be sustained for centuries. But if this wetland were to be significantly damaged, the three endangered species may be driven to extinction, and many of the other rare species may be extirpated. The wetland ecosystem which develops in that location after the perturbation will be significantly different from its predecessor.

The development of new technologies and social systems represents one kind of perturbation. Water availability to a city may be increased by decreasing the amount of water lost from the piping system that delivers its water supply. In a social context, the amount of water required to support a given population size may be reduced by developing a more water-conserving society. Conversely, new technologies (like faster home computers or larger cars) and/or social changes (the current population may decide that 60% of the land should be in a natural state; 50 years from now people may decide that 30% would be sufficient) may result in *increased* demands for resources, even if the population size were to remain constant. Whether a change has the potential to reduce or increase resource consumption, the point is that it will affect the function and ecological balance of the associated ecosystems.

1.8.3 SPATIAL AND TEMPORAL SCALE

Spatial and temporal scale also need to be taken into account when designing a sustainable system. For instance, most farms lose more soil every year than they gain. In some cases, however, it may take 200 years before the soil is severely depleted. Although initial short-term observations indicate that all is well with the soil system, the long-term result is that the soil is destroyed. Likewise, suppose the environment of a redwood (*Sequoia sempervirens*) forest is changed such that although the adult trees remain healthy, the redwood seedlings and saplings cannot survive. It would well be over 1000 years before the last adult redwood would die off. Their disappearance would be so gradual that most people would not even be aware of the change from one generation to the next.

An important aspect of spatial scale becomes obvious in cases where practices that are sustainable in one locality affect other areas negatively. The burning of coal releases large quantities of pollutants into the atmosphere, particularly sulfur dioxide. To minimize the local environmental impact of these pollutants, tall smokestacks have been designed and built. These allow for the dispersal of pollutants over large and distant areas. Although the tall smokestacks minimize air pollution problems in the local ecosystems, where the coal is actually burned, they create major environmental problems hundreds of kilometers away in the form of acid deposition. Likewise, development of land adjacent to New York City watersheds in the Catskill Mountains may not have any significant effect on the local environment, but could result in the pollution of the city's supply of drinking water.

1.8.4 CARRYING CAPACITY

Just how much life — and what kinds of life — an ecosystem can support depends on the environment associated with it. Environments richly endowed with the resources essential to life (water, minerals, light, stability, warmth, absence of toxicants, and so forth) have a greater capacity to support life than do environments with minimal resources.

No matter how well or how poorly equipped, in every ecosystem there is a limit to the amount of life that it can sustain. This limit is called a *carrying capacity,* defined as the maximum number of individuals of a species that can be sustainably supported. Because carrying capacity ultimately depends on the environmental complex associated with the ecosystem, it is not constant; it varies across time and space. For instance, the number of mosquitoes occurring in an area often varies significantly from one year to the next. The change in population could be due to a drought, the outbreak of a disease which kills mosquito larvae, or perhaps the spraying of a biocide. Whatever the reason, it is clear that a change in one or more environmental variables affects the ability of an ecosystem to support the populations of these tiny vampires.

Humans, like any other living creature, continually modify their environment. But because we are not restricted by our body's limited ability to alter the world around us (imagine trying to chop a forest down with your bare hands) and can create machines and tools, we have greatly augmented our ability to modify the environment. Whenever we encounter limits to our abilities, we swiftly put our brains to work to develop a technological innovation that will allow us to surmount the limitation. Relatively free of biotic limitations, humans modify the environment in a vast multitude of ways — many large scale and rapid. For better or worse, we are grand masters of environmental modification.

Therefore, through technology, humans can transcend the carrying capacities of the local ecosystems (local carrying capacity) in which they live. We can, in effect, switch to a carrying capacity based on a regional — or even global — ecosystem, with regional or global carrying capacities. For example, while the local carrying capacity for human population may be 1 person per square kilometer, we can design and develop systems capable of supporting 100 people per square kilometer in that same area. Of course, the additional food, water, etc., needed to support this larger population must come from other ecosystems; some nearby and others hundreds or thousands of kilometers away.

Consequently, it is up to us to decide how many humans should live in a particular area. On the one hand, we can design social and economic systems in which humans are a minor part of the landscape. We can also design systems where nature is virtually obliterated. But switching from a local to regional (or global) carrying capacity does not mean that we have avoided the consequences and limitations of carrying capacity altogether; it simply means that we have established limits based on larger scale environmental support systems.

Thus human ecosystems largely dependent on regional and global carrying capacities can only exist because of the support of ecosystems found elsewhere (to provide food, water, clothing, etc.). These "ghost acres" often cover many times more land than the area that they support and illustrate the importance of spatial scale. The watershed area (5066 km^2) needed to supply drinking water for New York City is more than six times greater than the area actually occupied by the city (787 km^2). When one considers the spatial requirements of a major city like New York, it is essential to factor in the extensive ghost acres that make it possible. All of these support ecosystems, which may look relatively empty or underdeveloped, are necessary, vital components of keeping such a city viable. If they are damaged or modified, there will be obvious impacts on the communities dependent on them. In fact, given the importance of preserving natural areas, extensive wilderness areas need to be set aside simply to compensate for the impact of human actions on all of the lands utilized for human-dominated ecosystems.

1.8.5 A REALISTIC APPROACH TO BUILDING SUSTAINABLE, HUMAN-BASED ECOSYSTEMS

When thinking about sustainability, one of the first questions that needs to be addressed is this: Just what is it that is being sustained? The focus could be on maintaining a large and stable human population in a small area via the support of regional and global carrying capacities or it could be on maintaining a small and variable population in the same area primarily using local carrying capacities. But there is more to life than quantity, so the dimension of quality also needs to be considered. A system may be able to sustain 1000 people, but will they have healthy and enriching lives? Thus the level of lifestyle desired and the maintenance of a healthy, nonpolluted physical environment must be factored into the equation. The time frame is also important — what is sustainable for the short term may not be sustainable on a long-term basis.

Thus, sustainable ecosystems must be based on complex requirements. In fact, almost every aspect of our lives and activities must ultimately be considered. For instance, a population cannot be sustained unless the agricultural system that produces its food is also sustainable. In turn, a sustainable agricultural system can only exist in an ecologically sound, economically viable, socially just, and humane context.

Any sustainable ecosystem we develop to support us will have to include components of natural ecosystems. But how much of nature should we preserve and build into our systems? This question is not easily answered and is beyond the scope of this brief essay. But it is clearly in the best interest of human kind to preserve as much natural landscape as possible. Doing so would maximize both the extent of *natural* control of ecosystems (as opposed to human management) and the preservation of the ecosystems with which we evolved — those proven to support us. Natural ecosystems have sustained life on earth for over three billion years; they provide the best models for the design and maintenance of sustainable systems. We certainly have no guarantee at present that a global-scale replacement of natural ecosystems with human-designed and -managed ecosystems would sustain us and the other species with which we share this planet. Until such proof is available, it makes little sense to destroy that which has a proven track record.

Living within the confines of a local carrying capacity provides direct feedback from the system that supports us, offering the greatest chance of achieving a sustainable system. If a human population's need for water were to be based only on a local carrying capacity, then a sustainable supply of water would result when water consumption by that population was less than, or equal to, the amount of water available in the local ecosystem. If water consumption were to exceed the local water supply, then the population's needs would not be met. Given the restriction of a local carrying capacity, two things could happen: (1) The population could maintain its size but use less water by developing of new technologies or social practices, or (2) the population could decline (due to emigration and/or death).

The restriction to a local carrying capacity requires that human population centers remain small; it also leaves a population at the mercy of local conditions. Regional and global carrying capacities, by providing resources far in excess of what local ecosystems can provide, allow for the development of larger cities and provide a buffer against local environmental restrictions. Currently, most human ecosystems are supported by a blend of local, regional, and global carrying capacities. In some cases, dependence on local carrying capacity predominates. In others, like modern major cities, dependence on regional and global carrying capacities predominates.

But regional and global carrying capacities do not provide immediate and direct feedback to local populations. The fact that one ecosystem is being sucked dry to provide a city several hundred kilometers away with water is not immediately obvious to the city dwellers. The average citizen may not become aware of the regional carrying capacity limitations of water until the very last minute, when it is too late to make easy adjustments to a significant reduction in water supply. Consequently, unless some way is established to make populations at the local level fully aware of how their use of resources affects other areas of the planet, it will be difficult, if not impossible, to achieve sustainable human ecosystems supported by regional and/or global level carrying capacities.

Each community must define the sustainable activities for its particular environmental settings. This is because what is sustainable in one environment may not be so in another environment. The shade created by the canopy of a large sugar maple (*Acer saccharum*) sustains the forest environment of an old-growth forest. If that tree were to be transplanted to an open field, however, its shade would greatly alter the environment, disrupting the lives of the species living there. Population size also needs to be considered. Although one person urinating behind a tree in a 1-acre wood lot is not a problem, 100,000 people peeing behind trees on the same lot is definitely not recommended.

The question of how many people a given area can sustain is thus complex. We first must decide on the balance of humanity and nature desired; this then helps to establish the relative blend of local, regional, and global carrying capacities required. It may be best to limit population to levels at or near the lower limits of the chosen carrying capacity model. This allows for the population to be sustained during periods of stress. If population levels are planned to be at or near the maximum limits of carrying capacity, no "give" is available when the carrying capacity declines in adverse conditions; there could be widespread suffering and death, as systems fail to meet the needs of a human population far too large to sustain at that point in time.

1.8.6 CONCLUSION

Sustainable ecosystems must be flexible and dynamic; capable of evolving in response to changes in the environment. They must also be sensitive and responsive to the feedback from the complex web of local, regional, and global carrying capacities that supports them. It is essential that feedback from regional and global level support systems be perceived and responded to at the local level.

Although social and cultural dimensions are beyond the scope of this essay, human societies will also have to adjust to living within a sustainable context, and this will not be an easy transition. For instance, if it is determined that only 100 people can be sustained in a given area, questions such as "Who gets to live there?" are certain to arise. And the right of property owners to develop their property as they wish must be meshed with the establishment of sustainable human-based ecosystems. In order for sustainable systems to work, it is necessary to have the cooperation of the people who will be a part of the system. Without a doubt, a new mind-set is required, perhaps one that views living with limits to be more creatively challenging than simply striving to surmount all limits.

It is interesting to note that a successful migration to space and other planets is predicated on our learning how to live sustainably on this one. We cannot leave earth without the support of an ecosystem-based, life-sustaining technology. Permanent human settlements on other planets will be possible only if we design and transport viable ecosystems off-world. Thus we will definitely not be alone when we migrate to other planets; we will be accompanied by a host of critters from the home planet. And it will be the combined actions of all of these creatures, great and small, which will create the ecosystems needed for the viable existence of earth's life on other worlds.

1.9 Two Cultures — Or Three Filters?[2]

BY GARRETT HARDIN
Professor Emeritus of Human Ecology
Department of Biological Sciences
University of California, Santa Barbara

In 1959 C. P. Snow (1905–1980), novelist and scientist, in his famous lecture *The Two Cultures* (Snow, 1993), sought to persuade contemplative people to reexamine the abstraction called "the intellectual." As one who had worked with both words and scientific apparatus Snow questioned whether competence with only the first of these tools was enough to justify the honorific title.

Snow's forceful attack evoked a counter attack. The well established literary critic F. R. Leavis fiercely defended the established view of intellectuality. Contemptuous of Snow's ability as a novelist, Leavis was unwilling to admit that the physical scientist was a true intellectual.

We now realize that Snow was examining the world as an anthropologist or a historian might, seeking to distinguish the principal varieties of actors in the human drama. Every animal faces a world that is rich in diversity; its survival depends on quickly distinguishing friends and foes. Discrimination must be made where there is never total certainty. In the deepest sense, speedy discrimination is an economic, sometimes even a vital, necessity. Among all active animals categorizing is the order of the day. Man is an animal.

But man is, as Aristotle said, a thinking animal. In recent times he has come to distrust his categorizations. All too frequently the motives of the categorizers are perceived to be contaminated by prejudice, bias, or bigotry. An external observer — the hypothetical Man from Mars, say — might be thoroughly objective; but where is the Man from Mars?

Like most people trained in the sciences I am uneasy about what I see as the excessive weight given to words by Leavis and his like. This emphasis has of course ancient roots. I can best state my position by repeating a passage I presented to a symposium of religionists several years ago:

> A scientist cannot accept the orientation of the first sentence of the book of John: "In the beginning was the Word, and the Word was with God, and the Word was God." No doubt this statement can be interpreted in terms of symbols, parables or myths, but all such substitutes for real propositions are ambiguous. Scientists are more attracted to the motto of the Royal Society of London: Nullius in verba. If I were charged with altering Scripture to conform with science I would say: "In the beginning was the World, which everywhere and forever envelops us; against this external reality all human words must be measured. (Hardin, 1990)

---

[2]This section has been reprinted with permission from *The Social Contract,* Vol. 9, No. 3, Spring 1999, pp. 139–145.

Adopting the Biblical position automatically generates a dispute with the Leavises of this world. The unreliability of words is well illustrated by a story from the life of Arthur O. Lovejoy, the great 20th century historian of ideas. Appearing as an expert witness in a court case he was asked to take the routine oath, "Do you . . . so help you God?" Lovejoy politely pointed out that before he could respond to the question he would have to know what kind of a God the questioner had in mind? Opposing counsel casually asked what kind of God he thought was possible. Lovejoy said he was so glad he had been asked, because this was a subject to which he had given considerable attention. There are many kinds of Gods, he said, even within the Christian religion. There is God the First Cause, God the Prime Mover, the God Who Answers Prayers, the God who . . . and so on.

Legal tradition dictates that an attorney, having raised a particular point, cannot keep the witness from flooding the record with his reply. Lovejoy went on until he had made (as I recall) 37 distinctions and then sweetly asked: "Which one of these Gods does the court have in mind?" The attorney threw up his hands in despair and urged that the trial move on to substantive issues.

The moral should be clear: there is no unambiguous relation of words to reality. Conventions rule; and we cannot expect that working with words-only will answer our questions about the world outside the words. Often non-verbal language is more intelligible than words. In a popular lecture given in 1883, the English physicist Lord Kelvin (1824–1907), discoursed on a language much used by scientists, namely the language of mathematics:

> I often say that when you can measure what you are speaking about and express it in numbers you know something about it; but when you cannot measure it, when you cannot express it in numbers, your knowledge is of a meagre and unsatisfactory kind: it may be the beginning of knowledge, but you have scarcely, in your thoughts, advanced to the stage of science. . . . (Thompson, 1891)

At this point we pause to name the intellectual stages discussed so far. The first stage, as illustrated in the quotation from the Bible, can be called the literate stage, using the word *literacy* and its derivatives to cover words transmitted by whatever medium (not just inked letters on paper). An intellectual of the Leavis stripe is said to be literate. The intellectuality of Kelvin (and other scientists) is now referred to under the rubric of *numeracy*, the ability to think and communicate with numbers. According to the *Oxford English Dictionary* the second word was coined by a royal commission in the very year of Snow's lecture. After 1959 the words numerate and numeracy appeared with ever greater frequency in general discussions.

The thrust of Snow's analysis makes us question the legitimacy of an important human activity: categorizing people into cultural groups. Doing so invites dissent because categories are only statistically valid. Moreover, the name of almost any category carries an emotional baggage: think of the uproar created by the chameleonlike terms "liberal" and "conservative."

Being pigeonholed is usually resented by the subject; but the passion for pigeonholing *other* people is a human failing. (Think of the joys of hurling at one's opponents the epithets *Bigot! Xenophobe! Racist!* and *Nativist!*)

Quite an amusing brouhaha resulted from the Snow–Leavis controversy; we can now see that the differences uncovered had more important functions than that of name-calling. But before going further we need to introduce one more thread into the argument.

In the last half of the 19th century the science of ecology was christened. Its definition was initially confined to the biological sciences; about a century later people interested in characteristically human problems broadened its meaning. The names "economics" and "ecology" are both derived from the Greek root *eco-,* meaning home, house, household. Despite the widespread prestige of economics, 20th century ecologists ultimately asserted that their specialty had a license to deal with all household problems of the human species, including both economic and environmental ones. That position made academic economics no more than a specialized corner of ecology. Needless to say, card-bearing economists did not agree.

That something was missing from official economics had been noticed at least as far back as *Gulliver's Travels.* In the fantasy land of Laputa, Jonathan Swift told how mathematical analyses produced ill-fitting clothes for the Laputans, while "their houses were very ill built . . . without one right angle in any apartment. . . . I have not seen a more clumsy, awkward and unhandy people. . . ." (Swift, 1726). Thus did Swift ridicule scholars who thought that purely numerical analyses could solve all human problems. That was in 1726.

Then in 1962 a biologist employed by the American government, Rachel Carson, gave an elegant presentation of human ecology in her book *Silent Spring,* which was both highly praised and passionately damned. In the generation that followed, well-established economists mounted one defense after another, the contents of which suggested that the thrust of Swift's satire was not mere fiction. A bouquet of their astonishing statements follows. (Unless otherwise noted, each quotation is by an established economist or a recognized spokesman for the group.)

> The world can, in effect, get along without natural resources.
> — Robert Solow (1974)

> There is no danger from the exhaustion of physical resources.
> — Peter T. Bauer (1981)

> In the received paradigm, economic growth can, in principle, continue indefinitely without resource constraints.
> — Allen V. Kneese (1988)

> The so-called Law of Diminishing Returns only operates in the absence, on the one hand, of significant technological changes, and on the other of significant social changes.   — Colin Clark (1985)

> On average, human beings create more than they use in their lifetimes. It has to be so or we would be an extinct species. This process

is, as the physicists say, an invariancy. It applies to all metals, all fuels, all food, all measures of human welfare. It applies in all countries. It applies in all times.   —Julian Simon, a merchandising and sales specialist (1994)

The United States must overcome the materialistic fallacy: the illusion that resources and capital are essentially things which can run out, rather than products of the human will and imagination which in freedom are inexhaustible.   — George Gilder, popularizer of economics (quoted in Daly and Cobb, 1989)

To these individual confessions should be added the testimony of Constance Holden, for many years a trusted reporter of the weekly *Science* magazine. In a planned encounter of economists and environmentalists at the World Bank, ". . . the economists at the meeting rejected the idea that resources could be finite. Said one: 'The notion that there are limits that can't be taken care of by capital has to be rejected.' Said another: 'I think the burden of proof is on your side to show that there are limits and where the limits are.' They were suspicious of well-worn ecological terms such as 'carrying capacity' and 'sustainability.' Said one: 'We need definitions in economic, not biological terms.' "

These objections by the economists take us to the heart of the economics/ecology controversy, the "burden of proof" issue. It is clear that economists assume that unlimited growth is normal and does not call for evidence. For an applied science that claims to be erected on a mathematical framework, it is astonishing to think that, since Adam Smith (1776), the evolution of this applied science has brought it into a domain where infinity is a legitimate operator in its equations. Early on, professional mathematicians forbad dividing by zero because it produces "infinity," the operational meaning of which has long been regarded as illegitimate. But not, apparently, in economics.

The frequent assertions of the economists quoted above cannot be reconciled with the traditions of the natural sciences. If Newton had denied the constancy of gravity he could never have successfully described our world. (The slight modification later called for by Einstein created a new constant entity with the dimensions of both mass and energy.) In biology, the variability of living things derives from a sort of constancy of hereditary elements.

When they wrestle with fundamentals scientists have to stop at some point, and say: "Thus far and no farther with our doubting." It is quite true that we have made mistakes in the past when we settled for "self-evident truths," "common sense," and the like, but unlimited doubt creates an infinite regress that leaves no firm foundation for further advances. Einstein expressed what may be the best way to terminate an infinite regress. His biographer wrote of him: "When judging a scientific theory, his own or another's, he asked himself whether he would have made the universe in that way had he been God" (Hoffman, 1972).

A modern term is needed for whatever mental entity we use to limit an infinite regress. Semantic coloring should not invest the chosen name with too much power. Of the suggestions made to date I like best the term, *the default position* (Hardin, 1991. After introducing "default" in the Abstract I then — unwisely, as I now think — used a different term in the body of the text.) This term implicitly lays the burden of proof on any contradictory position. Constancy (with no creation and no destruction) is generally taken as a default position in the sciences. The laws of thermodynamics are default positions. [Ecological applications of this new term are discussed at length in *The Ostrich Factor* (Hardin, 1999).]

Conflicts between economists and ecologists could have been aborted forty years before *Silent Spring* had the economists paid attention to the criticisms of Frederick Soddy (Kneese, 1988). The English chemist pointed out that the theory embedded in neoclassical economics in effect presupposes a perpetual-motion machine, a material form of infinity forbidden to science. Yet when, half a century after Soddy's critique, the multiply-authored work, *The Limits to Growth,* came out, Wilfred Beckerman expressed the opinion of his comrades in economics by calling it an "impudent piece of nonsense" (Perrings, 1987). Why "impudent"? Because it refused to make the economically fashionable assumptions of worldly infinities.

The standard conclusion of traditional economists pleases John Q. Citizen. Default positions, which tell us there are limits to what we can do, are seldom welcomed. ("How pessimistic!" says JQC.) After some ten thousand years of civilization, magic is still welcomed into millions of minds. Most daily newspapers in the United States still carry a regular column of astrological counsel; the flavor of the advice is certainly optimistic. So also are most of the orations given on ceremonial occasions: the speakers accept without question perpetual-motion machines proposed by the neoclassical economists. As evidence, consider a commencement address given by a philosopher at St. John's College:

> It is now common knowledge in the farthest corners of the world that hunger, sickness, nakedness, and homelessness — all those symptoms of the economy of scarcity under which we have all lived — can by the proper multiplication and distribution of science and technology be abolished from the earth. — Scott Buchanan (1982)

I have augmented C. P. Snow's two paths to a culture (the literate path and the numerate path) with a third one, the *ecolate* path. Should someone write a new essay entitled "The Three Cultures"? I think this would be counterproductive: we already have enough stereotypes to keep our repulsive passions well supplied. Our greater need is to become more aware of the different languages we use in our attempt to surround and express the truth. It is now clear that as the human mind processes the inputs from experience it uses three different filters, each connected with its characteristic question (Hardin, 1989).

Literacy: What are the appropriate words?
Numeracy: What are the operational numbers?
Ecolacy: And then what?

More consistently than the first two filters, the ecolate filter is focused on time and the probable consequences of a proposed action. Ecolacy presumes a consequentialist ethics (which is often at odds with the motivational ethics produced by earlier, and predominantly literate, intellectuals) (For views of consequential ethics in the pre-ecolate stage, see Scheffler, 1988). As human beings become ever more crowded together the need for a mastery of these three "languages" grows. Can we bring this about?

As one who was heavily involved in undergraduate education for more than three decades I have some suggestions to make. In the struggle to preserve the best of our civilization we have to take as our target those citizens who are most in need of a "general" education, namely the future teachers in primary and secondary schools, and the future journalists who will fashion the news reported in the press and in the electronic media. The informal fraction of the education of these two groups takes place outside the university and college curricula. Only the formal part is subject to academic influences. The principal emphasis of "general" courses should be on the needs of future citizens rather than those of students headed for academic specialties. In the past, successes in this area have been rather disappointing. Why? For the answer we need to plunge deeply into human behavior.

Reward determines behavior. In a world of living elements this is a universal default position. In any particular class of interactions, the determination need not be 100 percent effective: anything over 50 percent will make it determinative of the results. In the field of biology, natural selection is the overwhelming example of the operation of this default mode. Among domesticated animals and plants, rewards set up by animal and plant breeders determine what a strain will be like. In the economist's world of the free market, consumer preference determines the prosperity of human "producers."

What about the academic world? Many taxpayers are under the impression that teaching performance is the primary behavior that is rewarded in universities. Those who are close to the academic world know that this is not so — particularly not so in universities, which set the style more than do colleges. As far as abilities are concerned it is fortunate that great teaching ability and great research ability are not mutually exclusive. (This is shown by the numerous individuals who are very good at both.) But in universities generally, the criteria for filling academic slots result in research ability being rewarded more generously than teaching ability.

An academician's chance of advancement is largely determined by how his research accomplishment is perceived at other institutions. (Teaching reputations are less well known at a distance.) Since there are only 24 hours in the day, the perceptive and ambitious academician gives research more attention than teaching. *Rewards determine behavior* — Q. E. D.

Actually, rewards in academia are a mixed bag, and many professors exert themselves mightily in educating the next generation. "General" courses are devised to tell all the non-X students what they need to know about X, if they are going to be well balanced citizens later. But it is easier for a teacher to speak one language rather than several, and so the professor's specialty is apt to play the pirate in a course dedicated to more extensive goals.

The results of this unplanned selectivity is tragic: the neglect of the needs of the vast majority of students who stand most in need of a general education. This group includes those students who will one day be most responsible for training yet another generation of citizens. Tragedy compounded!

The two most important sub-groups are the primary and secondary teachers, and the journalists. The life stories of the two professions differ. For a long time teaching (in our part of the world) has been professionalized and esoteric. Administrators in the schools of education have a dictatorial clamp on the criteria for certification procedures. At a lower level the everyday workers, joined into teachers' unions, furnish the political power that keeps uncertified teachers out of the classroom.

University faculty who may take their advising duties seriously soon learn a sad fact: the education department or school at most universities has a reputation for general incompetence. Time after time, faculty advisors observe that the student who is no more than marginally competent at intellectual tasks ends up by becoming an "ed major." (Student scuttlebutt tells him that he will have an easy job there.) The final result was cruelly satirized by George Bernard Shaw almost a century ago: "He who can, does. He who cannot, teaches" (Shaw, 1903). Of course, this is only a statistical truth: a few schools of education give a quality education, but the profession suffers from the majority. Extensive reform would be frighteningly difficult, but the welfare of future generations is at stake.

The other sub-group, with perhaps even more influence on the future, is the fraternity of journalists. Our schools of journalism may be as bad as education departments, but journalists have not succeeded in producing a closed shop. Most practicing journalists did not come out of journalism schools, they just wandered into the occupation. Some are self-educated to a high level. But numeracy is still often mocked by the highly literate; the results show in the press.

As for ecolacy, this modern development is seen as threatening to many of the vested interests of society; and these, through the power of advertizing budgets, exert a powerful suppressive effect on ecolate analysis. The end result is that our civilization seems determined to get along with a single language, literacy — with numeracy and ecolacy being marginalized.

Can we survive? Well, who is we? If that great romantic dream of One World becomes — temporarily — a reality, the inevitable dysfunctions of scale probably preclude the possibility of universal reform. If, on the other hand, our world continues to be subdivided into many separate sovereignties (as it is now), there is at least a possibility that one or another of the semi-independent units may some day become truly tripartite in both thinking and communication. One such local success might then inspire others to follow the example.

Croplands are natural capital, which along with social capital and economic capital determine the quality of life on earth. Here, a farmer is shown spraying herbicides. Until a majority of consumers make a clear choice to pay more for organic produce, this lower-cost method of maintaining a crop will remain a common sight. PHOTOGRAPH COURTESY OF CORBIS.

CHAPTER 2

Business

2.1 Business Firms in the New Millennium

A **business firm** is an organization that acquires necessary resources and organizes those resources to produce and sell goods and resources. Firms continue to exist by getting the most out of their scarce resources by operating efficiently and effectively. Corporations are the best known and most used types of business organizations. Other types of organizations that are active in the marketplace are not-for-profit firms, partnerships, and government-owned firms.

The purpose of a business firm is to create value for all of its stakeholders. The firm strives to create new wealth for its shareholders, valuable products and services for its customers, and jobs and wages for its employees. At the same time it must strive to serve its community and create valuable benefits for the environment (or at least avoid creating negative effects and costs).

The **managers** of a business firm are persons who allocate the human and material resources and direct the operations of the organization.

The leading business firms in the new millennium are innovative companies that are doing what it takes to be a leader. They are companies that realize that staying ahead means successfully competing against and outperforming all other companies in their industry. They do not view the competitive market as a force beyond their control. They understand that leading companies create and manage their markets. They provide convenience and quality for their customers and suppliers. They recognize the importance of coordinating exchange and the need to develop effective supplier and distributor networks, and they provide their customers and suppliers with superior transaction services.

A firm leads its markets through effective innovation, sound management of resources, and a solid strategy. Being a market leader is attained by building and managing intellectual capital, excellent quality products, and continual attention to customer needs and wants.

The firm's performance is measured in terms of economic profit. This means the net present value of the company's cash flow, discounted at the appropriate rate of interest. This **discount rate** is the company's cost of capital, which reflects the riskiness of its business. Firms seek to maximize and grow profits. Profit maximization does not mean earning the highest possible profit in any given year, but rather earning the highest net present value of a stream of profits. This takes into account the time value of money. Moreover, looking at net present values of cash flows over a period of several years allows a company to take advantage of long-term opportunities.

2.2 Corporate Environmentalism

During the past 50 years, business corporations have adopted different strategies for addressing environmental concerns. During the 1950s and 1960s corporations tended to view environmental issues as a cost of doing business and used their internal units to address these issues as problems to be solved. This response was often reactive and hesitating. Many firms held the view that the role of business was appropriately in the seeking of economic value and viewed environmental impacts as externalities and beyond their purview. The tendency was for firms to be self-reliant based on technological self-confidence. Typically, environmental issues were handled by an operating (product) unit.

During the 1970s, regulatory standards grew in number and impact and business response tended toward compliance with the advent of the Environmental Protection Agency (EPA). The EPA became the issuer of rules and standards. Corporations became defensive; the EPA was perceived as being engaged in overregulation. Corporations established separate environmental departments during this decade. The goal was to minimize liabilities to the firm.

During the 1980s environmental advocacy and watchdog groups emerged as powerful voices. At this point, industry moved toward a goal of attaining social responsibility. Corporations were actively engaged in the development of responses that would reduce environmental impacts with the goal of retaining their reputations as good citizens (Hoffman, 1997). Corporations also have responded to the fact that there is a lack of equitable economic benefits, recognizing that one-fifth of humankind is well off, while the remaining four-fifths are almost as poor as ever.

During the 1990s corporations adopted a proactive stance by inaugurating recycling and waste reduction programs. The corporate environmental management department rose to new levels of power and leadership.

During the 1988–1998 period, a period of solid economic growth in the United States, carbon monoxide in cities' air was down 38%, nitrogen oxide down 14%, ozone smog down 19%, particulates down 26%, sulfur dioxide down 39%, and lead down 67% (Reilly, 1999).

As corporations move into the first decade of the new millennium, many expect them to work toward a full integration of economic and environmental goals by incorporating those goals in new products and resources as they are designed and built. The evolution of corporate environmentalism is summarized in Table 2.1.

Environmentalism is becoming a strategic issue and is being incorporated at the insurance liability, process and product design, and competitive strategy levels. Corporations will seek managers and engineers who knows the environmental implications of the job, not an environmental specialist who knows the implications of environmental management marketing and human resources. These managers and engineers will be called on to integrate environmental issues with physical design and process factors. Environmental issues cut across departmental lines and knowledge of environmental concerns will be a requirement for the management and technology departments.

Table 2.1 **History of Corporate Environmentalism**

Stage	Decade	Description
Industrial environmentalism	1960	Internal problem solving
Regulatory compliance	1970	Compliance with government regulations
Social responsibility	1980	Impact reduction and protection of reputation
Strategic environmentalism	1990	Recycling and waste reduction
Sustainable environmentalism	2000	Integration of business goals and environmental goals

2.3 The Three Factors of Quality of Life

The quality of life on earth is dependent on three factors as illustrated in Figure 2.1. The quality of life in a society depends on equity of liberty, opportunity, and health and the maintenance of community and households — this can be called **social capital** or social assets. The growth of the economy and the standard of living for people is a critical need for all people and we call this **economic capital** or economic assets. Finally, the environmental quality of a region or on a global basis can be called **natural capital** (as defined in Section 1.1). The interrelationship between these

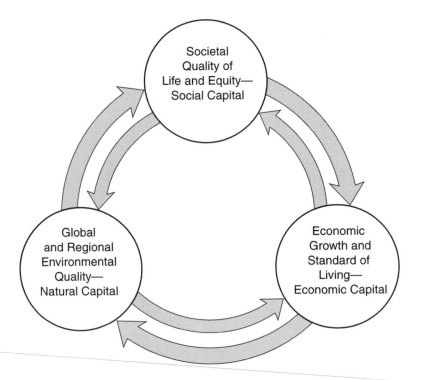

Figure 2.1 **The three factors that determine quality of life on earth.**

three factors adds up to the total quality of life. Quality of life includes such basic necessities as clothing, shelter, food, water, and safe sewage disposal. Beyond that, quality of life includes access to opportunity, liberty, and reasonable material and cultural well-being.

Business, government, and environmental leaders will need to build up capabilities for measuring and integrating these three factors and using them for decision making. One author aptly calls the sum of these factors the **triple bottom line** (Elkington, 1998).

As we strive to treat nature and society respectfully while enhancing people's quality of life, corporations strive to use nature only for what is needed and in balance with what can be recycled or replenished.

Recognizing the interconnectedness and interdepence of all living things, corporate leaders seek a balance using the triple bottom line concept. Economics, ecology, and society can be portrayed as a whole as shown in Figure 2.2. The four points of the compass are person, corporation, cultural values, and community. Decisions made by corporations or society need to account for all factors in the circle.

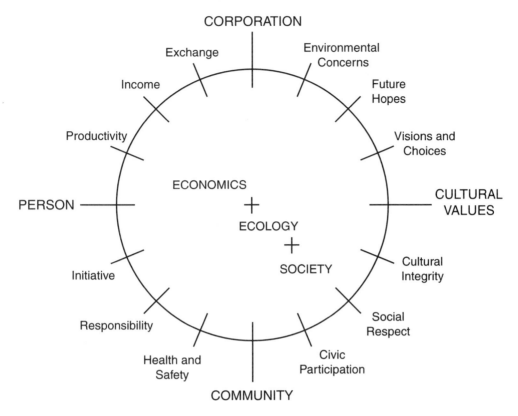

Figure 2.2 **The interdependence of economics, ecology, and society.**

2.4 It's Easier to Say Green Than Be Green[1]

BY LESTER B. LAVE* AND H. SCOTT MATTHEWS[†]

*Professor of Economics
Carnegie Mellon University

[†]Doctoral Candidate
School of Industrial Administration
Carnegie Mellon University

When it comes to environmental awareness, many big companies talk a good game. They issue glossy reports describing Earth Day celebrations and internal recycling activities and expressing strong commitments to preventing pollution. They endorse high-minded proclamations such as the Valdez-CERES Principles, which mandate that a company "sell products that minimize adverse environmental impacts," and the Business Charter for

---

[1]This section has been reprinted with permission from *Technology Review,* November/December 1996, pp. 68–69.

Sustainable Development, which entreats companies "to modify their operations to prevent serious or irreversible environmental degradation." But how does this expressed concern translate into willingness to raise product costs in order to improve environmental quality?

Not very well, it seems. To probe the real-world limits of corporate altruism, we surveyed 54 large U.S. companies — all of which have expressed environmental concerns through actions such as publishing an environmental report. The survey described the following hypothetical situation: Suppose that a material in one of the company's products were found to in some way harm the environment. Further suppose that a nontoxic substitute material were available. The substitute yields exactly the same product quality — but it costs more. How much more, we asked, would the company be willing to pay for this environmentally nontoxic material? Given the nonrandom nature of the survey, we urge caution in generalizing the results. Still, the answers offer insights into corporate priorities for the 25 companies responding.

We posed this question for several scenarios, in which the toxic material accounted for varying percentages of the product cost, and in which the substitute material carried a varying cost premium. When the toxic material constituted 1 percent of product cost, almost all respondents were willing to switch to a safer material that was only 1 percent more expensive. In other words, almost all companies were willing to raise product costs 0.01 percent (for example, $2 on a $20,000 car).

We were surprised to find how quickly this willingness to pay more to protect the environment fell off. Only two-thirds of the respondents answered that they would be willing to make a substitution that raised product costs 0.1 percent. One-third of firms said they were willing to raise costs 1 percent. Two companies of the 25 said they were willing to raise costs 5 percent. No company was willing to raise costs any further.

Admittedly, the hypothetical situation is vague; it doesn't spell out exactly how much damage to the environment might be averted by switching materials. The conclusion nevertheless seems clear: despite their PR, companies will take measures to reduce pollution only as long as they do not involve much effort or cost.

Remember — the companies surveyed all had publicly professed to be environmentally aware. If they are unwilling to increase costs more than a trivial amount to prevent pollution, what can we expect from firms that *don't* trumpet their ecological concern?

We are confident that more businesses would engage in environmentally friendly behavior if they perceived that such behavior would lead to favorable publicity with the potential of higher market share. Public opinion surveys, our discussions with company executives, and the massive business participation in efforts such as the President's Council for Sustainable Development indicate society's desire for pollution prevention.

In Northern Europe, in particular, consumers gravitate to products that they perceive as causing less pollution. Opel, for example, offered automobiles with U.S. tailpipe emissions-control at a price premium at a time when Germany did not require stringent controls. Opel's market share increased, apparently because German consumers desired a greener car despite the higher price. Managers on this side of the Atlantic believe that few U.S. consumers would respond similarly. In highly competitive industries, where profit margins are low and small differences in price are sufficient to take over the market, it would be suicidal for a firm to raise product costs unless consumers perceived the green product to be superior.

Companies respond to what they perceive consumers want. Some of our respondents, however, expressed frustration that they were unable to tell potential customers about the improvements they had made in their products. The Federal Trade Commission, Environmental Protection Agency, and some state regulators have concocted sufficiently narrow definitions of such environmentally related terms as "recycle" and "recyclable" that companies shy away from advertising the positive steps they have taken. IBM, for example, made a computer that was highly energy-efficient and designed to be easily recycled, but the company lawyers thought that advertising these advantages would violate existing rules.

2.4.1 WHAT MAKES A FIRM GREEN?

In many cases, the measures that prevent pollution also save companies money. Pollution is, after all, waste, which signifies inefficient use of resources. Wall Street should reward firms for steps that increase profit, but a firm should have to go beyond steps that raise its profit to earn environmental recognition. We propose five attributes that reflect a company's commitment to invest in pollution prevention and green products.

- *Green consumers.* A green firm invests in understanding the green demands of its customers and the market niches for green products.
- *Technology investment.* A green firm invests in technology to prevent and clean up the pollution that its facilities generate.
- *Materials choice.* A green firm is willing to increase product costs a small amount — perhaps 5 percent — to prevent toxic discharges.
- *Redesign.* A green firm is willing to invest a portion of its product revenues — again, 5 percent seems reasonable — toward a redesign that lowers pollution.
- *Sustainability.* A green firm attempts to lower its use of resources to leave more for future generations.

In an increasingly competitive marketplace, firms must be able to benefit from their green actions. At a minimum, businesses must be able to tell their customers, salespeople, regulators, and other groups about their green actions. Unfortunately, some firms will claim they are green without taking the actions. Thus some mechanism must be put in place to certify the claims. One possibility is to create an organization to establish criteria and then to certify that products meet these criteria. The quickest way to begin these certifications would be for a government agency to consult with the affected companies, then promulgate agreed-upon criteria, and delegate product testing to a non-governmental organization.

One good model is the Energy Star ratings, created by EPA, for products' energy efficiency. A similar program could regulate claims that a product is recyclable or is made from recycled material, that it is free of toxic materials, and that the manufacturing process produced little or no hazardous waste.

Our proposed attributes are realistic. Some companies define themselves as being on the leading edge of technology, and so commit themselves to be on top of new technology and to performing R&D to develop a stream of state of the art products. Others define themselves as a source of the highest quality products, where quality is in the eyes of the consumer. Companies of both types make investments that have uncertain payoffs, but generally are successful.

By analogy, we define a green company as one that is committed to informing its engineers about green technologies and tools; to R&D aimed at lowering the discharges of toxic substances; to experimenting with the greenness of its products to elicit consumer response; to lowering the use of nonrenewable resources; and to using renewable resources at sustainable levels. We challenge companies to back up their expressions of environmental concern by making these commitments.

2.5 Natural Capitalism

In Chapter 1 we defined **natural capital** as those features of nature such as minerals, fuels, energy, biological yield, or pollution absorption capacity that are directly or indirectly utilized or are potentially utilizable in human social and economic systems. Natural capital includes the natural resources (NR) of the earth's ecosystems as well as the ecosystem services (ES) to provide habitat and regulation of atmosphere, water, and climate. Thus, natural capital (NC) is

$$NC = NR + ES.$$

The value of ecosystem services does not appear as an asset on the balance sheets of governments or corporations. Yet, one estimate of the value of the earth's ecosystem services is at least $33 trillion a year (Lovins *et al.*, 1999). Amory Lovins, L. H. Lovins, and Paul Hawken propose an approach called **natural capitalism** in which natural capital is properly valued and included in the economic analysis and determinations of business and government (Lovins *et al.*, 1999).

Natural capitalism, as envisioned by Lovins *et al.*, is based on four major shifts as outlined in Table 2.2. Examples of reducing waste are the reduction of the use of paper for a company's office communication or the use of more efficient pumps for fluid flow in a factory. The adoption of more efficient, innovative technologies will result in the use of fewer natural resources.

Table 2.2 **Four Major Shifts in Business Practices**

1. Increase the productivity of natural resources:
   - Reduce waste
   - Increase output while reducing resource input
   - Adopt innovative, efficient technologies
2. Shift to biologically inspired production models:
   - Shift to closed-loop production systems
   - Use output (waste) as input for another production system
   - Return unavoidable waste harmlessly to ecosystem
3. Move to a solutions-based business model:
   - Consider value to be a flow of services
   - Consider value to be quality, utility, and performance
   - Improve life-cycle quality
4. Do not allow waste emissions to exceed the assimilative capacity.

Lovins *et al.* (1999) summarize the opportunity for productivity improvement as follows[2]:

> Whether through better design or through new technologies, reducing waste represents a vast business opportunity. The U.S. economy is not even 10% as energy efficient as the laws of physics allow. Just the energy thrown off as waste heat by U.S. power stations equals the total energy use of Japan. Materials efficiency is even worse: only about 1% of all the materials mobilized to serve America is actually made into products and still in use six months after sale. In every sector, there are opportunities for reducing the amount of resources that go into a production process, the steps required to run that process, and the amount of pollution generated and by-products discarded at the end. These all represent avoidable costs and hence profits to be won.

A shift to biological models of production systems is based on the concept of reusing or using initially all the natural resources productively within a closed-loop system. Lovins *et al.* (1999) state:

> Every output of manufacturing should be either composted into natural nutrients or remanufactured into technical nutrients — that is, it should be returned to the ecosystem or recycled for further production. Closed-loop production systems are designed to eliminate any materials that incur disposal costs, especially toxic ones, because the alternative — isolating them to prevent harm to natural systems — tends to be costly and risky.

---

[2]Copyright © 1999 by the President and Fellows of Harvard College; all rights reserved. Excerpts reprinted by permission of *Harvard Business Review,* "A Road Map for Natural Capitalism," by A. Lovins, L. H. Lovins, and P. Hawken, June 1999.

The third stage of the shift to natural capitalism is the shift of the business model toward valuable solutions that incorporate quality, utility, and performance through a continuous flow of services.

Finally, in step 4, natural capitalism will involve efforts toward restoring, sustaining, and expanding natural capital by investing in habitat and biological resource base. An example is a lumber company that invests in its forest habitat and its reforestation practices. Reinvesting in natural capital is more profitable, in the long run, for farmers, fisherman, forest managers, and semiconductor manufacturers.

The four steps for shifting the business system can be summarized by the flow diagram shown in Figure 2.3. The first step works toward increasing the productivity of the business processes. The second step focuses on recycling waste back into the production process and the natural decomposition processes. Step 3 involves showing customers and vendors that the value of a service is the good it provides. Last, step 4 involves reinvesting in natural capital.

What are the deterrents to the development of natural capitalism? Lovins *et al.* (1999) point to lack of good measurements of performance, financial and government distortions through subsidies, and the methods of capital allocation. Many companies pay less attention to saving resources because they are a small percentage of their total costs.

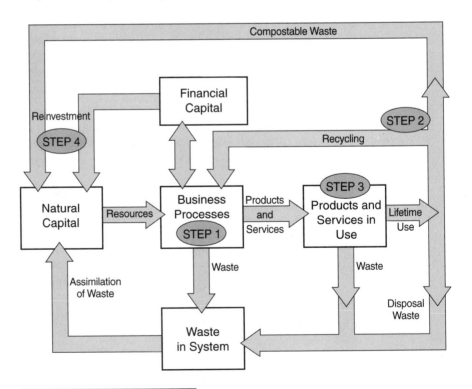

SOURCE: Adapted from "Natural Capitalism" by Peter Senge, *Whole Earth,* Winter 1999.

Figure 2.3 **The four major shifts in the business system.**

In the following summary, Lovins *et al.* (1999) point toward a recognition of the necessity of paying attention to natural capital:

> The logic of economizing on the scarcest resource, because it limits progress, remains correct. But the pattern of scarcity is shifting: now people aren't scarce but nature is. This shows up first in industries that depend directly on ecological health. Here, production is increasingly constrained by fish rather than by boats and nets, by forests rather than by chain saws, by fertile topsoil rather than by plows. Moreover, unlike the traditional factors of industrial production — capital and labor — the biological limiting factors cannot be substituted for one other. In the industrial system, we can easily exchange machinery for labor. But no technology or amount of money can substitute for a stable climate and a productive biosphere. Even proper pricing can't replace the priceless.
>
> Natural capitalism addresses those problems by reintegrating ecological with economic goals. Because it is both necessary and profitable, it will subsume traditional industrialism within a new economy and a new paradigm of production, just as industrialism previously subsumed agrarianism.

2.6 Green and Competitive

Corporations operate in the world of dynamic competition. They are constantly creating innovative solutions to concerns raised by customers and regulators while responding creatively to competitive pressures and actions. As Porter (1995) notes:

> Properly designed environmental standards can trigger innovations that lower the total cost of a product or improve its value. Such innovations allow companies to use a range of inputs more productively — from raw materials to energy to labor — thus offsetting the costs of improving environmental impact and ending the stalemate. Ultimately, this enhanced *resource productivity* makes companies more competitive, not less.

Often, companies can attain productivity increases from an innovation that simultaneously uses fewer resources and emits less waste. When environmental factors are internalized within a firm's business strategy, the firm can strive to generate positive organizational and technological benefits that accrue privately to the firm, thus, increasing its competitive advantage. As a proactive approach to environmental factors takes hold, we would expect it to redesign its product, production, and service delivery processes. Such a redesign would likely involve the acquisition and installation of new technologies. We expect that improved environmental performance to enhance economic performance. Physical resources can be a source of competitive advantage if they outperform equivalent assets within competitors.

Companies are addressing the concept of pollution (waste) prevention and resource input reduction or materials substitution. Corporations view as attractive the idea of moving toward high quality by means of reducing defects. In the same way defects can be reduced by quality product and process design. Using input resources more efficiently results in less waste.

Many companies are using innovations to command price premiums for "green" products. Many ask: If a company can increase resource productivity and reduce wastes, why have regulations at all? In a sense, regulation sorts out the priorities and provides a push toward the high-priority issues.

Few companies have fully analyzed the true cost of wastes, toxicity, and resource use. As they trace the full cost up and down the supply chain they will find opportunities to reduce costs and environmental effects. The first step, as summarized in Table 2.3, is **pollution prevention** (Hart, 1997). Pollution prevention focuses on reducing or eliminating waste before it is created. Pollution prevention pays off with reduced costs and better processes. **Product stewardship** focuses on all environmental impacts associated with a product's life cycle. Step 3 is concerned with the **design for environment**, which is a method for making products easier to recover, reuse, or recycle. Step 4 is concerned with **clean technology**, which focuses on creating technologies that are less impacting and more sustainable.

Table 2.3 Steps for Environmentally Responsible Corporations

1. Pollution prevention — eliminate waste.
2. Product stewardship — reduce environment impacts throughout a product life cycle.
3. Design for environment — design products that are easier to recover, reuse, and recycle.
4. Clean technologies — use sustainable, less impacting technologies.

As an example of an industry that can follow the four steps shown in Table 2.3, consider the automobile industry. Step 1 involves a production system with lower wastes such as scrap steel and paint. Step 2 involves cars with fewer materials and lower pollution emissions over the life of the car. Step 3 involves designing cars that may be disassembled after full use so that parts can be reused or recycled. Step 4 encourages the use of lower impact technologies such as more fuel efficient engines or alternative power devices such as fuel cells.

Another excellent example of step 4 is the concept of the Monsanto new leaf potato, a new type of potato that eliminates the need for insecticides to help it resist the potato beetle (Magretta, 1997). Monsanto has used biotechnology to create a product that wastes less and is more productive. Robert Shapiro, CEO of Monsanto, has stated (Magretta, 1997):

I offer a prediction: the early twenty-first century is going to see a struggle between information technology and biotechnology on the one hand and environmental degradation on the other. Information technology is going to be our most powerful tool. It will let us miniaturize things, avoid waste, and produce more value without producing and processing more stuff. The substitution of information for stuff is essential to sustainability.

Monsanto has seven teams working on their process, products, and opportunities as summarized in Table 2.4. This is an excellent example of a corporation attempting to do good and do it well.

Ten principles of sustainable business practices are provided in Table 2.5. This is a summary of the Coalition for Environmentally Responsible Economics (CERES) principles endorsed in 1992 by several leading global corporations (see *www.ceres.com*).

Table 2.4 **Monsanto's Green Teams**

1. Eco-Efficiency Team — Maps and measures the ecological efficiency of processes
2. Full Cost Accounting Team — Responsible for life-cycle costs including environmental costs
3. Index Team — Establishes criteria for measuring moving toward the triple bottom line
4. New Business/New Products Team — Examines value in the market and ecological stress of products
5. Water Team — Looks at global water needs
6. Global Hunger Team — Examines the development of technologies for alleviating world hunger
7. Communications and Education Team — Trains all employees about a framework for sustainable business

Table 2.5 **Ten Principles of Sustainable Business Practice**

- Protect the biosphere and biodiversity through reducing pollution
- Promote sustainable use of natural resources
- Reduce waste
- Conserve energy
- Reduce environmental, health, and safety risks
- Design safe products and services
- Restore environmental damage
- Inform the public of environmental conditions
- Ensure management commitment
- Self-audit annually

Corporations can cooperate with nonprofit organizations to mitigate the effects of environmental damage. In a recent article, the late John Sawhill of the Nature Conservancy provided an example of a cooperative effort with a corporation to find a solution to offset environmental damage. In 1992, the Walt Disney Company wanted to build a wild animal park near Orlando, Florida. Florida regulators were concerned with damage to wetlands. The Nature Conservancy, Disney, and local, state, and federal agencies worked toward a mitigation solution that helped the environment while enabling Disney to develop its theme park. Sawhill described it as follows (Howard, 1995):

> In exchange for permission to develop the Orlando site, which will affect about 340 acres of wetlands over a 20-year period, Disney agreed to purchase, protect, and restore 8,500 acres of wetlands and wilderness in central Florida. Disney will donate this land in phases to the Conservancy and provide an endowment to make sure that we can continue to operate it. By the way, these kinds of mitigation agreements will become more and more popular in the future.

Business is wisely considering a shift from antagonism toward environmentalism and is attempting to find new alliances for joint progress. Table 2.6 provides a summary of a new paradigm for sustainable business.

2.7 How Much Environmental Damage Is Too Much? Economic Perspectives[3]

By Eban Goodstein
Associate Professor
Department of Economics
Lewis and Clark College

Global warming will pose a serious threat to many natural ecosystems, and to the welfare of hundreds of millions of people around the planet — especially those in low-lying countries, and poor countries, where agricultural systems are dependent on a predictable and stable climate (IPPC, 1996). At the same time, fighting global warming by switching to nonfossil fuel energy sources will be costly. If we stay with business as usual, over the next century CO_2 levels in the atmosphere will rise to double their preindustrial level — 560 parts per million — and temperature is expected to rise by between 1.5 and 6.5°F. Given this, what should we, as a global society, do?

---

[3]This section has been excerpted with permission from *The Encyclopedia of Global Environmental Change* (Ted Munn, ed.), New York: John Wiley and Sons, 1999.

Table 2.6 **The New Paradigm for Sustainable Business**

Old paradigm	*New paradigm*
1. Financial bottom line	Triple bottom line
2. Improvement	Innovation
3. Shareholders	Stakeholders
4. Resource exploitation	Resource use reduction and recycling
5. Rights of the corporation	Responsibilities
6. Limited responsibility	Stewardship through the life cycle
7. Wasteful	Reducing waste
8. Control of resources	Stewardship
9. Economic growth	Sustainability and growth
10. Environmental compliance	Design for the environment
11. Environmentalists as antagonists	Collaborate with environmentalists

Accept the business-as-usual scenario? Attempt to stabilize carbon dioxide levels at their current level of 360 parts per million? Or seek a target somewhere in between?

Confronted with any environmental problem — whether toxic emissions from a local incinerator, or the vast conundrum presented by global warming — this "How much is too much?" question necessarily forms the starting point for analysis. Since pollution and resource degradation are a by-product of material production, and human welfare depends on both material production and a livable environment, what is the right trade-off? Of course, such a question is a normative one; as such it has no objectively right or wrong answer. As we will see below, although economists generally agree both on a common ethical framework, and a particular definition of sustainability as a desirable social goal, there is significant disagreement about how much pollution and resource degradation society should tolerate if we are to achieve a sustainable economy. Some support the use of benefit–cost analysis to guide resource development and pollution decisions; others prefer a stricter precautionary principle.

2.7.1 THE ETHICAL FRAMEWORK OF ECONOMICS: UTILITARIANISM

As a starting point, economic analysts are concerned with *human* welfare or well-being. From the economic perspective, the environment should be protected for the material benefit of humanity and not for strictly moral or ethical reasons. To an economist, saving the blue whale from extinction is valuable only insofar as doing so yields happiness (or prevents tragedy) for present or future generations of people. The existence of the whale independent of people is of no importance. This human-centered (or anthropocentric) moral foundation underlying economic analysis, which has as its goal human happiness or utility, is known as utilitarianism.

Some environmentalists are hostile to utilitarian arguments for protecting the environment. Indeed, an economic perspective on nature is often viewed as the primary problem, rather than part of the solution. The philosopher Mark Sagoff (1995) puts it this way:

> . . . the destruction of biodiversity is the crime for which future generations are the least likely to forgive us. The crime would be as great or even greater if a computer could design or store all the genetic data we might ever use or need from the destroyed species. The reasons to protect nature are moral, religious and cultural far more often than they are economic. (p. 618)

The focus on anthropocentric, utilitarian arguments by economists is not meant to discount the importance of other ethical views. Indeed, over the long run, nonutilitarian moral considerations will largely determine the condition of the planet which we pass on to our children and theirs. But economic arguments invariably crop up in short-run debates over environmental protection, often playing pivotal roles.

Within this utilitarian framework, sustainability takes on a particular meaning. Sustainable outcomes require that we reduce pollution (or stop depleting resources) if doing so, on balance, prevents the decline of living standards below their current level for the typical (median) member of any future generation (Pezzey, 1992).

2.7.2 MARKET SYSTEMS AND SUSTAINABILITY: EXTERNALITIES

Economists largely agree that pure free-market systems are unlikely to generate sustainable outcomes (Goodstein, 1999, Chap. 3). Laissez-faire systems fail because of the existence of negative externalities: costs generated in the production or consumption of a good which are not borne by the producer or consumer of the good. For example, when I buy a gallon of gasoline and burn it in my jet ski, I impose a variety of costs on society as a whole, ranging from the emission of local air and water pollutants to global warming gasses. Many of these costs are not reflected in the price of gasoline. They are thus "external" to the buyer and seller.

If all resources in an economy were privately owned, and environmental damages could be proved easily (with low "transactions costs"), then externalities would be internalized through private negotiation or litigation. For example, if Bill Gates owned the Mississippi River, and my jet ski was polluting it via emissions of soluble hydrocarbons, then Bill could sue me and force me to "internalize" the external costs imposed on his river. However, many important resources are not privately owned — the air, rivers and streams, oceans, most forests, deserts, and other natural habitats. Moreover, environmental damages are often spread among many parties, and difficult to prove in specific cases, creating high transactions costs for private resolution of the externality problem. Given these two features — the widespread presence of common property resources, and the high transactions costs associated with pressing court claims for damages — economists agree that free market systems will generate excessively high levels of pollution and resource degradation.

The conventional solution to this problem is to force companies and consumers to internalize externalities through government regulation. Regulation can be either prescriptive (command-and-control), in which firms are required to install particular types of clean-up technologies, or it can be incentive based. The latter approach includes both pollution taxes, and marketable permit systems. These incentive-based methods internalize externalities by forcing firms to pay for pollution, but leave the specific method of pollution reduction up to the companies themselves.

If government forces firms to pay for pollution through regulation, how much "should" firms pay? The conventional view is that businesses should pay an amount per unit of pollution equal to the damage done by the last unit of pollution emitted (the marginal unit of pollution). Thus, if the last ton of sulfur dioxide coming out of a power plant smoke stack leads to economic damages of $200, than the plant should pay $200 for every ton that it emits. In competitive markets, setting pollution prices equal to marginal economic damage in this way balances the costs of pollution control (higher prices for products like electricity) against the benefits (less damage from pollution like acid rain), leading to maximum total monetary benefits to society.

Economists have developed a variety of techniques to determine monetary values for pollution damages (Goodstein, 1999, Chap. 8). These range from the mundane (fewer sick days for workers) to the highly controversial (placing a dollar value on stroke deaths prevented, reduction in child IQ avoided, or biodiversity preserved). How can these latter types of valuations be done?

As one example, in benefit–cost studies, the value of a life saved ranges from $2 million to $10 million. These numbers comes from studies which examine the wage premium for risky employment. Put overly simply, economists have found that workers like police officers receive a wage premium of around $500 for accepting a 1 in 10,000 increase in the risk of premature death on the job. Ten thousand police officers thus exchange one of their lives for about $5 million.

One can quarrel with this line of reasoning on many grounds. And clearly, if considerations of this type dominate, then uncertainty in dollar valuations of the benefits of environmental protection can loom large. Nevertheless, economists have developed a number of tools for monetizing important components of quality of life traditionally left out of measures like GDP. This allows society to put a monetary value on pollution prevention: For example, global warming damage estimates from carbon dioxide emissions from coal-fired power plants range from around 0.5 tenths of a penny per kWh to three cents per kWh (Krupnick and Burtraw, 1996). And while numbers in this range may turn out to be the "wrong" price estimates, defenders of this kind of estimation of the dollar benefits of environmental protection respond that even the wrong dollar value is better than the default price of zero.

2.7.3 MARKET SYSTEMS AND SUSTAINABILITY: SHORT TIME HORIZONS

While the problem of externalities can, in theory, be dealt with via regulation, free market economies have another feature which may lead to unsustainable outcomes. In modern market systems, firms expect profit rates of return on the order of 15% to 20% in order to induce them to either conserve resources or to invest in new technology. With such high discount rates, projects with payoffs occurring much more than 5 years in the future seldom look profitable. Thus, from a business perspective, it seldom makes sense to preserve rain forests on the grounds that in 50 years the biodiversity it contains will be highly valued; neither are energy companies much interested in solar power as long as cheap oil and coal have at least 5-year lifetimes.

It is easy to show that these short time horizons create a situation in which future generations are not as well off as we could possibly make them — if we invested more today in conservation of certain resources or in basic R&D, then we could with a high probability raise the welfare of our descendants. This is the general rationale for government support of science and technology. However, it is not clear whether our short-term bias is unsustainable — that is, whether it actually *reduces* the welfare of future generations below the level that we enjoy today. Even with their short-term biases, market systems are incredibly dynamic in terms of the development of new technologies. It may be that the 5-year time horizon — coupled with government support for longer term research — can still insure that the well-being of future generations will not decline. This depends in turn on the degree to which new technologies can substitute for depleted or degraded natural resources and ecosystem services (Goodstein, 1999, Chaps. 6 and 7).

2.7.4 SUSTAINABILITY: THE NEOCLASSICAL VIEW

On this point, broadly speaking, two camps have developed: neoclassical and ecological. Neoclassical economists (in this context) are technological optimists. This means that they believe that in "well-behaved" market economies — those in which negative externalities have been internalized via incentive-based regulations like pollution taxes or marketable permit systems — substitutes will emerge for scarce (and increasingly expensive) natural resources and environmental waste sinks. System sustainability is thus assured by assumption.

Neoclassicals often use the example of whale oil or copper telephone wires — as these resources became relatively scarce and their prices rose, petroleum and fiber optics emerged as subsititutes. How about a harder case, like the loss of climate stability due to global warming? Neoclassicals respond, first, that as greenhouse gas emissions are regulated, the externality is internalized, and the price of CO_2-based services rise, then new low CO_2 technologies will emerge. Moreover, advances in agricultural techniques will insure adaptation to a changing climate, and dikes can be built to hold back sea level rise. Clearly there will be losers from climate change, but on balance, the argument goes, living standards for the typical person on the planet will continue to rise even in the face of "moderate" climate change.

Technically, this means the economy will display a positive rate of growth of per-capita net national welfare, the familiar GDP measure of economic output adjusted downward to reflect the costs of economic growth. If net national welfare, which balances the benefits of increased material consumption against the attendant environmental and social costs, is growing on a per-capita basis, then by definition, the well-being of future generations is also rising (Goodstein, 1999, Chap. 6).

While system sustainability is thus assured by assumption in the neoclassical framework, sustainable decisions at the micro level should be made using benefit–cost (BC) analysis. For example, when considering whether or not to build a coal plant, one should compare the discounted stream of measurable monetary costs (material and environmental) over the expected life of the plant, against the discounted stream of measurable monetary benefits (cheaper power). If the benefits exceed the costs, then building the plant would be sustainable, because doing so would raise the welfare of the typical member of future generations.

Future costs and benefits are discounted in benefit–cost analysis to reflect the opportunity cost of productive foregone investment. For example, the value of $100 received in 10 years (ignoring inflation) is a lot less than the value of $100 received today. If the interest rate were 10%, I could invest that $100 today and have $260 in ten years. Conversely, to have $100 on hand in 10 years I would need to bank only $39 today. This figure of $39 is called the present discounted value of $100 received in 10 years, if a 10% interest or "discount" rate is available. While discounting makes sense for individual financial decisions, at the social level it can generate perverse outcomes.

To illustrate, suppose we are considering installing a filter on our coal plant that will generate a one-time-only benefit of $100 million 100 years from now. At a discount rate of 10%, a $100 million benefit gained 100 years from now is worth only $7,200 today. The logic of benefit–cost analysis thus suggests that we should not spend a measly seven thousand dollars and change today to yield a benefit of $100 million to our great grandchildren. Why? Because if we put the money in the bank, future generations would have more than $100 million on hand.

The problem is that, for real-world projects subject to benefit–cost analysis, we don't intend to put net savings in a trust fund; the alternative is instead to spend it elsewhere in the economy (say, to reduce energy prices from the coal plant) where it may or may not yield a 10% rate of return to society as a whole. While discounting makes obvious sense on a personal level for short time horizons, it breaks down for actions yielding a stream of social benefits far into the future.

On the flip side, not discounting at all also generates perverse outcomes. Consider a project that yields $1 worth of benefits every year, forever. Such a project has infinite value, and would thus seem to be worth sacrificing the entire planet's output to finance. Given these two extremes, there is much debate over the proper approach to discounting future benefits and costs (Portney and Weyant, 1999).

If indeed the analysis is sensitive to the choice of a discount rate, or else tries to weigh controversial benefits such as lives saved or biodiversity preserved, then uncertainties can easily grow large enough to render determination of a (neoclassically) sustainable outcome at the project level difficult, if not impossible.

2.7.5 SUSTAINABILITY: THE ECOLOGICAL VIEW

In contrast to neoclassicals, ecological economists are technological pessimists. Tracing their intellectual lineage explicitly to Malthus, ecologicals have no faith that technological substitutes for important natural resources, such as climate stability, will be forthcoming from real-world market systems. Thus, ecologicals would reject a coal plant following the logic that CO_2 emissions are already destabilizing the global climate. Essentially, ecologicals argue that for important resources such as fresh water, UV protection, biodiversity, and climate stability, the future consequences of current resource degradation are too uncertain to countenance the use of BC analysis. Instead ecologicals argue for what they call the precautionary principle. To preserve the welfare of future generations, therefore, these unique forms of natural capital should be protected *unless the costs of doing so are prohibitively high* (Ciriacy-Wantrup, 1968; Daly, 1996). Moreover, the argument is often simultaneously advanced that the cost of protection will not be as high as is frequently claimed (Laitner *et al.*, 1998).

Along this line, ecologicals point to survey evidence suggesting that, in developed countries, broad growth in material consumption in fact buys very little increase in societal happiness. This occurs, it is argued, because beyond a basic level, utility from consumption depends on relative rather than absolute levels of consumption. To the extent that this is true, environmental quality is being sacrificed only to feed a "rat race" in which human welfare does not rise with increased material consumption — clearly an unsustainable state of affairs (Mishan, 1968; Howarth, 1996).

Which view is correct? Neoclassicals argue that the ecological position is too extreme. They insist there are trade-offs, and that we can pay too much for a pristine environment. Resources and person-power invested in reducing small cancer risks or preserving salmon streams, for example, are resources and people that cannot be invested in schools or health care. Benefit–cost analysis is needed to obtain the right balance of investment between environmental protection and other goods and services. Moreover, in the sustainability debate, neoclassicals argue that history is on their side: Malthusian predictions have been discredited time and time again.

Indeed, in recent years, prominent natural scientists (though not economists) in the ecological economics community have made stunningly bad forecasts. The problem with all these early predictions, as with Malthus's original one, was that they dramatically underestimated the impacts of changing technologies. Neoclassicals point to these failed predictions to support their basic assumption that natural and created capital are indeed good substitutes — we are not "running out" of natural resources or waste sinks.

Ecologicals respond that history is not a good guide for the future. One hundred and fifty years after the beginning of the industrial revolution, the argument is that accumulating stresses have begun to fundamentally erode the resilience of local and global ecosystems upon which the economy ultimately depends. Indeed, ecologicals have largely shifted their 1970s concerns about running out of nonrenewable minerals and oil to other resources: biodiversity, fresh water, environmental waste sinks, productive agricultural land. While ecologicals stretching back to Malthus have indeed done their share of crying wolf, this does not, of course, mean the wolf is not now at our door.

The neoclassical and ecological perspectives differ dramatically in their assessment of the likelihood of sustainable outcomes from real–world market economies. Ecologicals prefer that resource development or (environmental degradation) be regulated according to a precautionary principle; neoclassicals prefer a benefit–cost test. Nevertheless, both perspectives are "economic"; both accept the common definition of sustainability offered above; both are grounded in utilitarian philosophy; and both deal in the currency of trade-offs.

2.8 Profit[4]

BY PAUL HEYNE
Professor
Department of Economics
University of Washington

The word **profit** is regularly used in everyday speech and writing with different and even contradictory meanings. Since usage determines the meaning of words, there is no single correct definition of profit. The meaning most commonly employed today by economists was developed largely by Frank H. Knight in the 2nd decade of the 20th century as part of his effort to refine the theory of a "free-enterprise" economy (Knight, 1921).

Profit is a residual: what is left over from the undertaking of an activity. The profit from an activity or enterprise is what remains out of total revenue after all costs have been paid. Costs are payments promised by the

---

[4]This section has been reprinted with permission from *The Technology Management Handbook* (Richard Dorf, ed.), pp. 4-10–4-12, Boca Raton, FL: CRC Press, 1998. Copyright CRC Press, Boca Raton, Florida.

undertaker of an activity to the owners or effective controllers of whatever productive resources the undertaker deems necessary for the successful completion of that activity. Costs include wages promised to obtain labor services, rent promised for the use of physical facilities, interest promised for the use of borrowed funds, payments promised to purchase raw materials and other inputs, as well as taxes paid to secure the cooperation or noninterference of government agencies. The undertaker of the enterprise is usually known today by the French term **entrepreneur** because we have surrendered the English word to the undertakers of funerals. When entrepreneurs themselves own some of the productive resources employed, the cost of using them is what those resources could have earned in their best alternative opportunity, or what economists call their **opportunity cost**. Thus, what the entrepreneur could have earned by working for someone else is a cost of production, not a part of the entrepreneur's profit.

In the absence of uncertainty, there would be no profits. Suppose it is generally known that a particular activity will certainly generate more revenue than the total cost of undertaking that activity. Then more of that activity will be undertaken. This will either reduce the price obtainable for the product or raise the cost of obtaining the requisite resources or do both, until expected total revenue exactly equals expected total cost. Profit, consequently, cannot exist in the absence of uncertainty. It is a residual that accrues to those who make the appropriate decisions in an uncertain world, either because they are lucky or because they know more than others. Losses accrue to those who make inappropriate decisions.

The extraordinary productivity of free-enterprise economies has evolved as entrepreneurs have sought out and discovered procedures for producing goods for which people are willing to pay more than their cost of production. This evolution has been largely a process of ever finer specialization, both in the productive activities undertaken and in the goods produced.

Critics of the free-enterprise system have generally failed to recognize what an extraordinary quantity of information is required to coordinate a highly specialized economic system and have vastly underestimated the uncertainties that permeate a modern economy. They have viewed profit for the most part as a simple surplus obtained by those who were in the fortunate position of being able to purchase the services of resources, especially labor, at prices less than their value or worth as measured by the revenue obtainable from the sale of the products of those resources. From this perspective, private profit and the private hiring of labor have no useful social function. The state or some other representative of society can simply take over the control of productive activity and apply the ensuing surplus to public rather than private purposes, thus ending exploitation and special privileges. The organization of production is a technical administrative task in a world without uncertainty.

However, in the actual and highly uncertain world, no one knows exactly what combination of productive activities will generate the largest "surplus" under current circumstances. What precise mix of goods should be produced and how should they be produced? No one knows! In a free enterprise system, however, entrepreneurs decide. Entrepreneurs are persons who act on their belief that a particular rearrangement or reallocation will extract a larger surplus or profit from available resources and that they themselves will be able to appropriate a significant share of this increased surplus or profit. To implement their beliefs, they must obtain command of the requisite resources. Entrepreneurs do this by promising to meet the terms of those who own the resources they want to use and then claiming for themselves the residual, that is, the profit.

The entrepreneurs' promises must be credible, of course. People will not work for someone else unless they confidently expect to be paid. Owners of physical resources will not let others employ the services of their resources without assurance that they will obtain the promised rent. Lenders of money want guarantees of repayment. Insofar as entrepreneurs cannot provide satisfactory assurances, they either cannot undertake the projects they have in mind or must surrender a measure of control over their contemplated projects. Lenders who have been given adequate security don't care what the borrower does because they know that their terms for cooperating will be met. However, lenders who cannot be certain they will be paid in a timely manner the amounts contracted for by the entrepreneur will want some kind of voice in the business, to "protect their investment," and will want to share in any profit that results, as a reward for accepting the uncertainty associated with their participation.

Thus, entrepreneurs obtain control over the projects they want to undertake, or become "the boss," and claim for themselves (or shoulder by themselves) the entire residual profit (or loss) by providing credible guarantees to all those whose cooperation they require.

How can one provide credible guarantees? The most obvious way is to grant the cooperator a "mortgage" on assets of well-established value. Entrepreneurs who don't themselves own such assets must employ other arts of persuasion and will often be compelled to accept some kind of co-entrepreneurship by agreeing to share both control and the residual with other members of the producing "team."

Marxists and others who wanted to do away with "capitalism" effectively outlawed the free-enterprise system when they obtained political authority by prohibiting the private hiring of labor and private profit. However, they never found an effective alternative way of coordinating productive activity in a world characterized by uncertainty. The central planning organizations they established reduced some uncertainties by curtailing the ability of consumers to influence what would be produced and by removing from the hands of individuals much of the power to decide how resources would be employed. However, this turned out to entail not only a radical reduction in individual liberty but also a drastic crippling of productivity. Central planners simply could not bring together under their oversight and control the information necessary for the coordination of a modern, highly specialized economic system (Hayek, 1945).

No one designed the free-enterprise or profit system *in order to* solve the information problems of modern economies. The system, including the social institutions that provide its framework and support, evolved gradually in response to the initiatives of entrepreneurs of many kinds who, in their efforts to profit from the projects in which they were interested, "were led by an invisible hand to promote an end which was no part of [their] intention" (Smith, 1776).

Defining Terms

Entrepreneur: The person who acquires control of a project by assuming responsibility for the project through guaranteeing to meet the terms of all those who own the resources required to complete the project; the **residual claimant**.

Opportunity cost: The value of all opportunities forgone in order to complete a project.

Profit: Total revenue minus total opportunity cost.

Residual claimant: The person who is entitled to everything that will be left over, whether positive (profit) or negative (loss), after a project has been completed and who thereby acquires the incentive to take account of everything that might affect the project's success; the entrepreneur.

2.9 Managing the Business Interest

Managers of a firm are employed to maintain superior returns for their shareholders and balance the needs and concerns of the other stakeholders. At the same time they must respond responsibly to social demands for environmental performance. When managers balance the triple bottom line — economic, environment, and equity (social) — they will strive to maintain their economic returns while balancing the environmental and social equity issues. As stated by one author (Reinhardt, 2000):

> I take as a premise that companies are in business to make money
> for their owners, and they can pursue the satisfaction of other
> stakeholders only insofar as it serves this basic purpose.

Business managers will perform best for all when they view the environmental issues as a business problem amenable to solutions that can be obtained using the skills and tools they possess. Basically, a business manager examines and, to the extent possible, controls revenues, costs, and profits. We will assume that the manager wants to improve the environmental performance of the firm while maintaining profitability.

The manager normally assumes that costs will rise due to environmental improvements. Profits can be maintained if the extra costs can be passed on to the customers. A second option is to find other internal cost savings that more than offset the cost of the environmental benefits. A third option is to secure an environmental standard set by government or agreed on by an industry-wide association so that all competitors pay the additional cost and thus the cost can normally be passed on. The structure of the business architecture will embody the means of provision of the environmental benefit while retaining or enhancing profitability. As firms compete internationally, they may face competitors who can avoid the cost of abating the environmental issues.

Investments in environmental benefits normally have to be justified when a firm is creating new benefits to the customer or reducing risk of hazards or loss to the firm. Managers may best be able to access the creativity of their employees when they pose an environmental challenge as a business problem and obtain a good solution that in large part solves the problem while maintaining the profit performance of the firm. They need to formulate a solution using the resources of the firm and seizing the real opportunities while avoiding unlikely pathways. Most managers strive to balance the short-term profits and costs with the long-term profits and cost. The manager looks for a trajectory that leads to long-term strength while maintaining shorter term flexibility and resiliency. The creation of a realistic business organization that is responsive to environmental needs while attentive to shareholder interest is challenging to any manager.

Thomas A. Edison (1847–1931) realized in 1878 that the development of an electric lighting device was of great commercial importance. His success was both a technical and a commercial achievement. Edison presided over a well-staffed laboratory in Menlo Park, New Jersey, for many years.

Edison was on the right track, seeking a direct replacement for the gas light that would be easily replaceable and easy to control. Edison eventually found the "ideal" filament for his light bulb in the form of a carbonized thread in 1879. Other materials that could be used for a thin high-resistance thread were later substituted. Edison also undertook a program to improve the vacuum in a light bulb, thus reducing the effects of occluding gases and improving the efficiency of the filament and bulb.

FROM RICHARD C. DORF AND JAMES SVOBODA, *INTRODUCTION TO ELECTRIC CIRCUITS*, NEW YORK: WILEY, 2000. PHOTOGRAPH COURTESY OF THE U.S. DEPARTMENT OF THE INTERIOR, NATIONAL PARK SERVICE, EDISON NATIONAL HISTORIC SITE.

CHAPTER 3

Science, Technology, and Progress

3.1 Science and Technology

The history of technology is longer than and distinct from science. Science is the systematic attempt to understand and interpret the physical and biological world. While technology is concerned with the fabrication and use devices and systems, science is devoted to the conceptual enterprise of understanding and describing the physical and biological world. By the 15th century the emergence of a commercial structure caused some merger of purpose of science and technology. Francis Bacon, in the 17th century, recognized three great technological innovations — the magnetic compass, the printing press, and gunpowder — and he advocated experimental science as a means of enlarging man's dominion over nature. By emphasizing a practical role for science in this way, Bacon implied a harmonization of science and technology. Bacon, with Descartes and other contemporaries, for the first time saw man becoming the master of nature, and a convergence between the traditional pursuits of science and technology was to be the way by which such mastery could be achieved.

The role of Thomas Edison is illustrative of the growing relationship between science and technology, because the trial-and-error process by which he selected the carbon filament for his electric light bulb in 1879 resulted in the creation at Menlo Park, New Jersey, of what may be regarded as the world's first industrial research laboratory. From this achievement the application of scientific principles to technology grew rapidly. It led easily to the engineering rationalism applied by Frederick W. Taylor to the organization of workers in mass production. This was not just a one-way influence of science on technology, because technology created new tools and machines with which the scientists were able to achieve an ever-increasing insight into the natural world. Taken together, these developments brought technology to its modern highly efficient level of performance.

3.2 Technology

The Western world, and increasingly the Eastern world, are technological worlds. Man lives with the tools he has developed; he depends on them for his livelihood. **Technology** is science plus purpose. While science is the study of the laws of nature, technology is the practical application of those laws toward the achievement of some purpose or purposes. One may define technology as the organization of knowledge for the achievement of practical purposes. A more expanded definition of the term is a use of devices and systematic patterns of thought and activity to control physical and biological phenomena in order to serve man's desires with a minimum of resources and a maximum of efficiency.

Many describe technology as the application of science to industrial and commercial objectives. Others consider technology to be the machines, processes, methods, materials, tools, and devices applied to industrial and commercial objectives. All descriptions of technology note that the primary uses of technology are industrial and commercial. Other uses of technology are for military and health and safety objectives. Clearly, technology is used for practical purposes. Many see technology as the means or activity by which mankind seeks to change or manipulate the environment.

The contemporary world is to a great extent influenced by technology. Major technological changes can set population shifts in motion, determine development patterns, and create or solve pollution problems. For example, population throughout the world is shifting from rural to urban areas because farming technology allows fewer people to produce more food. Our complex energy systems have given us a wide array of conveniences, but at the cost of the degradation of the air, water, and land.

Perhaps no change has had an influence as great as the general introduction of the automobile in the United States. With the introduction of the automobile, new industries were no longer limited to locating along a river or railroad line. People were able to work in the city and yet live far away from it. The scale and speed of technological change may well have outstripped the ability of our institutions to control and shape the human environment.

3.3 Technological Systems and Innovation

Technologies are not confined to a specific business, application, or industry. They are potentially applicable to several businesses. For example, the integrated circuit was originally developed for the electronics industry but is now widely used by the auto industry. Furthermore, technologies can be combined by linking them into larger systems. We can think of the steam engine of the 19th century as an example of a technology arranged into many different systems for use in railroads, steamships, steel making, and textiles.

Innovation is the creation of new technology that satisfies a commercial purpose. The innovation process is illustrated in Figure 3.1. A new need or want is identified in the marketplace and restated as a problem to be solved. Then, using existing technologies and scientific knowledge the research and development (R&D) process is used to create a new technological innovation. There is a spectrum of R&D activities. At one end is fundamental research — the pursuit of knowledge without regard to practical application. At the other end is development — the translation of scientific knowledge into products or processes. In the middle of the spectrum is applied research — studies undertaken with the aim of contributing to the solution of practical problems.

An example of the application of the innovation process is the current R&D activity that seeks to commercialize a fuel cell for automobiles. The need is described as the requirement to reduce air pollution from autos while improving miles driven per unit of energy input. The R&D process is currently involved with examining the existing fuel cell technology and using the science of chemistry and materials to develop a new fuel cell that is capable of powering an electric vehicle.

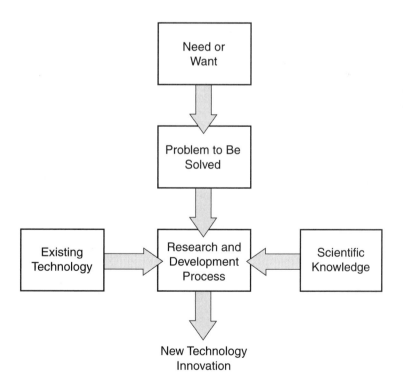

Figure 3.1 **The innovation process.**

A supportive social environment is receptive to new ideas and is one in which the leaders are prepared to support innovation. The existence of important groups willing to encourage innovators is crucial to support new technologies. This condition exists in the United States and many developed nations.

The development of new technologies is a cumulative matter in which each generation inherits all the preceding science and technology on which it can build. Some societies advance, stage by stage, from primitive to moderate and then to advanced technologies. The fact that many societies have remained stagnant for long periods of time even at quite developed stages of technological evolution, and that some have actually regressed and lost the accumulated techniques passed on to them, demonstrates the ambiguous nature of technology and the critical importance of its relationship with other social factors.

The evolution of innovation is illustrated by the list shown in Table 3.1.

Table 3.1 **The Evolution of Innovation**

Corrective lenses	1268	Roger Bacon recorded that glass lenses could correct vision
Printing press	1448	Johann Gutenberg
Steam engine	1712	Thomas Newcomen
Lightning conductor (rod)	1750	Ben Franklin
Electric telegraph	1838	Samuel Morse
Telephone	1875	Alexander Graham Bell
Incandescent lamp	1879	Edison's carbonized cotton thread inside a bulb
Zipper-hookers fastener	1893	Whitcomb Judson invented the clasp locker
Paper clip	1899	Johan Vaaler of Norway used small curved lengths of metal
Radio	1899	Marconi transmits radio signal
Television	1927	Farnsworth and Zworykin
Velcro	1948	A fastener with tiny hooks on one fabric and loops on the other
Transistor	1948	Shockley, Bardeen, Brattain
Alkaline battery	1959	Lew Urry of Eveready Co. used an alkaline cell
Integrated circuit	1959	Jack Kilby and Robert Noyce

3.4 Thomas A. Edison: Innovator, Businessman, and Engineer

Electrical energy can be converted into heat, light, and mechanical energy. Nature exhibits the first two conversions in lightning. When the battery was invented in 1800, it was expected that the continuous current available could be used to produce the familiar spark. This continuous current was ultimately used in the arc light for illumination.

During the first two-thirds of the 19th century, gas was used to provide illumination and was the marvel of the age. By 1870 there were 390 companies manufacturing gas for lighting in the United States.

In 1878, in Ohio, Charles F. Brush invented an electric arc lighting system. Later, a search was undertaken to find a suitable light that would give less glare than an arc light. Thus, research was centered on finding a material that could be heated to incandescence without burning when an electric current was passed through it.

Thomas A. Edison (1847–1931) realized in 1878 that the development of an electric lighting device was of great commercial importance. His success was both a technical and a commercial achievement. Edison presided over a well-staffed laboratory in Menlo Park, New Jersey, for many years. In an interview, Edison stated:

> I have an idea that I can make the electric light available for all common uses, and supply it at a trifling cost, compared with that of gas. There is no difficulty about dividing up the electric currents and using small quantities at different points. The trouble is in finding a candle that will give a pleasant light, not too intense, which can be turned on or off as easily as gas.

Edison was on the right track, seeking a direct replacement for the gas light that would be easily replaceable and easy to control. As we may recall, Edison eventually found the filament for his light bulb in the form of a carbonized thread in 1879. Other materials that could be used for a thin high-resistance thread were later substituted. Edison also undertook a program to improve the vacuum in a light bulb, thus reducing the effects of occluding gases and improving the efficiency of the filament and bulb.

With his early engineering insight, Edison saw that a complete lighting system was required, as he stated:

> The problem then that I undertook to solve was . . . the production of the multifarious apparatus, methods, and devices, each adapted for use with every other, and all forming a comprehensive system.

By 1882 the system had been conceived, designed, patented, and tested. The equipment for the lighting system was manufactured by the Edison companies. In his prime and during the development of his electric lighting system, Edison depended on investment bankers for funds. By 1882 he had built a system with 12,843 light bulbs within a few blocks of Wall Street, New York. How consistent he was in his approach can be seen by the way he finally marketed the electric light. To emphasize to customers that they were buying an old familiar product — light — and not a new unfamiliar product — electricity — the bills were for light-hours rather than kilowatts. Today, nearly one-quarter of the electricity sold in the United States is devoted to lighting uses.

3.5 Social Goals

For good or ill the contemporary world is and will continue to be substantially shaped by technology. As Professor Elting Morison observed (Morison, 1966),

> All earlier history has been determined by the fact that the capacity of man had always been limited to his own strength and that of the men and animals he could control. But, beginning with the nineteenth century, the situation had changed. His capacity is no longer so limited; man has now learned to manufacture power and with the manufacture of power a new epoch began.

In the middle of the 20th century, two technological projects of the United States demonstrated to its citizens that technology could be harnessed to national goals. The two triumphs of American technology were produced by large government-funded enterprises. The first was the successful development of atomic weapons and the second was the conquering of interplanetary space and setting foot on the moon. Both were American conquests of "the impossible." Perhaps the most nearly comparable American achievement was the building of the railroad across the United States. Charles Francis Adams, Jr., called the transcontinental railroad a new

> enormous, incalculable force . . . exercising all sorts of influences — social, moral, and political; precipitating upon us novel problems which demand immediate solution; banishing the old before the new is half matured to replace it; bringing the nations into close contact before yet the antipathies of race have begun to be eradicated; giving us a history full of changing fortunes and rich in dramatic episodes.

The railroad, he stated, might be "the most tremendous and far-reaching engine of social change which has ever either blessed or cursed mankind."

The sense of social need must be strongly felt, or people will not be prepared to devote resources to a technological innovation. The thing needed may be a more efficient automobile or a means of utilizing new fuels or a new source of energy. In modern societies, needs have been generated by marketing activities. Whatever the source of social need, it is essential that enough people be conscious of it to provide a market for product or service that can meet the need.

3.6 The Benefits of Technology

The **welfare** of a society refers to the quality of life that exists in a society. Technology enables new goods and services to be delivered to a society. Technology may increase the efficiency of production as well as the quality of the goods. An example of a new technology is the sewing machine, which was invented in 1846 by Elias Howe. This machine enabled people to make better garments in less time and thus created the garment industry. The airplane was demonstrated by the Wright brothers in 1906. The improvement of aircraft technology now provides travel at 500 mph for about 2.5 cents per seat-mile (Boeing 747).

A good telegraph operator could send 25 bits per second of Morse code in 1890. The telephone (single line) could send 4,000 bits per second in 1910. The typical computer modem now receives about 56,000 bits per second. Each strand of optical fiber can carry about 10 billion bits per second. As communication technologies improve, performance improves and costs decline. The capacity of telecommunication channels is illustrated in Table 3.2.

Table 3.2 **Capacity of Telecommunication Channels**

Means	*Year*	*Capacity (bps)*
Telegraph	1860	20
Telephone (early)	1900	4000
Telephone (multiline)	1920	9×10^4
Coaxial cable	1950	2×10^6
Microwave	1956	9×10^6
Satellites	1980	10^8
Optical fibers	1990	10^{10}

In 1000 B.C. the average person relied on a costly oil lamp, which provided a low level of light. Using lumens as a measurement of light (refer to a light bulb package) the person in 1800 paid $100 per lumen (in 1990 dollars). By 1900, with electric lighting a person paid $10 per lumen. By 1990, the typical cost was 10 cents per lumen.

Technology improved the efficiency of agricultural production. Farmers and others in the agricultural industry promoted the use of new machines and processes to improve the productivity of their estates. Under the same sort of stimuli, agricultural improvement continued into the 19th century and was extended to food processing. The steam engine and later the internal combustion engine were adopted for agricultural purposes such as threshing and plowing. The McCormick reaper and the combine harvester were both developed in the United States, as were barbed wire and the food-packing and canning industries, with Chicago becoming the center for these processes. The introduction of refrigeration techniques in the second half of the 19th century made it possible to convey meat from ranch to city.

Technologies are adopted by users over time. Consider the car: The basic patent for an internal-combustion engine capable of powering a car was filed in 1877. By the late 1920s — 50 years later — more than half of all American households owned a car. The first electricity generating station was built by Edison in 1881 but it was only in the 1920s that electricity was widely used in the home and industry throughout the United States and

Europe. The speed of adoption of several technologies is recorded in Table 3.3. The percent of U.S. households with three modern communication technologies is given in Table 3.4.

Table 3.3 **Speed of Adoption of a Technology**

Technology	Year invented	Years until 25% of the U.S. population adopted it
Electricity	1873	46
Telephone	1875	35
Auto	1885	55
Television	1925	26
Microwave oven	1953	30
Personal computer	1975	15
Cellular phone	1983	13

Table 3.4 **Percent of U.S. Households with the Technology in 1999**

Television	98%
Cordless phone	72%
Personal computer	46%

The benefits of improved productivity and health can be, in part, attributed to the global diffusion of technologies. For the period from 1950 to 2000, global economic output has increased sixfold and life expectancy has increased from 47 to 64 years.

3.7 Technology and Social Progress

Many observers see the Renaissance as the period in which there first emerged the restless, technical inventiveness that so strongly characterizes today's Western culture. They point out that it was at this time that the pursuit of innovation and technology was first recognized and honored. This period marked a new awareness of invention.

Invention is the act of creating a new device, mechanism, or artifact. **Innovation** is the act of creating a new commercial or social useful device, process, or artifact. Thus, an invention might be a new idea for space travel, while an example of an innovation is the introduction of the integrated circuit or the Internet. We distinguish an invention from an innovation by noting that innovation is the actual introduction of the new method or product to social use. Thus, fusion reactors are inventions until they become practical and can be economically introduced for use by utilities and power companies.

Technology involves the application of reason to techniques that are useful within society. Seeking to increase knowledge and to apply it to the problems or needs of the world follows faith in the notion of progress. Sustained material progress is a tenet of the technological society. Optimism and progress underlie the modern enthusiasm for technology and its wide application. For example, many believe that the Internet and related innovations will lead to a new burst of material and social progress. Benjamin Franklin wrote in 1780: "It is impossible to imagine the height to which may be carried, in a thousand years, the power of man over matter."

A recent book challenges this optimism by raising the issues of externalities such as pollution and congestion (Bronk, 1998). Bronk challenges what some call techno-optimism. The idea of progress penetrates society widely. However, our increasing anxiety concerning the depletion of resources and the increase of pollution causes many to assert that there are limits to progress. However, most will not give up hope of further improvement in the overall conditions of life. Thus, we have the "sustained progress dogma." An excellent example is the statement by Schwartz (1999) in a recent book:

> The Long Boom (the years 1980 to 2020) is a period of remarkable global transformation. No other age possessed the tools or knowledge to do what we can do today.

Not only do we expect sustained progress, but we expect it faster. The length of time between the discovery of a fundamental process and its application has become progressively shorter. For example, 112 years (1727–1829) passed between the discovery of the physical phenomenon used in photography and application of photography itself. In contrast, it took only three years (1948–1951) in the case of the transistor.

Many current technologies are complex combinations of science and techniques, which makes the actual "source" of the first discovery difficult to trace. Moreover, one technology tends to lead to another through a snowball effect.

Technology is seen as improving productivity of society as measured by output per unit of weighted labor and capital input. For example, in the household mobile robots are predicted to be carrying out home and workplace chores such as delivering packages, cleaning, and other manual activities by 2010. By 2040, some believe we will have a mobile robot with the "intellectual capabilities of a human being" (Moravec, 1999).

On the other side of this issue of progress, one must recognize that energy in combination with machines has been substituted for human labor in many sectors (Ayres, 1998). For example, as shown in Figure 3.2, energy is used as a substitute for labor on U.S. farms.

Daniel Bell first published *The Coming of the Post-Industrial Society* in 1999. In his book, Bell described the role of technology in driving social change and the importance of knowledge (intellectual capital). Bell described the idea of political institutions directing technological change and thus driving toward positive, less harmful, change. We now have a world where time, not raw material, is the scarce resource and where intellectual capital is paramount.

3.8 Positive and Negative Outcomes of Technology

The outcomes of new innovations are expected to be beneficial, but may turn out to actually be negative. A shorter workweek and reduced physical labor are beneficial, but intense urbanization may be less so. Material well-being and improved health and longer lives are examples of beneficial outcomes. New technologies such as the automatic elevator displace human operators but improve productivity (*Economist,* 2000).

Our new computer technology has moved us into an information society. Using industrial output as a measure of prosperity may be outdated. We produce and manipulate information to increase the productivity of services. Since the invention of the wheel or the lever, man has applied technology to overcome physical limitations. In the future, by prudent use of improved technology, land can be made to grow more food, substitutes can be developed for scarce natural resources, and devices can be made for controlling pollution. However, new technology can also create many new and often unanticipated problems. Automotive air pollution, persistent pesticides, and nondegradable solid wastes, for example, are the consequences of technological innovations.

The law of **unintended consequences** states that technological solutions to problems may yield predicted positive outcomes but often result in undesired, often negative, outcomes. We use pesticides to improve crop yields but may cause pesticide poisoning of farm workers or farm neighbors.

Often costs incurred by third parties to an action may be outside of the realm of understanding of the problem solver. Civilization is often thought to have started in the lower valleys of the Tigris and Euphrates, and its basic technology, which permitted development, was irrigation. Yet the very

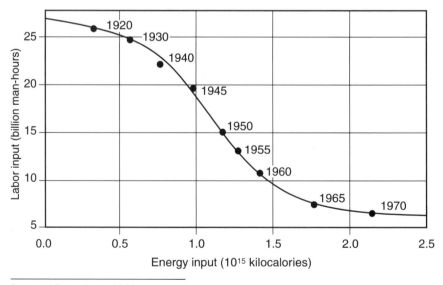

SOURCE: From Ayres, 1998.

Figure 3.2 **The substitution of energy for human labor on U.S. farms.**

process of irrigation, through evaporation, left salt in the topsoil. The salt eventually destroyed the fertility of the land and the civilization based thereon. Again, modern civilization is based on some technologies that may lead to adverse consequences.

As an example, consider the relatively benign use of cellular phones. Some 80 million Americans use cell phones and 100 million Europeans use cell phones. The unintended consequence of cell phone use is people using them while driving. It is estimated that less than 1% of the accidents on the highway are caused by drivers on a cell phone. Nevertheless, some countries such as England and Italy ban their use in a car by a driver.

Edward Tenner (1996) in a recent book, *Why Things Bite Back*, explains the occurence of malfunctions that led to the formulation of Murphy's law: "If anything can go wrong, it will." As Tenner states:

> The technological dream of a self-correcting world is an illusion no less than John von Neumann's 1955 prediction of energy too cheap to meter by 1980. The social goal of a new Athens, of machine-supported leisure, has proved a noble mirage. Technology demands more, not less, human work to function. And it introduces more subtle and insidious problems to replace acute ones. Nor are the acute ones ever completely eliminated; in fact, unless we exercise constant care and alertness, they have a way of coming back with new strength.

Tenner distinguishes a side effect from a revenge effect. **Side effects** are trade-offs, whereas **revenge effects** are unintended negative consequences. Revenge effects happen because new devices or systems react with people in ways we could not foresee. Consider the *Titanic,* which was designed to be unsinkable with a side effect of larger tonnage and poorer fuel economy. The unintended, revenge, effect was the increased speed of operation since the ship was perceived as nondestructible. When a safety system or modification encourages additional risk-taking such that accidents occur, that is a revenge effect. Another example of a revenge effect is flood-control systems which encourage settlement in flood-prone areas, increasing the risk to life and property.

Revenge effects do not always come from using technology; they can also come from adopting deceptive solutions in place of costlier ones (as part of the trade-off process). An excellent example of a proposed technological advance is an intelligent vehicle highway system for increasing highway capacity (see Chapter 8). Individual vehicles would be operated under coordinated electronic control with tight spacing between the vehicles. The potential revenge effects include a tire failure or computer error, which could cause a large pile-up. Most technological solutions to problems lead to new problems and so on. All designers of technological solutions need to be prepared to encounter unexpected consequences — both good and bad.

Society, in the form of government, intervenes on behalf of its citizens to attempt to obtain solutions to their problems. Yet safety and reliability are difficult to achieve at an optimum level. Safety is never absolute and the actions of man must be accounted for. For example, after five decades of experience in improving the safety features of farm machinery, manufacturers are frustrated to find that owners of the safer equipment simply push it to a higher level of performance, accepting a constant level of hazard. A tractor redesigned to lessen the likelihood of tipping over is then driven on a steeper hillside, for instance.

3.9 Technology and Developing Nations

Wide disparities of wealth and well-being exist across the world. Many programs exist with the goal of addressing inequities and creating more sustainable development paths for the less wealthy nations, including the use of technologies to improve conditions. Perhaps the most important program is to ensure worldwide availability of birth control technologies.

The idea that efforts should be concentrated on a search for technological solutions to environmental limits is more likely to seem attractive and feasible to a developed nation than to an underdeveloped nation because the less developed nations may see these actions as limiting their economic growth. Some of them see the current attention being paid to ecological issues as a new form of economic imperialism, in that it appears to be an attempt to deny to them the development pathways already pursued by the economically developed nations. Furthermore, the attempt by the developed nations to develop technological substitutes can look like an attempt by the natural capital consuming nations to extend their control over commodity prices by reducing demand. Any reduction in demand is particularly serious for a less developed economy that is heavily dependent on a few exported commodities.

Developed nations use more resources per capita as shown in Table 3.5. It is feared that as the affluence of developing nations increases, they will use resources at the same levels as developed nations. Conventional wisdom holds that developing countries cannot hope to reduce pollution of air and water until they reach a level of affluence similar to that of developed nations. Many believe that economic growth of less developed nations can only increase their level of pollution.

Another prevalent belief is that growing global trade and open borders are encouraging dirty industries to move to developing countries, which cannot afford to curb environmental abuses. Many observers believe that this outcome will not occur. Poorer nations will act to curb pollution because they have decided that the benefits of abatement outweigh the costs. Environmental regulators in developing countries are trying fresh approaches and finding new allies in the battle to curb pollution.

Technology, like any other social element, will only exist as long as it is useful as a means to the ends of society. Just as the various economic systems disappeared as conditions changed, so will a given technology give way to another when the old is no longer capable of accomplishing the goals of the society in the face of current conditions.

Table 3.5 **Per-Capita Resource Use in 1996**

	United States	India	World
Fossil energy use (GJ/year)	300	6	60
Fresh water (m³/year)	1,900	600	650
Population density (persons per 1,000 hectares)	294	3,177	442

SOURCE: Data from World Resources Institute.
Key: GJ, gigajoules; m³, cubic meters.

An excellent example of an appropriate response to the needs of a developing nation is the Grameen Bank of Bangladesh (Yunus, 1998). Grameen ("Village") Bank strives to release villagers from the clutches of the moneylender by making small loans (the average is $30) to the poorest people. This bank builds a network of borrowers by lending to women who sell cell phone service acting as "human pay phones." Another venture is the village Internet program using "cyber kiosks." Technology may aid the poor by leveling the playfield and offering opportunities not otherwise available.

3.10 Limits to Technological Solutions

Numerous examples of technological successes are evident in our daily lives. Perhaps the most glamorous and widely hailed successes are in the fields of medicine and space technology. However, most notably, technology is criticized for its failure to solve the problems of transportation and to ameliorate the serious urban and environmental problems.

Indeed, if one examines a variety of national programs of the 1990s that have been developed to attack major social problems, one is tempted to draw the conclusion that a technological solution to such problems is destined to be incomplete. In the area of education, for example, problems yield only in part to technical solutions such as the use of computer-aided instruction and educational television. Such can be ameliorated, slowly, with the proper application of technological approaches merged wisely with social and political awareness of the many dimensions of the issue.

Our society is increasingly accustomed to calling on technology to solve problems, whether they be problems of war, poverty, or disease. However, many realize that massive applications of technology can be ineffective, often counterproductive, in the absence of effective social goals.

Why do people fear technology? There are many reasons why people fear technology and its consequences, but some are predominant at this time. Technology appears to many to have too much momentum; it is difficult to "control." Technology produces change too fast and without effective opportunities to debate its effects and the desirability of its introduction into use. As mentioned in Chapter 2, the span of time from discovery of a new technology to practical application is constantly decreasing. Whereas 35 years elapsed between discovery of the vacuum tube and its use in a practical radio, only 3 years elapsed from discovery of the transistor to practical application. With increasing rapidity, knowledge is converted to useful products that produce social change. Hence, the problem of anticipating social change becomes increasingly difficult.

Many believe that civilization has now advanced to a level where the rate of increase in technology by far exceeds the rate at which we dare create new demands on the available environment as a result of increases in population or in its demands made on the environment by rising standards of living. If the optimization of living conditions is the standard of judgment, then many say, "From here on out, that population growth must be minimized."

Furthermore, critics warily view the modern corporation together with modern technology and the power of financial capital. The scientist's and engineer's self-perceived moral responsibility is limited to advancing objective instrumental knowledge, and the corporate executive's self-perceived moral responsibility is limited to maximizing corporate profits. The result may be a system in which power and expertise are delinked from moral accountability, instrumental and financial values override life values, and what is expedient and profitable take precedence over what is nurturing and responsible (Korten, 1999).

Many critics attack the technocratic society, which they may have come to see as coldly rational and inhuman — a world of technology in which human ends are forgotten in the search for rational techniques and industrial efficiency. In some sense society is dominated by technology, and some people contend that technology determines not only means, but also ends. Technology is seen as the search for the one most efficient way, with no room for human choice or judgment about values.

Some critics are overtly antitechnology and regularly express antipathy to the high technology-based global industrial system emerging today. They see globalization as destructive of small-scale communities and of local cultures. Their view of the new economic order as destructive of the social and psychological benefits associated with meaningful work places their political and social goals in proximity to those espoused by isolationists and protectionists. Their deep distrust of technology is shared not only by many environmentalists but by groups as diverse as the Freemen, survivalists, and militiamen.

Criticisms of technology have been vocal since the mid-19th century. Ralph Waldo Emerson said that "Things are in the saddle and ride mankind." Aldous Huxley noted in *Brave New World* (1932) a future society in which technology was firmly enthroned, keeping human beings in bodily comfort without knowledge of want or pain, but also without freedom, beauty, or creativity, and robbed at every turn of a unique personal existence.

The theme of technological tyranny over man's individuality and his traditional patterns of life was expressed by Jacques Ellul (1973) in his book *The Technological Society*. Ellul asserted that technology had become so pervasive that man now lived in a milieu of technology rather than of nature. He characterized this new milieu as artificial, autonomous, self-determining, and, in fact, with means enjoying primacy over ends. Technology, Ellul held, had become so powerful and ubiquitous that other social phenomena such as politics and economics had become situated *in it* rather than influenced *by it.* The individual had come to be adapted to the technical milieu rather than the other way around.

A major problem area of modern technological society is that of preserving a healthy environmental balance. Though man has been damaging his environment for centuries by overcutting trees and farming too intensively, and though some protective measures, such as the establishment of national forests and wildlife sanctuaries, were taken decades ago, great increases in population and in the intensity of industrialization are promoting a worldwide ecological crisis. Nevertheless, it can be said that the fault for this waste-making abuse of technology lies with man himself rather than with the tools he uses. For all his intelligence, mankind in communities behaves with a lack of respect for his environment that is short sighted.

The forces of technological change cause us to focus our attention on how to adapt to change as well as shape it. The concept of a zero-sum system permeates our analysis. In a **zero-sum game**, a gain by one party necessitates a loss for the other party. It is the assumption of a fixed pie that can be redistributed but not enlarged that underlies this viewpoint. Thus, environmental problems, such as global warming, are seen as zero-sum in which increasing wealth is seen as declining environment. Will technology create new wealth or declining ecologies?

Postrel (1998) in a recent book discusses the future and asks:

> Do we search for *stasis* — a regulated, engineered world? Or do we embrace *dynamism* — a world of constant creation, discovery, and competition? Do we value stability and control, or evolution and learning? . . . Like the present, the future is not a single, uniform state but an ongoing process that reflects the plenitude of human life.

Postrel goes on to state:

> Contrary to the scare stories of reactionary greens, many innovations substitute ingenuity for physical resources, reducing environmental impacts. That is particularly the case as technologies are refined and enter the phase of incremental improvement. . . .
>
> Technology, art, and culture are based not on uncovering a few fixed facts but on coming up with new combinations of ideas, testing them, finding their faults, trying possibly better combinations, and so on.

Technology is used to control nature for the betterment of life, but often goes awry. Peter Huber in a recent article outlines the benefits and costs of the uses of technology. In the following section, he argues forcefully that technology, wisely used, can improve mankind's conditions.

3.11 Fear Nature, Not Technology[1]

BY PETER HUBER
Senior Fellow
Manhattan Institute

Yesterday's devastating earthquake in Turkey killed more than 2,000 people and injured at least 10,000. The worst damage apparently occurred around the industrial city of Izmit, where an oil refinery caught fire. Turkey will respond, and with it the rest of the world, rushing in cranes and bulldozers to clear the rubble, along with food, clean water, diesel generators, fuel to power them, antibiotics and vaccines. Experience teaches that often many more people die in the aftermath of a natural disaster than directly from it — for want of simple necessities and the infrastructure of pipes and roads, food, water supplies and power lines that make possible their timely delivery.

According to early reports, the Turkish quake measured 7.8 on the Richter scale, just a shade less powerful than the 7.9-magnitude San Francisco quake of 1906. That one killed 700 people. What human toll would a repeat exact in the U.S. today? It would be far costlier in dollar terms than Tuesday's quake, for the simple reason that we have more in the way of economic assets to destroy. But notwithstanding our nuclear power plants, highways and high-rises, the death toll from a comparable U.S. quake and its aftermath would likely be lower.

3.11.1 LESS VULNERABLE

We are less vulnerable to natural disaster than we used to be, and less vulnerable than most of the rest of the world. Less vulnerable notwithstanding all our high-rises and high technology. Indeed, less vulnerable *because* of them.

The multibillion-dollar natural disaster in the U.S. today — hurricane, tornado, earthquake or flood — typically costs dozens of lives, or very occasionally hundreds. But rarely more. And because our emergency forces are able to respond so quickly and effectively, those lives are almost always lost as a direct consequence of nature's violence, rather than in its aftermath. For all its pesticides and nukes, its fossil-fuel engines and its bio-engineered crops, Western capitalism has become very robust and resilient indeed.

The technologies and resources Turkey now needs the most are the ones humanity has been repeatedly advised to abandon. In the days to come, Caterpillar's earthmovers will roll in to dig the living and the dead out from Istanbul's rubble. Engineers will deploy heavy machinery to rebuild the water pipes, sewer systems and roads. If Turkey needs emergency food supplies, it will call on the Iowa farmer, and the abundance of his bio-engineered grains, broad-spectrum pesticides and combine harvesters — not on an organic vegetable garden lovingly tended by the Prince of Wales. The food will be airlifted to Turkish airports by fuel-guzzling transports. Nobody will waste time flying solar collectors to Istanbul: solar is a hobby

[1]This section has been reprinted from *The Wall Street Journal,* Aug. 18, 1999. Reprinted with permission of *The Wall Street Journal.* © 1999 Dow Jones & Company, Inc. All rights reserved. Mr. Huber is the author of *Hard Green,* New York: Basic Books, 2000.

for comfortable people who can plug into a conventional grid if they need to. When electricity is powering the rescue workers' lights, drills and pumps around the clock, you burn diesel.

For many years we have been advised that the real perils lie not in nature's fault lines but in high technology's. This is the sand-pile theory of technological advance. Its most prominent proponent is Vice President Al Gore. The story runs something like this: The technology that subdues nature so effectively will subdue man, too. Pesticides, nukes and transgenic mice are inherently unstable. They are like so many sand piles — so many avalanches just waiting to go critical. Sooner or later, as Mr. Gore concludes in his 1992 manifesto, "Earth in the Balance: Ecology and the Human Spirit," they are bound to crash down upon us.

All the evidence bearing on accident rates and mortality statistics is directly to the contrary. Most of capitalism's technological creations, it turns out, are not sand piles at all — they behave, instead, more like stable gobs of honey. If they were getting more brittle, catastrophic accidents would become more numerous year by year. In fact, catastrophic failures of dams, power plants, jumbo jets and chemical factories are growing less frequent.

And the wealth and power that this resilient technology delivers have proved to be the best defense we have against nature's catastrophes. Technology can't stop an earthquake — not yet, though in time it may do that too. But it has produced stronger buildings and highways, more-robust power systems and the capabilities of instant, material response that can save thousands of lives in the aftermath.

Capitalism's high technology has proved even more essential in staving off bio-catastrophes — not killer tomatoes concocted by Monsanto, but killer viruses that emerge from swamps and rain forests. The biosphere, like the geophysical environment, doesn't love us or hate us, it just wholly lacks an attitude. The rainforest periodically releases Ebola, typhoid, cholera, malaria and leprosy. There may be a cure for AIDS somewhere up there in the trees, too, but trees are also where HIV came from in the first place, by way of monkeys. The totality of nature has not been mystically perfected for human benefit. Darwin teaches otherwise.

Free markets, by contrast, do have an attitude — a life-affirming one. Safety, resilience and stability sell better than the alternatives. Capitalism ends up crafting things accordingly. Technology, it turns out, is less fickle, less unstable, than nature. It can do us harm, and sometimes does, but for the most part it is designed to do us good. Nature isn't.

3.11.2 SHELTERING US

For several decades now, green oracles of catastrophe have been trying to pin earthquake scenarios on capitalism, and its technological fruits. Return to nature, they insist, and we reclaim the better life — cleaner, safer, more stable. They have it completely wrong. Capitalism and its technology don't end in catastrophe; they distance us from it. They don't cause catastrophe, they stand as our main shelter and defense against it.

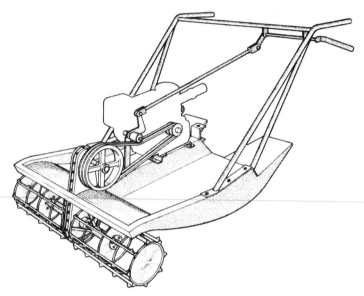

An example of an appropriate technology for developing nations: a small floating rotary tiller for flooded rice fields commonly used by farmers in the Philippines. DRAWING REPRINTED BY PERMISSION OF THE INTERNATIONAL RICE RESEARCH INSTITUTE (IRRI).

CHAPTER 4

Sustainable and Appropriate Technologies

4.1 Sustainability and Technology

Our ignorance about the nature, cause, and significance of perceived ecological risks is compounded by ignorance about how to rationally manage the growth and technological evolution of market economies and simultaneously contain both the sociological and ecological hazards they engender. Unless notable progress is made soon to increase our understanding, it is possible that, within this century, humankind will seriously jeopardize critical natural processes, ecosystems, species populations, and resource stocks. This unnecessary ecological scenario is likely to be accompanied, or preceded, by accelerating economic, social, and political instabilities.

Almost everyone in modern societies has a subjective understanding of technologies and networks of technologies as production and consumption processes. Engineers develop and patent them, entrepreneurs and economists promote their novelty and virtues, Wall Street investors speculate on their commercialization, and environmentalists frequently condemn their destructive powers. Almost everyone has a subjective understanding of the biophysical interactions between economies and the natural abiotic processes of the atmosphere, hydrosphere, and geosphere and the biotic processes of the biosphere as natural ecosystems. Furthermore, most people have an informal understanding of what market economies are. However, from a social perspective, the important thing is not to know what technologies, ecosystems, and economies are, but how to measure their effects and control them for favorable outcomes.

Many hold a favorable vision of long-term economic growth that creates jobs while improving and sustaining the environment. Reconciling these goals requires a technology strategy that helps industry shift from waste management to pollution prevention, efficient resource use, and industrial ecology. Many believe that companies will become more beneficent by lowering their energy and resource needs while reducing or eliminating their waste cleanup and disposal costs. This is part of a long-term goal of sustainable development and technology's role in enabling or impeding it. Whether technology is used to protect or merely exploit the natural environment will be one of the most challenging policy choices of the 21st century. The decision of which path to pursue rests in the hands of no one government or sector of society. Rather, the decision lies in the collective actions of individuals, corporations, and governments throughout the world. Recognizing this distinction may be the key to the appropriate development and application of technology.

In the follow section, Professor Robert Ayres provides an insightful analysis of the potential for sustainability and the role of technology, corporations, and government.

4.2 Sustainability versus Unsustainability[1]

By ROBERT U. AYRES
Sandoz Professor of Environment and Management, Emeritus
Center for the Management of Environmental Resources
INSEAD

In 1985 the UN General Assembly created the World Commission on Environment and Development (WCED), more widely known as the Brundtland Commission after the name of its chairwoman, Prime Minister Gro Harlem Brundtland of Norway. The commission, and its staff, labored for two years, met dozens of times in different cities, and listened to hundreds of experts. Its report was published in 1987 under the title "Our Common Future." The theme of the commission report was "sustainable development," which was defined as economic growth to meet the needs of the people living today without compromising the possibility for future generations to support themselves. It recognized that there is a very real conflict between meeting the needs and desires of the five and a half billion people then alive (already more than six billion) and the possibility of satisfying the nine to ten billion people expected to be living on earth by the middle of the next century.[2]

---

[1] The section has been taken in large part from Chapter 9 of the author's book, *Turning Point: An End to the Growth Paradigm,* London: Earthscan Publications, 1998.

[2] The impact of this report triggered the creation of a UN Commission on Environment and Development (UNCED), chaired by Maurice Strong of Canada. This finally culminated with a Global Environmental Summit, which occurred in Rio de Janeiro in June 1992 and produced many documents, including a global plan of action entitled *Agenda 21* (Strong, 1992).

A large number of alternative definitions of "sustainable development" have been proposed. To read such a list and attempt to sort out the subtle differences among them is a mind-numbing academic exercise. One common problem is that most definitions, including the one originally proposed by the Brundtland Commission, use words that are deliberately open to a variety of interpretations, so as to achieve the widest possible consensus. This is understandable, given the political realities. But a definition that merely specifies that several incompatible objectives (e.g., environmental protection, alleviation of poverty, economic development, and intergenerational equity) should be reconciled and achieved simultaneously, has no operational significance. In fact, restated as above, it is probably an oxymoron.

The current interpretation of *sustainable* development from mainstream institutions like the World Bank is virtually indistinguishable from the notion of *continuing* economic growth as measured in the usual way, i.e., in terms of consumption of market-tradeable goods and services. The requirement of assuring future generations an equal chance at the "good life" is supposed to be met automatically by accumulating man-made capital and (if necessary) substituting man-made capital for natural capital. In short, to address issues of sustainability, mainstream economics relies on the standard tools and assumptions of neoclassical growth theory, which tends to assume that the only needed inputs are capital and labor.

In contrast, what has been called the "ecological" criterion for sustainability admits the likelihood that some of the important functions of the natural world cannot be replaced by man-made capital or human technology within any realistic time frame—if ever. In other words, man-made capital and engineering technology, no matter how sophisticated, cannot substitute for all of the essential services provided by natural capital, such as topsoil, the hydrological cycle, the temperate climatic zones, the protective ozone layer of the stratosphere (which screens out harmful UV radiation from the earth's surface), or biological waste assimilation and detoxification. It is fairly obvious that these environmental services, not to mention others including natural scenery and biodiversity, cannot be either recreated by human technology nor replaced by man-made equivalents.

Nevertheless, it is a lot easier to identify *unsustainable* activities and trends than it is to define sustainability in a precise and operational manner. I use the word "unsustainable" here in the most literal sense: Unsustainable trends and activities cannot continue indefinitely (or even for very much longer, in some cases) without running into a natural limit. For instance, continued dependence on hydrocarbon fuels is unsustainable because the resulting "greenhouse gas" emissions to the atmosphere are apparently warming the global climate, changing rainfall patterns in harmful ways, increasing storm damage along coasts, and raising the sea level.

Population growth is also unsustainable. The slums of the world's megacities are holding zones for people who can no longer support themselves from the land and whom the cities can neither employ nor feed. It might be tempting to think of these slums as a potential source of bright ideas and future market for goods [as Julian Simon (1980) does in *The Ultimate Resource*]. But, at the same time, one must recognize that slums breed crime and disease, not to mention social instability. Most other problems of the global environment cannot be solved if the world's population is not stabilized soon.

Other trends are also unsustainable by any reasonable criterion. The loss of biodiversity is one example. The present rate of disappearance of species has no historical precedent, and is still accelerating. It has been called "the sixth mass extinction." The last major extinction of species, probably triggered by an asteroidal collision, ended the reign of the dinosaurs 65 million years ago. The current extinction is caused by a combination of human-induced land use changes, reduced barriers to species migration, eutrophication (i.e., excessive fertilization) of rivers and lakes, acidification, buildup of toxic elements in soils, and climate warming. All of these trends, in turn, are unsustainable.

The notion of "sustainable level" with regard to pollution, toxification, acidification, greenhouse gas buildup, and so on, is widely assumed but seldom defined. The existence of such a level is predicated on the idea that natural processes can and will compensate for some of the environmental damage. For instance, natural weathering of rocks generates some alkaline materials that can neutralize acid rain. Similarly, some of the excess carbon dioxide produced by combustion processes may be absorbed in the oceans or taken up by accelerated photosynthetic activity in northern forests. Some plant species grow faster in the presence of higher levels of carbon dioxide. The deposition of sulfates and, especially, nitrates from combustion products can also increase growth rates in some forests where these substances are initially scarce. This eutrophication (fertilization) phenomenon already seems to be stimulating forest growth in the Northern Hemisphere. This is removing somewhat more of the excess carbon dioxide from the atmosphere than we could have expected based on earlier data, although by no means enough to compensate for the added inputs.

Eutrophication is an example of a negative feedback process that could partially abate the impact of increased emissions of carbon dioxide. On the other hand there are also significant possibilities of positive feedbacks that could further accelerate climate warming. One is that warming in the far north is beginning to melt the permafrost — frozen subsoil — resulting in enhanced activity by anaerobic bacteria and the release of methane into the atmosphere. (Methane is hundreds of times more potent as a greenhouse gas than is carbon dioxide.) Another possibility, recently discovered, is that increased rainfall in Canada, from global warming, may cause increased runoff of freshwater via the Saint Lawrence River. This can reduce the salinity of northern Atlantic waters, which, in turn, can cause the warm Gulf Stream water to sink before it reaches the Arctic and cools off (by warming the air and the land). This would not only result in a significant cooling of northern Europe (e.g., the "little ice age"), it would also cause a very slight warming of the deeper ocean waters. But the consequence of even a slight warming of the deeper ocean waters could be that frozen methane crystals (methane clathrates) that are known to exist in large deposits under the silt along continental shelves may melt and release the methane. While it would initially be dissolved in seawater, some of it would inevitably be released to the atmosphere. This could accelerate the overall climate warming dramatically.

4.2.1 THE ENVIRONMENT/DEVELOPMENT DILEMMA

As suggested above, population growth and agriculture are the most critical problem areas. Some technological optimists have projected that early 20th-century rates of increase in agricultural productivity can and will continue into the indefinite future.[3] However, most agricultural experts are much less optimistic. The alarm has been raised once again, certainly on better grounds than in Malthus's time. Today there are no more "new lands" waiting for cultivation (except land cleared by cutting and burning the last of the tropical rain forests). The total cultivated area in the world has remained virtually constant for the last two decades. Some new lands are being cleared — as mentioned above — but this is roughly compensated by the loss of formerly fertile land in other areas by erosion and desertification. Most tropical forest soils are not very fertile to begin with, and they lack organic (humus) content. Once cleared, they are rapidly exhausted of soluble nutrients by leaching as well as cropping. Eventually they become degraded to a sort of hardpan.

Moreover, experts say that the potential increases in agricultural yield available from the use of chemical pesticides, fertilizers, and plant breeding have already been largely exhausted in the industrialized countries. The potential gains from wider application of these "industrial" methods, for instance in China and India, are surely worth seeking. But they are definitely limited. Erosion is carrying off topsoil four or five times faster than it is being created by natural processes, even in Europe and America where soil conservation is practiced. In Asia the problem is more acute, due to deforestation, overgrazing, and cultivation of upstream hillsides that should have been left forest covered.

Some important rivers are now virtually dry part of the year because of excessive withdrawals. The Yellow River in China and the Colorado River in the United States are two examples. Moreover, freshwater (for irrigation) is rapidly becoming a source of conflict between upstream and downstream users. This conflict is already acute in the case of the Jordan River and the Euphrates River (which starts in Turkey and flows through Syria before reaching Iraq). It is likely to be a major problem soon for the Nile. Moreover, nonrenewable groundwater is already becoming seriously depleted

---

[3]For a super-optimistic view see *The Next 200 Years* (Kahn *et al.*, 1976), *The Resourceful Earth* (Kahn, 1984), and *The State of Humanity* (Simon *et al.*, 1995).

and/or contaminated in many regions of the world — including America's Great Plains — where intensive irrigation *cum* chemical agriculture have been practiced for only a few decades.

With regard to the possibility of continuing to increase the productivity (yield) of existing arable land, there is a continuing push to develop improved varieties and higher photosynthetic efficiencies. Biotechnology was — until recently — beginning to be harnessed to increase food production. There was optimistic talk of a "second green revolution." But, consumer activists have raised fears of genetically modified (GM) foods. Some of these fears may be exaggerated but others might be legitimate. In any case, for some years past global grain production per capita has actually been declining. Thus, significant incremental improvements in the productivity of the available farmland and crop types will be needed just to keep up with inevitable population growth.

With regard to renewable resources, such as forests, fisheries, and groundwater, it has long been known that there is a level of exploitation that can be sustained indefinitely by scientific management. However, beyond that level, harvesting pressures can drive stocks down to the point where recovery may take decades, or — in the case of fisheries — may never occur at all. Many fisheries appear to be in this situation at present, though admittedly it is difficult to establish precisely what fish populations are self-sustaining.

The point of this discussion is that modern industrial technology, as applied to agriculture, is approaching limits. This is not to deny that some new technological possibilities, especially through genetic engineering, may be capable of increasing output of food and fiber significantly without requiring more land or water than is now available. For instance, methods of artificially accelerating tree growth may compensate for some net decrease in the area devoted to plantation crops. Marine farming can be expanded, in some areas. But these are only marginal improvements. The only remaining really big potential for increasing the human food supply would be to cut back sharply on the global consumption of grain-fed red meat, especially beef and pork.

Notwithstanding the foregoing, it has been suggested, especially by the World Bank, that economic growth, in general, is actually good for the environment.[4] Proponents of this proposition go on to assert that economic growth *per se* has been a significant factor in the improvements that have occurred in North America, Europe, and Japan over the past two decades. The empirical backing for this argument, as recently articulated,[5] is based on the so-called "environmental Kuznets curve," or EKC. This is the inverted U pattern that one sees if certain environmental pollution mea-

---

[4]This argument has been used to support the free trade agenda, for instance, on the grounds that free trade is good for economic growth and therefore good for the environment.
[5]Notably in the 1992 World Bank "World Development Report" (World Bank, 1992).

sures (such as smoke, sulfur dioxide, and organic water pollutants) are plotted on a scatter chart against national income (GNP) per capita. The poorest countries, without any industrial base, exhibit low levels of these types of pollution. The pollution level rises as more fuel and other resources are processed. As GNP per capita increases, the level of pollution also increases for a while because higher GNP is correlated with more intensive industrial development. However, as countries become still more prosperous, services begin to predominate over manufacturing, and the curves turn down again — hence the inverted U.

However, the empirical correlation between prosperity and environmental improvement at higher levels of GNP per capita does not actually prove causation. For example, lower rates of sulfur oxide and particulate pollution in the industrialized countries is probably due mainly to the substitution of natural gas for coal as a domestic fuel in Europe and North America. This was only partly attributable to investment in gas pipelines; it also depended on the availability of nearby and convenient sources of the gas itself. Most of the less developed countries are not well endowed with natural gas.

Also, the inverted U phenomenon only applies to a relatively small subset of indicators, notably the ones mentioned above. In many other cases (e.g., nitrogen oxides, chlorofluorocarbons, methane, toxics) the relationship doesn't hold.

The inverted U argument is nevertheless taken seriously by many economists for a different, more theoretical, reason: namely that, as people get richer they will value the environment more and pay more to protect it better. For instance, people living in slums will drink contaminated water if they must choose between feeding their children and buying fuel to boil their drinking water. However, while this is doubtless true, it doesn't prove that economic growth is good for the environment, in general.

To be sure, there is some evidence that the environment of the United States is in better shape now than it was two or three decades ago. For instance:

- Forests are regrowing in parts of the Appalachians (eastern U.S.), where farmers and charcoal-burners once cut.
- Eroded cotton plantations in America's "Old South" have given way to tree farms that feed the pulp and paper industry. The soil erosion has now all but stopped.
- Most kinds of air pollution have decreased in the U.S. Lead pollution is the best example: It is down 90 percent. Particulate emissions are also down significantly, along with carbon monoxide. Pittsburgh, Cleveland, Chicago, and other industrial cities of the Midwest — and the same is true in Japan and western Europe — are much less smoky than they once were. Sulfur dioxide emissions overall are down slightly in the nation as a whole, and somewhat more sharply in the big cities.

- The greatest environmental improvements since the 1960s have been in water pollution. The major rivers and lakes of North America, and also of western Europe and Japan, are now much cleaner than they were two decades ago. Lake Erie is no longer a "dead" lake. There are fish — even salmon — once again in the Hudson, and the Thames.

The first big environmental success story was the regrowth of the Appalachian forests. It is true that forests now cover the hills that were once farmed or cut over for charcoal burning. The same phenomena are visible, on a reduced scale, in Europe. This occurred mainly due to the substitution of iron for wood in shipbuilding and coal in place of charcoal for iron smelting, brick-making, and other industrial purposes. The regrowth of the forests also occurred because the opening of transportation routes to the fertile Midwest made agriculture uneconomic in the hills. Farms were abandoned in the Appalachians because they could no longer compete. The new land opening up in the Ohio River Valley was much more productive and, with the opening of the Erie Canal, and the railways, it became accessible.

The shift from cotton to tree plantations, and consequently reduced soil erosion in the southern states of the United States, occurred only because the original rich topsoil had been washed into the sea. "King cotton" abandoned the South, because the productivity of the land had declined drastically due to massive erosion. Also, thanks to government subsidized irrigation projects, it became cheaper to produce cotton in Texas and further west. Tree farms replaced cotton fields in Georgia, Alabama, and Mississippi because the eroded land was — and is — literally worthless for other agricultural purposes. Monocultural tree farms are better than nothing, but they do not harbor diverse biological communities.

I should add that reforestation is a local phenomenon. As regards the global picture, the World Wildlife Fund (WWF) has compiled the first detailed database. Global forest-cover (defined as 40% crown cover) has fallen from 50% to 31%, of which 10% is tropical, 13% is northern coniferous, and 7% is temperate broadleaved. Only 6% of this total forested area is protected and the annual loss rate from deforestation is now 1% per year. In the United States, only 2% of the original old growth forest remains uncut, and cutting has accelerated.

In the case of air pollution, the reduction in particulates and sulfur occurred partly as a consequence of the phasing out of coal as a fuel for domestic and most industrial purposes. This occurred after World War II throughout the eastern states of the United States as soon as pipelines to carry natural gas from the Louisiana and Texas oil fields were built. In Europe, the availability of oil and natural gas from the North Sea, and from Russia, may have been the single biggest factor in reducing air pollution. The building of nuclear power plants was another factor. Air pollution from automotive exhausts — except for lead — is only slightly lower in American cities than it was in 1970 and in most of the world it is much worse. The decline in lead pollution was entirely due to the federal ban on use of gasoline with tetraethyl lead by new passenger cars, which began in the early 1970s. However, nitrogen oxide emissions have not fallen; they have increased.

Further environmental progress was due to regulation, not economic growth. The improvement in water quality in North America came about for two reasons. First, government regulation forced communities to borrow large sums of money (subsidized by the federal authorities) to build sewage treatment plants. The story is similar in Europe and Japan, though salmon do not swim, as yet, in the Rhine or the Elbe due to industrial and mine wastes. Second, government regulation forced industry to install water treatment facilities and/or to otherwise dispose of many wastes and pollutants that had formerly been dumped indiscriminately into rivers. In many cases, such as the chemical and metallurgical industries, the major change has been that pollutants formerly dumped without treatment are now concentrated into "sludges" and buried in landfills. In a few cases industrial processes have been changed significantly to convert wastes into fuel or into salable by-products. The pulp and paper industry, once the most notorious of all polluters, is perhaps the best example of real progress in this area. Nowadays most of the so-called "black liquor" (lignin waste) is burned to recover heat and chemicals. But this would not have happened without regulation.

The so-called Clean Air laws of the 1970s, which have been amended and tightened several times, have forced the use of so-called catalytic converters to reduce emissions from motor vehicles. Europe has followed with similar legislation. Regulation has also forced many large industrial polluters to install pollution control equipment and (in some cases) to improve their operations so as to reduce emissions into the atmosphere. The banning of lead, DDT and other dangerously long-lived pesticides, PCBs, and CFCs was also entirely due to actual regulation or (to a lesser extent) voluntary changes by industry in the anticipation of future regulation.

To summarize, the favorable changes in our environment have occurred either (1) because of technological developments (such as the availability of natural gas and increasing electrification) that had no relation to environmental problems or (2) because of regulation. I know of very few cases — if any — where totally unregulated competition in the private marketplace solved an environmental problem, for its own sake. Technological alternatives sometimes emerge to solve environmental problems as a result of purely private responses, in *anticipation* of future regulatory or economic incentives imposed by government. Technology is not autonomous. It does not respond to social needs unless there is a customer, with cash in hand, willing to pay the price. In the case of environment protection, the only paying customer (apart from a few foundations) is government itself or industry anticipating government regulation. That rule is unlikely to be repealed in the future.

Now, I return to the question that was posed earlier: Is economic development good for the environment? In brief, the answer is "Sometimes but not always." Consider four stylized facts:

1. Economic growth is historically correlated closely with commercial energy consumption (not to mention other resources); although the E/GDP ratio does tends to decline in richer countries, per capita consumption nevertheless tends to increase (Figure 4.1).
2. Most environmental problems (and many health problems) arise from pollution.
3. Harmful pollution is, for the most part, directly traceable to the use of fossil fuels and/or other materials, such as toxic heavy metals or chlorinated chemicals. In short, every material extracted from the environment is a potential waste, and it usually becomes an actual waste within months or a few years at most.
4. Most wastes that are disposed of, either in air or in water, are capable of causing harm to the environment, if not directly to human health.

Many environmentalists, and some environmental economists too, argue from these facts that "sustainable development" is literally a contradiction in terms. This may be putting it too strongly. On the other hand, a majority of business and political leaders appear to assume that the problems are real but minor and we should defer the more serious cleanup efforts until our society is richer, or until we know more, which will be sometime in the future. (The question: how rich must we be, and how much must we know before tackling the tough problems is never addressed). According to this view,

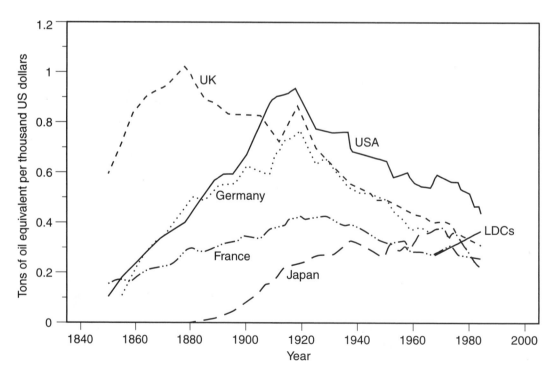

Figure 4.1 **Long-term trends in commercial energy intensity.**

growth is the key to everything and only minor changes in current technology and/or regulatory policy would suffice to overcome any known environmental threat.

In fact, many so-called environmentalists appear to believe that the most serious environmental threats we face are those directly related to human health, namely, contaminated water or food or skin cancer. Lesser or minor problems that get attention include forest die-back ("Waldsterben") due to acid deposition in forest soils, oil spills, dirty beaches, litter, haze, bad smells, noise, and so on. Few people worry about disturbances to the large scale bio-geo-chemical cycles, for instance. Yet earth's habitability rests on the integrity of these cycles.

In short, I don't fully agree with either of the two standard positions with respect to sustainability. On the one hand sustainable development is not a contradiction in terms. On the other hand, it is not a second-order problem that can be put off until we have more money or knowledge. Environmental protection is essential, not optional. But, at the same time, economic growth need not be antithetical to environmental protection. Ecologically sustainable growth is theoretically possible. But it is surely not what is happening now; nor is it inevitable.

4.2.2 THE SHARP END: POLLUTION, DEPLETION, HEALTH, AND WEALTH

People living in the wealthy and (relatively) underpopulated OECD countries[6] haven't experienced the sharp end of environmental degradation. It is relatively easy, sitting in a window overlooking a well-tended park, or a pasture with cattle grazing in the distance, to think that surely all these problems are being exaggerated by "doomsayers." Besides, isn't nature perfectly able to recover from a little localized damage? The answer to that is usually "yes." There is another side of the story, however. Most people live in crowded cities or tiny rural villages that are anything but park-like. These people are dying — very slowly but in enormous numbers — of disease, poverty, or pollution. They are just out of sight of the big new luxury hotels where the business elite gather. This phenomenon is being experienced today in many countries. In fact, it would be more accurate to say *most* countries.

This topic is more complicated than it seems, just at first. The problem, in a nutshell, is that wealth for the elite in developing countries is now being produced, in large part, not by any creative process but by "spending" natural capital. Throughout the so-called "developing world" natural wealth — such as rich agricultural land, groundwater, mineral deposits, and forests — is being appropriated (if not expropriated) by government agencies for mega-projects and/or by multinational corporations. The immediate profits are typically redistributed to the wealthiest and least deserving. Meanwhile, dispossessed tribesmen and landless rural peasants — by the millions — are being evicted and forced to work in mines or migrate to cities where they are

---

[6]Obviously this does not apply to Japan, nor to parts of Europe, especially the low countries, western Germany, and northern Italy.

unwanted and unfitted to work productively. There they are starved and exposed to water and air pollution on a scale we in the West can scarcely imagine. The inequity of natural resource exploitation and environmental degradation deserves closer scrutiny.

I cannot summarize the environmental miseries of Afghanistan, northeastern Brazil, Central America, China, Egypt, Haiti, India, Mexico, Nigeria, North Korea, Peru, or the former USSR in a few paragraphs. I can only cite a few tell-tale statistics and try to make one overwhelmingly important point. The point is that apparent economic growth — even where it exists — is largely being "paid for" by degrading natural capital. The natural capital that is being degraded includes quantifiable components like minerals taken from the ground, virgin forests being cut and burned for charcoal or pulped into paper, topsoil being lost from farmland, and fisheries whose productivity has been depleted by overfishing. To put it in domestic terms, the national account statistics, as currently structured, assume that if you sell your furniture and your house, the proceeds are counted as part of current income. This is exactly what countries are doing.

All of these kinds of depletion can and should be accounted for in official national accounts statistics, but they are not (yet). As a consequence, gross national product for many countries includes a large component of capital consumption — capital that is being depleted but not being replaced.[7] It means, of course, that GNP growth rates are being overstated by failing to allow for depletion of natural capital. Quantitative studies of the extent to which apparent GNP growth rates for some countries have been exaggerated by this omission have been carried out by Robert Repetto and his colleagues at the World Resources Institute in Washington, DC, in collaboration with scholars in several countries. On the basis of detailed quantitative analysis of Costa Rica, for instance, it was found that the asset value of forests and fisheries declined by $4.1 billion (1985 dollars) from 1970 to 1989, more than one year's GDP. Topsoil loss from erosion amounts to 13% of the value added by livestock production, 17% for annual crops, and between 8% and 9% for agricultural production overall. The traditional fishery has essentially been wiped out by overfishing and pollution. Profits were negative in 1988, even valuing fishermen's time at the subsistence wage level. The Philippines and Indonesia exhibit similar phenomena. Depreciation of natural capital in Indonesia was in the neighborhood of 14% of GDP in 1985, and rising steadily.[8] Recent studies suggest that deple-

---

[7]The UN Statistics Office and several national governments are working on a new system of "green" national accounts that is supposed to rectify some of the omissions in the existing statistics. Meanwhile, governments (and the World Bank) continue to use the old statistics and to judge their relative performance thereby.

[8]Indonesia is a major oil and gas producer; these accounted for much of the country's income during the 1970s and 1980s. But, the statistical problem is complicated by the fact that new discoveries are treated as additions to resource stocks—which they are not. In 1974 reported Indonesian oil reserves increased very sharply (apparently due to changes in U.S. tax laws) resulting in a statistical "blip."

tion of forests, farm and grazing lands, and water shortages amounted to between 5.5% and 9.5% of China's GDP in 1990.[9]

But depletion of forests, fisheries, and underground mineral resources is only a part of the problem; perhaps the least part. Another sort of depletion is occurring in those parts of the world where rapid industrialization is occurring without adequate environmental protections. Just to consider one example: People who live and work in heavily polluted air, as coal miners in Wales and West Virginia once did, and coal miners, in northern Czech Republic, East Germany, Poland, the Donetz basin of Russia, and coal mining regions of India and China still do, sacrifice years of potentially productive working life to silicosis and "black lung" disease, not to mention bronchitis, emphysema, lung cancers, and other debilitating illness. The economic cost in terms of lost working man-days is very considerable. Workers in many coal mining regions of the world have an average life expectancy in the low fifties. Workers in some Chinese coal mines cannot work after the age of 36 (Smil, 1996).

While large new electric power plants nowadays generally incorporate electrostatic precipitators (ESPs to remove the fly ash from the effluent stream, flue gas desulfurization is almost unknown in these countries. Yet all hydrocarbon fuels average 1% to 3% sulfur by weight, and some of the coals in southern China, for instance, average 4% or more. Worse, most of this low-grade coal from local mines is not burned in large efficient electric power plants (which get the best quality coal), but in small factory boilers or household stoves which burn the fuel inefficiently and remove none of the ash or the sulfur. Thus several hundred million Chinese — not to mention Russians, Poles, Czechs, Indians, Turks, and others — are constantly exposed to a dense pall of sulfurous smoke. It is very hard to attach realistic economic values to this and other forms of pollution. However, based on very conservative valuations of hospital costs and lost work days by economically productive adults only, Smil (1996) has estimated the cost at between 1.7% and 2.5% GDP for China in 1990. Estimating costs more liberally would increase these totals considerably.

4.2.3 THE ROLE OF MARKETS

I now want to make another fundamental point here, namely that the present economic system is institutionally unlikely to cure itself. Here I am taking direct issue with the assumption, shared by most business and financial writers and some conservative economists, that free competitive markets will automatically create the necessary incentives (via price signals) to end unsustainable practices. This faith arises from a central thesis of neoclassical economics: namely that, given the right incentives (prices) and time enough, technology is capable of finding a way to avoid essentially *any* physical resource bottleneck.

---

[9]See Smil (1996) and other studies cited therein.

To be sure, markets generally do function quite well within their domain. For instance, alarmists since Thomas Malthus have warned of the forthcoming exhaustion of natural resources, from land and water to minerals and fuels. Yet, with the sole exception of forest products (and land itself), prices of most classes of commodities and exhaustible natural resources have actually tended to decline continuously over the decades. This fact stands as a continuing reproach to those who underestimate the power of efficient markets to call forth innovative activity.

However, the market is never *perfectly* competitive, nor does it encompass *all* of the environmental goods — or services — that need to be preserved. The last point is the crucial one. Consider the environmental services such as air to breathe, rainfall, the ozone layer, or the carbon cycle. These services are essential to the continuation of all life on earth. They are "indivisible" in that they cannot be subdivided into small pieces that can be individually owned or exchanged. They are what is called "common property." Everybody — hence nobody — owns them. They cannot be bought or sold as such in any marketplace. Scarcity of environmental services such as these does not raise their prices (because there are no prices) and hence does not inhibit demand. Anyhow, there are no producers who could or would respond to price signals. When demand exceeds capacity, in the case of a common property resource, the result is degradation and destruction. This is the "tragedy of the commons."[10] In fact, when a common property resource is scarce, as is now the case for a number of fisheries as well as many of the large mammals such as great whales, tigers, and rhinos, there is likely to be an intensified competition to capture the last few that remain.

For all these reasons, the standard economic model of market-driven resource allocation does not apply to environmental services. In principle, when market transactions result in damage losses to "third parties" external to the main transaction, those who suffer the damage should be compensated by those who gain by the transaction. The market price of whatever good is exchanged must be high enough to allow for such compensation, in addition to production costs and profits. The higher price would also inhibit demand for such goods, creating economic incentives to minimize emissions or damages. This notion has become popularized as the "polluter-pays-principle" or PPP. However, the principle would not have had to be formalized if it were built-in to the market system. In fact, PPP is an ideal, not an actuality.

For instance, a number of quantitative studies, using a variety of methodologies, have suggested that the real social (e.g., health) costs of using coal are several times higher than the current market price. The same is true for gasoline, in the United States at least, and probably also in Europe. Hence the existing market price is far below what it should be to optimize the balance of benefits and damage costs from use. It is difficult to determine the

---

[10]The reference is to a famous article by Garett Hardin (Hardin, 1968).

"right" prices in cases like this where the market price does not reflect all the social and environmental costs of use. Practical difficulties abound. Damage costs are much greater in densely populated urban areas, for instance, than they are in rural areas. Worse, a number of governments — including Germany and China — actually subsidize the mining of coal, rather than taxing it heavily, as they should to implement the PPP.[11] The reason, of course, is that the Chinese government wants to encourage industry, and coal is the main source of energy for industry in China. In Germany, the coal miners unions are politically powerful and they use their power to protect their own jobs regardless of the public interest or other considerations.

To summarize, it cannot be assumed technology will come to the rescue even in the public domain of common property resources, where normal market mechanisms of profit and loss don't function effectively. It follows that, to achieve long-run eco-sustainability, governments will have to intervene much more vigorously to create the missing economic incentives. They must correct the distortions that decades of growth-oriented policies have introduced into markets. However, I do not share the conclusions of some environmentalists that the only answer is an end to growth itself. In fact, it seems to me that a "no-growth" policy is a non-starter. It would be politically unacceptable to the developing world, even if Americans, Europeans, and Japanese could be persuaded to adopt it. However, as I have said, I am convinced that economic growth and a clean and healthy environment can be reconciled.

The point is that current economic and environmental trends, supported and encouraged by current tax and trade policies, are definitely antithetical to ecological sustainability. As far as trade is concerned, the first-order effects of reducing barriers to trade are clear. They are (1) increasing goods traffic and (2) continued exploitation of primary extractive activities in remote areas at the expense of secondary and recycling activities in the importing countries that might otherwise compete with them. Both of these effects, in turn, are antithetical to the environment.

To take one example, consider the impact of goods traffic. It is hard to believe (but true) that German potatoes are currently shipped, by truck, across the Alps to Italy for washing, and then shipped back to Germany for frying! Dutch pigs, fed on manioc and other feeds imported from Thailand, are also shipped in trucks across the Alps to Italy where they are slaughtered and processed into "Parma" ham, which is then sold all over the world. The biggest fishing port on the Adriatic coast of Italy is now totally dependent on imported fish brought by refrigerator ship from the South Pacific, the local fisheries having long since been depleted. Meanwhile, Austrian attempts to restrict heavy truck traffic across the Brenner Pass — the only all-weather pass currently capable of carrying such traffic — because of

---

[11]The first to suggest the use of taxes to correct for market failures was the Cambridge economist, Cecil Pigou. Such taxes are often referred to as "Pigovian" taxes.

local noise and pollution problems, have been strongly opposed by the Germans and Italians as "restraint of trade." One of the reasons the Swiss elected not to join the European Union (EU) was for fear of being forced by EU rules to allow more such trans-Alpine traffic, thus permitting heavy trucks to pollute their ecologically fragile high valleys. The Swiss are building a major new railway tunnel, instead, to accommodate future north–south goods traffic.

With regard to the second environmental effect of trade, one consequence of reducing trade barriers is that it is getting easier for rich countries to export their industrial (and other) wastes. This is a rapidly growing business, despite international agreements restricting it. Somewhat surprisingly, perhaps, the "green" Germans are the world's biggest exporters of wastes, partly to Poland and other parts of eastern Europe, and partly to more distant countries like Indonesia where German packaging wastes, for instance, are sold as raw materials — thus undercutting local scavengers and reducing the incentives for German industry to develop uses for these materials, as was originally intended.

The lowering of trade barriers and transport costs, jointly, have favored large centralized producers with good access to local raw materials and ocean shipping. By the same token, these factors have reduced the effectiveness of environmental protection laws and reduced incentives to develop efficient ways of reuse, repair, renovation, remanufacturing, and recycling materials in a local region. Obviously, environmental protection regulations are harder to enforce and "take back" legislation that encourages manufacturers to be responsible for the final disposal of their products, is harder to justify when manufacturers are located far away.

To summarize this section, economic development along standard lines is not good for the environment. The "invisible hand" of the market does not necessarily have a green thumb. On the contrary, increasing labor productivity, as a response to perceived needs for increasing competitiveness, means decreasing labor intensiveness by further increasing capital intensiveness, materials intensiveness, and energy intensiveness. Increasing resource productivity, on the contrary, would imply a reverse of all of the above.

4.2.4 SUMMARY AND IMPLICATIONS

This section makes several important points. The first of them is that the natural environment provides essential services that cannot be replaced by man-made capital or technology. Critical environmental problems — global warming, ozone depletion, acidification, soil erosion, deforestation, exhaustion of underground water supplies in many areas, toxification of soils, fishery depletion, and loss of biodiversity — cannot be prevented or "solved" by technological intervention alone. True, technology can make a difference, mainly by reducing the rate at which we approach the point of planetary no-return. Renewable energy technologies, such as photovoltaics would help, and eliminating the use of fossil fuels would also help a lot in reducing the rates of acidification and toxification.

But many of the projected environmental damages are irreversible — or, at least, very long lasting — and neither known nor imaginable technology can reverse any of them. There is no conceivable technology to lower the sea level once it has risen. No technology can restore the glaciers or the ice-caps once they melt. There is no technology to remove greenhouse gases from the atmosphere, to manufacture ozone in the stratosphere, to detoxify soils, to replace fossil groundwater or eroded soils. Above all, extinct species cannot be recreated — except in Hollywood.

In short, technology is still a minor player, at best, in the area of environmental repair and rehabilitation. Even where we know technology can help (solar energy, for instance), it will not come to the rescue of the environment automatically. Established interests may well gang up to oppose and delay it. As always the potential losers know who they are, and they have financial power and political clout. The potential winners are disorganized and some of them do not know who they are. Anyhow their potential gains are in the future. The contest is uneven. A technology like photovoltaics, that could be competitive even now on a level economic playing field, may be delayed for many years by powerful forces that combine to tilt the playing field (called the "free market") in their own favor.

A second major point is that environmental protection and economic growth, along current trajectories, are indeed antithetical. It is not true — as has been claimed by the World Bank, among others — that uninhibited growth is always good for the environment. It is true that there have been some environmental gains in the last three decades, especially in the richest countries. But, on closer examination, there is no serious case for arguing that these environmental improvements occurred because of economic growth. In the United States they are mainly the result of one of three things: (1) the substitution of coal for wood as a fuel industry in the 19th century, (2) the increased availability of natural gas (to replace coal) as a domestic fuel in the postwar period, or (3) direct regulation. The two technological substitutions in question may have contributed to economic growth, but they were not consequences of it.

In developing countries, like China, the relationship between economic growth and environmental damage is far more negative. Depreciation of natural capital, which ought to be deducted from GDP (but is not) typically ranges from 3% to 10% of GDP, depending on circumstances. Health costs, paid for by reduced life expectancy and lost working time, due to exposure to air and water pollution, add to this total. These costs are directly related to industrialization without adequate attention to environmental protection.

The third important point is that unfettered market forces have not, and will not, create technology to solve environmental problems except (as in the above examples) by accident. It is hard to find any reason to think that "painless" market-driven technological progress will eliminate greenhouse gases such as CO_2 except, and unless, some other nonpolluting source of energy can be shown to be cheaper and more profitable. For instance, the oil industry is currently secure in its oligopolistic dominance. This is partly the result of the bounty of nature, partly a legacy of past and present subsidies to producers of oil and to users of motor vehicles,[12] and partly a consequence of the sheer financial muscle of the combined oil and auto industries. These industries are well protected, by economies of scale and the costs of market entry, from potential competitors, such as electric vehicles or photovoltaic hydrogen-powered fuel cells.

The oil industry expects to go on indefinitely, drilling more and more, deeper and deeper wells, even in the last Arctic wilderness areas, in the last tropical jungles, and under the ice. The industry has no serious plans to develop technological substitutes for fossil fuels or fossil fuel burning vehicles. Its current strategy in fending off critics is to insist that any change would be enormously costly, and that no change is warranted, pending "scientific proof" of need. The energy industry is much more likely to resist and obstruct the introduction of promising new technologies, such as renewables and fuel cells, than to promote them. If obstruction fails at last, Exxon, BP, Mobil, and Shell can always buy out the competition. (*Note added in proof:* Since this was written, both BP and Shell have begun to invest in nonfossil energy sources. Exxon and Mobil have merged.)

4.3 Selecting an Appropriate Technology

One movement concerned with the appropriate use of technology and its suitability of smaller scale settings is known as intermediate or appropriate technology. *Appropriate technology* can be defined as the identification, transfer, and implementation of the most suitable technology for a set of conditions. Typically the conditions include social factors that go beyond routine economic and technical engineering constraints. Identifying them requires attention to an array of human values and needs that may influence how a technology affects the novel situation. Thus, "appropriateness" may be scrutinized in terms of scale, technical and managerial skills, materials and energy, physical environment, and social and human values.

Often, appropriate technology also implies that the technology should contribute to and not distract from *sustainable* development of the host country by providing for careful stewardship of its natural resources and not degrading the environment beyond its carrying capacity. Proponents of appropriate technology often focus on the development and transfer of technologies to less developed countries that is labor intensive, easy to understand and use, and appropriate for improving the lot of the poor rural masses.

---

[12]Subsidies, direct and indirect, to automotive transportation have been conservatively estimated at $300 billion in the United States alone (WRI, 1993). They include direct subsidies extraction (the depletion allowance), subsidies to local and state road building and maintenance, military expenditures to protect the "lifeline" to middle Eastern sources, costs of health care to uninsured road accident victims, free parking, health damage (and costs) due to air pollution, and so on.

A major criticism of appropriate technology is that it is often preoccupied with developing low-tech products that may not foster the development of technology capabilities in emerging nations.

In the following section, Barrett Hazeltine provides an excellent summary of appropriate technology.

4.4 Appropriate Technology

By Barrett Hazeltine
Professor
Engineering Division
Brown University

4.4.1 DEFINITION OF APPROPRIATE TECHNOLOGY

What is "appropriate technology"? Every type of technology might be appropriate somewhere, so in a trivial sense, all technology is appropriate. The term **appropriate technology** is used here in a narrow sense. The U.S. Congress's Office of Technology Assessment (OTA, 1981) characterizes appropriate technology as being small scale, energy efficient, environmentally sound, labor intensive, and controlled by the local community. The Intermediate Technology Development Group in London, an organization that works toward the betterment of developing countries, uses nearly the same description in its journal *Appropriate Technology,* but adds that the technology must be simple enough to be maintained by the people using it.

A central concept of appropriate technology is that the technology must match both the user and the need in complexity and scale. Some examples of how these two can be mismatched follow. A village in Botswana received ten combines (grain harvesting machines) as part of an aid project. When these machines broke down there were no trained mechanics and no spare parts. The combines lie rusting at the edge of the fields as the farmers use traditional methods to harvest their grain. The underpinnings necessary to support this equipment do not exist in the community. A small fishing community in the Marshall Islands receives a space age photovoltaic system to provide electricity for lighting. Only after installation is it determined that there is no significant need for a system so sophisticated. The well-known tragedy caused by the explosion of a fertilizer factory in Bhopal, India, is another example of technology not matching local conditions, in this case presence of trained staff. In the United States, people point to the use of electric heating in houses, especially if such systems require new power plants, as a mismatch between a fairly simple need and an elaborate solution. One could say the same thing about one person using a full-sized automobile for commuting.

The proponents of appropriate technology believe it is applicable to many situations both in the United States and in the so-called Third World. Certainly, it is an alternative approach that should be seriously considered. On the other hand, it does not make sense everywhere. Like other alternatives it needs critical and serious evaluation.

4.4.2 HISTORY OF APPROPRIATE TECHNOLOGY

Small-scale technology has been used for a long time, of course. The people who colonized America, for example, had no other options. The modern appropriate technology movement is attributed, however, to E. F. Schumacher, a British economist. Schumacher (1973) was an economist for the Coal Board and also an advisor to the government of Burma — now called Myanmar — and later that of India. He wrote several papers from 1955 to 1963 entitled "Economics in a Buddhist Country," "Non-Violent Economics," and "Levels of Technology" which were eventually incorporated in a book called *Small Is Beautiful*. His ideas inspired the creation of the Intermediate Technology Development Group referred to earlier. The approach gained attention during the 1960s coincident with the social responsibility movements of those years. Several world leaders, including President Carter, became active supporters and many projects and support groups were formed. Their methods are now being used in many places. In the United States, the National Appropriate Technology Center offers information by telephone or mail, and most states have either government or private groups doing relevant research or actually building projects. U.S. AID, the foreign aid division of the State Department, now supports appropriate technology projects in many parts of the world.

4.4.3 EXAMPLES OF APPROPRIATE TECHNOLOGY

Appropriate technology can be used for energy, for food, for health care, for sanitation, for transportation, and for meeting other needs. Water wheels, wind generators, and photovoltaics produce electrical energy. In the United States photovoltaic devices are the cheapest way to provide electricity to certain remote sites. Solar hot water systems are also less costly than oil or gas powered systems in many places in the northeast. Efficient cook stoves use one-third the fuel of ordinary stoves. Vegetables are grown intensively with hand tools and without chemical fertilizers, herbicides, and pesticides. Aquaculture can be done on a small scale. Oral rehydration therapy is a simple technology that has saved many children struck with diarrhea. Composting toilets and pit latrines are effective and can be constructed by a nonprofessional. Methane digesters, which can also be built and operated by a nonprofessional, convert manure and other wastes to a combustible gas and an odorless fertilizer. A bicycle can be adapted to carry much baggage and more than one person. Many houses in the Third World are built by the eventual owner of pressed earth bricks.

4.4.4 BENEFITS OF APPROPRIATE TECHNOLOGY

Schumacher's focus was on jobs — people in the Third World desperately need employment. He argued that the capital investment required to create western, high-technology jobs is too large to be practical. A modern foundry requires, perhaps, $20,000 for each job created. Existing technologies, such as those used by a village blacksmith, correspond to an investment of $2 per job. Something in between, intermediate, is needed, say, $200 per job, similar to a small machine shop in the United States. This intermediate technology could be much more productive than what now exists and could serve to move a country out of poverty without requiring huge capital investments.

Schumacher further argued that the $200 technology would be close to existing methods and so would lead to improvement in skills for many local people. The $20,000 technology would be so technical that highly trained specialists would be required. The cost of training is high, so it would not be possible to train many specialists. The conclusion is that intermediate technology increases the knowledge level for most of the population, whereas high technology produces a class of experts separated from the rest of the economy. If technology only slightly different from that existing is introduced, for instance, an improved farming method, then even those not directly involved with the new method can learn about it. More people therefore benefit from the improvement.

A final difficulty with $20,000 technology, like an integrated circuit fabrication plant, is that it uses imported supplies, produces mostly export items, and needs expertise and components from outside the country for maintenance and improvement. Thus, the high-technology factory is not an effective way for the country as a whole to develop.

A major reason for using appropriate technology, as Schumacher argues, is that it provides goods, services, and jobs that will not be provided any other way. No company or organization will be able to invest enough in high-technology factories in developing countries to provide sufficient jobs. An analogous argument is that the cost of tearing down old houses in U.S. cities and replacing them with modern high-rises is more than the government can afford. If a limited amount is of money available for low-cost housing, an appropriate technology approach, such as training people to rehabilitate their houses themselves, will have much greater impact. In the Third World few farmers can afford the equipment and chemical supplies to make farming there similar to that in the United States, but for a reasonable cost many farmers could be given the small machines and training necessary to improve their productivity.

Introducing a new technology related to an existing one has two other advantages over a completely new technology. It is less disruptive to the social structure and it can be adapted. A factory or mine, which takes young people from villages to a city, affects village life much more than a small workshop located in the village. Augmenting the training of traditional doctors — so-called "herbalists" — often improves health care more than building Western-style hospitals, because people feel more comfortable using herbalists and the cost is less. If a new technology is similar to the existing one, the user can adapt it most effectively to the local situation. The user can adapt high-efficiency wood stoves to burn straw, or whatever fuel is available, while gas or electric stoves are much less flexible. Even methods of producing sophisticated equipment can be adapted. In one assembly plant in Taiwan, for example, small groups working in teams, not on assembly lines, produce pocket calculators, just as people worked together in traditional society. The process is more cost effective than U.S. production methods.

Appropriate technology benefits a society more than high technology because it fosters self-reliance and responsibility. Compare a person who has renovated her own house with one who has been assigned an apartment. Who is probably more prepared to take responsibility in a job? Similarly, an experienced owner of a small farm is probably better prepared for a leadership role in the community than a worker on a mechanized farm. Appropriate technology not only teaches skills, it also gives people experience in solving problems and getting things done. Many people would much prefer to be their own boss, to be responsible for themselves, than to work for another. Appropriate technology is more likely than high technology to give this opportunity, because smaller organizations mean more organizations, with more leaders. A trend in health care is to involve the patient more in both decisions and treatment — kidney dialysis at home is an example. This trend resonates with the self-reliance aspect of appropriate technology.

This same quality, of being responsible for one's own success, applies to groups also. Appropriate technology is small scale and thus the people directly involved can have significant control. A small machine shop, with general-purpose machines, can adjust its products and meet market needs more effectively than an automated factory, partly because the tools are more adaptable, and partly because fewer people have to be convinced in a small group. Community self-help groups are often more effective at meeting housing needs than bigger, city-wide organizations because the community groups are closer to the people and their needs. Of course, situations do exist where large size is essential; it is hard to think of an economic car factory which is small, and a small group of artisans making pottery may need a large organization to do their marketing, because it may be difficult for a small company to gain access to a market.

Advocates of appropriate technology point out that not only does it foster self-reliance and responsibility, but it also fosters other desirable attitudes, including cooperation and frugality. If machines are not available, then people's strength must substitute, and this usually means that groups of people must work together. In the colonial United States the neighboring farmers gathered to help put the roof on a new house. Neighborhood associations or cooperatives have done many of the successful appropriate technology urban projects. The reason appropriate technology promotes frugality may be simply that the users have less to waste or it may be that hand work encourages care and concern, which translates into less waste.

A related reason why some people are attracted to appropriate technology is the *type* of job it produces. It is not surprising that a person operating a self-sufficient farm is willing to work 55 hours a week while an assembly line worker feels 40 hours are too many. The farmer has more interesting, challenging, and rewarding work. The satisfaction one gains from seeing a farm succeed is, for some people, worth much more than the money one could earn in another job. The types of jobs promoted by appropriate technology are easy to integrate into a lifestyle that many people aspire to — self-reliance, fulfillment through one's profession, and concern for others.

A final set of advantages for the appropriate technology approach is related to the environment. Appropriate technology farming is an example. It minimizes the use of chemical fertilizers and pesticides. It uses manure and other waste products as fertilizer. It avoids the use of heavy machinery, which damage the ground. It entails growing several different crops at once. All these help to preserve the soil and the surrounding groundwater. Solar collectors reduce the amount of fuel required to heat a home, so the air is less polluted. Craftpersons working in small groups are often more careful than unskilled workers on an assembly line, so small-scale hand production may generate less waste than mass production. Appropriate technology can often be less of a burden to the environment than other technologies. One must, however, be careful with this argument. Use of a wood stove would seem to offer many of the advantages of appropriate technology, but if most people used wood stoves, air pollution would be serious and the forests would be severely depleted. Few people recall that a major reason automobiles were originally encouraged was to reduce the serious pollution problem caused by horses in city streets. Now most people would consider automobiles much more of an environmental threat than horses.

A related environmental concern is diversity. So-called "modern" agriculture consists of planting a single crop using the most productive seed variety. A plant disease endemic to that variety then can wipe out a whole season's yield for an entire region. Such came close to happening with corn in the United States in 1974. The appropriate technology approach uses a variety of crops, so it is less likely a disease would spread widely, and a diversity of species, so at least some should survive a blight.

4.4.5 CONCERNS ABOUT THE USE OF APPROPRIATE TECHNOLOGY

A major concern is whether appropriate technology can provide sufficient goods and services. It does not appear that small-scale hydroelectric systems would by themselves give sufficient energy for the United States. Could small farms provide enough food? Could small factories produce at sufficiently low cost or are they inherently always inefficient compared to mass production? Questions like this have to be answered case by case but examples do exist where appropriate technology provides more at less cost.

A related concern is whether people will accept the appropriate technology approach or the inventions designed by appropriate technology. Even if oil-heating costs were higher than solar, would people prefer to use oil? Do many people really prefer a high-technology solution such as a complex kitchen gadget to a more effective lower tech one? (The Defense Department is accused of this bias toward high tech.) Would most people really prefer to work 40 hours a week at a boring job and then pay someone else to improve their house when they have leisure time, or would they prefer to do their own work if they could?

A concern for Third World countries has been whether appropriate technology does in fact lead to national development, in the sense of a trained workforce. Will a nation following the appropriate technology approach end up with a population generally conversant with technology, but with a technology that is of no use in the modern world? The question is: Will people with a background in appropriate technology have an easier time learning newer technologies compared to people with no background at all? An analogous question in the United States is whether people trained to rebuild their own homes can use those skills elsewhere.

Some Third World leaders are understandably suspicious of people from the industrial countries and wonder if appropriate technology is a way to discourage the Third World from industrializing and becoming competition. One answer to these leaders is that there may be no other way for the nations of the Third World to industrialize on their own besides through appropriate technology.

Another area of concern is that the appropriate technology approach seems to be difficult to plan or manage. Would it not be better to take established technologies from the United States, say, and spend one's time and effort making them work? By definition appropriate technology is specific to a locality, so transferring expertise from one country to another is difficult. Because appropriate technology is small scale and done by many independent people, it is difficult for a government official to understand what is happening and take action when needed. People who object to appropriate technology for this reason, however, may be too optimistic about the problems in transferring an established technology and too pessimistic about the problems in transferring appropriate technology.

4.4.6 THE FUTURE OF APPROPRIATE TECHNOLOGY

A reason we should care about appropriate technology is based on what we see in the future. Certainly, we see more people on the earth and thus greater threats on resources and the environment. Appropriate technology can use fewer resources and be easier on the environment than the alternatives, for the same output.

Appropriate technology may be beneficial for intangible reasons also. Technological changes will certainly take place. Better communication and transportation will mean people, ideas, music, and so forth will travel further, faster. The risk is loss of cultural diversity. Unless care is taken, communities all over the world will look the same, have the same music, grow the same food, and so forth. Another risk of better communications is the temptation to build larger organizations, to centralize more decisions so individual members of the organizations have less control. As technology becomes more complex, a threat develops that fewer people will understand it. It is dangerous when only a few understand the essential industries of a society and the rest must proceed on trust. A complex technology is also a threat if it results in a split society, with a very few people having rewarding, challenging jobs and the great majority having meaningless, dull jobs. Appropriate technology may be a way to make people's lives richer and safer.

A more general reason we should care about what happens to appropriate technology is that appropriate technology represents a choice. Appropriate technology may not be the best choice in all situations but cases do exist where it is best. Because appropriate technology is often well matched to the needs of the Third World and the underdeveloped parts of the United States and Europe (e.g., blighted urban neighborhoods) it may offer there a possibility of relative parity in standard of living without unrealistic investment. We should care about the future of appropriate technology because the hardware promoted can be more effective than the alternatives, because it promotes valuable social attitudes, and because the approach reminds us to take a full range of factors into consideration when making a choice.

IMCO Recycling, Inc., will soon have an important opportunity to further increase its participation in the auto component market because of an unexpected rise in the amount of aluminum scrap available for recycling. This increase will occur because more autos are being produced, the amount of aluminum content in all types of vehicles has moved steadily higher over the past decade, and the processes of vehicle dismantling and aluminum scrap recovery are becoming more efficient. PHOTOGRAPH COURTESY OF **IMCO** RECYCLING, INC.

CHAPTER 5

Business and Technology Methods

5.1 Sustainable Business Practices

During the Industrial Revolution resources seemed inexhaustible and nature was viewed as something to be tamed and civilized. Recently, however, some leading industrialists have begun to realize that traditional ways of doing things may not be sustainable over the long term. Now companies are shifting toward a cleaner model for their business and technology processes. The hope is that this cleaner model will transform human industry from a system that takes, makes, and wastes into one that integrates economic, environmental, and ethical concerns.

The goal of this cleaner model of industrialization is to release less pollution, think long term, reduce regulations, produce fewer toxic materials, reduce wastes, and improve the condition of the earth's ecosystem.

In Chapter 2 we described the four steps for improving business practices (see Table 2.2). They are summarized in Table 5.1. Business would focus on creating an economy that favored the more cyclic use of materials. Waste would be redefined as those by-products that had no useful application within the entire industrial system, rather than, as now, simply the discarded by-products or emissions from individual products, processes, or service operations. In such an economy, by-products and recycled materials would be bought and sold like any other goods. Translating the concept into practice is a challenge. Indeed, in absolute terms, an ultra-efficient industrial ecosystem is not achievable because no single cycle is actually a totally closed loop. Recycling processes produce wastes such as wastewaters or contaminated residuals. Energy recovery produces residual ash. Disposal will always remain part of the chain of management for some by-products of industrial processes. Nevertheless, it is hoped that these remaining wastes will be compostable or can be used in other products.

Table 5.1 The Four Steps for Improving Business Practices

1. *Increase productivity:* Increase the productivity of the business process in order to reduce use of natural resources and reduce waste.
2. *Recycle or use elsewhere:* Recycle used materials and components within the production system or use them as inputs to another production process.
3. *Improve quality and extend life:* Improve the product or service, its quality, and its utility, and extend their lives.
4. *Reinvest in natural capital:* Invest in natural capital by recycling compostable waste and investing financially in the natural resource base and the maintenance of natural capital.

5.2 Eco-Efficiency

Eco-efficiency is achieved by delivering competitively priced goods and services that satisfy human needs and provide quality of life, while progressively reducing ecological impacts and resource intensity throughout the life cycle, to a level at least in line with the earth's carrying capacity. Fussler (1996) proposes a goal of cutting the current energy and material flows by one-half. Overall the goal is to produce more from less material and energy.

The idea is to design products that lead to an industrial system that eliminates almost all wastes. Customers would buy the *service* of such products, and when they had finished with the products, or simply wanted to upgrade to a newer version, the manufacturer would take back the old ones, break them down, and use their complex materials in new products. This system should be so powerful and productive that it eliminates the need for most regulations. Wastes will be reused and will regenerate the natural capital by benignly composting. An energy system will be designed that is sustainable, using renewable sources for many processes and fossil fuels only for necessary activities.

Eco-efficiency is a business response to the challenge of sustainable development. Businesses that do this may create profitable benefits by creating products and services that meet the aspirations of most of the world's consumers for prosperity coupled with a clean and healthy environment. In addition, eco-efficiency should reduce the costs and liabilities associated with resource consumption, waste, and pollution (DeSimone and Popoff, 1997). The proactive approach is to see every environmental challenge as an unmet need for more effective processes and products and to then fill it.

In the following section, Joseph Fiksel provides an introduction to eco-efficiency.

5.3 Achieving Eco-Efficiency through Design for the Environment[1]

By Joseph Fiksel
Senior Director, Strategic Environmental Management
Battelle Memorial Institute

"Eco-efficiency" is a term that does not yet appear in dictionaries but has already gained considerable force in shaping the environmental policies and practices of leading corporations. The Business Council on Sustainable Development (BCSD) sounded a trumpet call with their 1992 manifesto, "Changing Course." Due to the credibility of the companies that constitute BCSD's membership — including Dow Chemical, 3M, Northern Telecom, Ciba-Geigy, Volkswagen, Nissan, Mitsubishi, and many others — their message has had a substantial influence on the strategic thinking of company executives around the world. BCSD's concept of eco-efficiency suggests an important link between resource efficiency (which leads to productivity and profitability) and environmental responsibility.

Eco-efficiency makes business sense. By eliminating waste and using resources wisely, eco-efficient companies reduce costs and become more competitive. As environmental performance standards become commonplace, eco-efficient companies will be at an advantage for penetrating new markets and increasing their share of existing markets. This article describes the business practices companies are adopting to increase their eco-efficiency, and improve their competitive advantage.

Corporations that achieve ever more efficiency while preventing pollution through good housekeeping, materials substitution, cleaner technologies, and cleaner products and that strive for more efficient use and recovery of resources can be called eco-efficient.
— Declaration of the Business Council
on Sustainable Development, 1992

It is helpful to group the approaches to eco–efficiency suggested by BCSD into three successively broader categories, as illustrated in Figure 5.1:

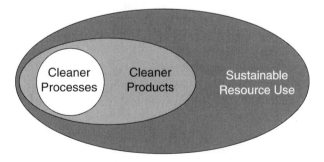

Figure 5.1 **Approaches to achieving eco-efficiency.**

[1]This section has been reprinted with permission from *Total Quality Environmental Management*, Summer 1996, pp. 47–54. Copyright © 1996 John Wiley & Sons, Inc. The author is a senior director in the Strategic Environmental Management practice of Battelle Memorial Institute.

- *Cleaner processes* — modifying production processes and technologies so they generate less pollution and waste. This approach assumes that the product definition is fixed.
- *Cleaner products* — modifying the design and material composition of products so they generate less pollution and waste throughout their life cycle. This approach includes development of cleaner processes, but also allows for more fundamental changes in the product itself.
- *Sustainable resource use* — modifying the production system as a whole, including relationships with suppliers and customers, so that fewer material and energy resources are consumed per unit of value produced. This approach includes cleaner products and processes, but also allows for broader technical and economic innovations.

To achieve these changes, leading companies are adopting a new practice called Design for Environment (DFE). Here, we define DFE as *systematic consideration of environmental performance during the early stages of product development* (Fiksel, 1996). DFE embraces all three of the BCSD approaches. It also addresses the traditional concerns of health and safety management, to the extent that they are important product or service considerations. In short, DFE is the design of safe and eco-efficient products and processes.

Some practitioners use the term "life-cycle design" instead of DFE, since awareness of life-cycle considerations is vital to this practice. There are many interpretations of life cycle depending on the viewpoint taken — process or product, physical or commercial. Figure 5.2 shows an example of the physical life cycle for durable goods such as refrigerators or personal computers. The arrows represent the flows of materials and energy, and the boxes represent economic agents or processes responsible for these flows. From this perspective, the purpose of DFE is to find profitable ways to minimize the flow of energy and materials from the center of the diagram (i.e., depletion of resources), minimize the flow of wastes into the center of the diagram, and maximize the cyclical flow of materials around the perimeter of the diagram.

5.3.1 CLEANER PROCESSES: POLLUTION PREVENTION

Pollution prevention activities preceded DFE as the first wave in voluntary proactive environmental management. The best-known corporate pollution prevention program is the Pollution Prevention Pays (3P) program at 3M Corporation, which has eliminated over 600,000 lbs. of air emissions, effluents, and solid waste, with a savings of over $500 million in capital and operating costs. Many other similar programs have also achieved impressive results, including Chevron's SMART (Save Money and Reduce Toxics), Dow Chemical's WRAP (Waste Reduction Always Pays), and Coors' SCRAP (Save, Conserve, Reduce, and Profit).

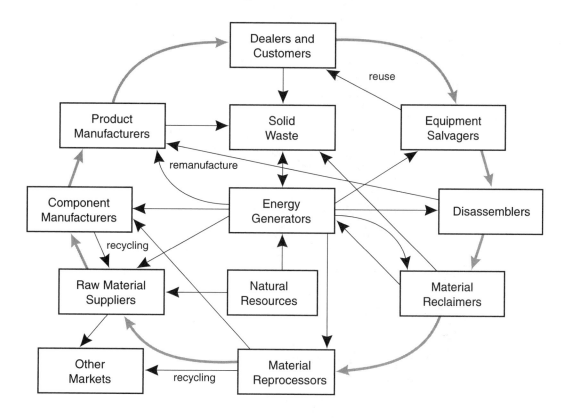

Figure 5.2 **A physical life-cycle model for assembled durable products.**

There are four major categories of pollution prevention:

- *Good housekeeping* practices ensure that resources are used efficiently and that preventable material losses are not occurring.
- *Chemical substitution* reduces or eliminates the presence of undesirable substances such as heavy metals, carcinogens, or CFCs.
- *Manufacturing process changes* simplify production technologies, reduce process emissions, or introduce closed–loop recycling.
- *Resource recovery* captures waste materials and reuses them as inputs to other manufacturing processes or for secondary applications.

5.3.2 CLEANER, SUSTAINABLE PRODUCTS THROUGH DFE

Pollution prevention, by its nature, can produce only marginal improvements. By the time pollution prevention is initiated, the investments in capital equipment have already been made, and the basic process parameters are already established. If pollution prevention thinking can be shifted into the design cycle, before the products are specified and the plants are constructed, it can have an order of magnitude greater impact. This is, essentially, the purpose of DFE.

DFE seeks to discover product innovations that will meet or exceed cost and performance objectives while reducing pollution and waste at any or all stages of the life cycle. DFE cannot be practiced in isolation — it must be balanced against other design considerations to optimize a design. To effectively integrate DFE into a new product development process, the following key elements are required:

- *Eco-efficiency metrics* are driven by fundamental customer needs or corporate goals and support environmental performance measurement.
- *Eco-efficient design practices* are based on in-depth understanding of relevant technologies and supported by engineering guidelines.
- *Eco-efficiency analysis methods* assess proposed designs with respect to the above metrics and analyze cost and quality trade-offs.

In addition, it is important for design teams to have an information infrastructure that supports the ongoing application of DFE metrics, practices, and analysis methods.

5.3.3 METRICS

The choice of high-level environmental performance metrics determines what types of signals are sent to engineering and manufacturing staff responsible for meeting operational goals. In addition, it determines the available options for communicating company performance to outside audiences. Once eco-efficiency goals have been expressed in terms of specific metrics, the next step in product development is to decompose these metrics into quantitative parameters that can be estimated and tracked for a particular product design. Examples of environmental performance metrics that can be used to establish design objectives include the following (note that these can be measured in absolute terms or normalized with respect to product units):

- energy consumed during the product life cycle
- power used during operation (for electrical products)
- total fresh water consumed during manufacturing
- toxic or hazardous materials used in production
- total industrial waste generated during production
- hazardous waste generated during production or use
- air emissions and water effluents generated during production
- greenhouse gases and ozone-depleting substances released over life cycle
- product disassembly and recovery time
- percent of recyclable materials available at end of life
- percent of product recovered and reused
- purity of recyclable materials recovered

- percent of recycled materials used as input to product
- product mass
- useful operating life
- percent of product disposed or incinerated
- fraction of packaging or containers recycled
- average life-cycle cost incurred by the manufacturer
- purchase and operating cost incurred by the customer.

5.3.4 DESIGN PRACTICES

There are a variety of specific DFE practices associated with eco-efficient design. Ideally, a single design innovation may contribute to achieving several different goals. For example, reducing the mass of a product can (1) reduce energy and material use, which contributes to resource conservation, and (2) reduce pollutant emission, which contributes to health and safety.

The following are some of the more common DFE practices in industry today:

- *Material substitution* — replacing product constituents with substitute materials that are superior in terms of increased recyclability, reduced energy content, etc.
- *Waste source reduction* — reducing the mass of a product or its packaging, thus reducing the resulting waste matter per product unit.
- *Substance use reduction* — reducing or eliminating undesirable substances that are either incorporated into a product or used in its manufacturing process.
- *Energy use reduction* — reducing the energy required to produce, transport, store, maintain, use, recycle, or dispose of a product and its packaging.
- *Life extension* — prolonging a product or its components' useful life, thus reducing the associated waste.
- *Design for separability and disassembly* — simplifying product disassembly and material recovery using techniques such as snap fastening components and color-coded plastics.
- *Design for recyclability* — ensuring both high recycled content in product materials and maximum recycling at end of life.
- *Design for disposability* — assuring that nonrecyclable materials and components can be disposed of safely and efficiently.
- *Design for reusability* — enabling components of a product to be recovered, refurbished, and reused.
- *Design for remanufacture* — enabling recovery of post-industrial or post-consumer waste materials or components for recycling as input to new product manufacture.
- *Design for energy recovery* — enabling extraction of energy from waste materials, for example, through incineration.

5.3.5 ANALYSIS METHODS

Finally, to "close the loop" in the development process, analysis methods are needed to assess the degree of improvement expected from a new design with respect to the eco-efficiency metrics of interest. Analysis methods may range from focused estimation of parameters (e.g., market surveys of expected recycling rates) to full-scale life-cycle assessment of environmental impacts. They may also include relationships between environmental metrics and other cost or performance metrics such as reliability and durability.

Analysis methods are used by design teams in at least four ways:

1. *Screening* methods are used to narrow design choices among a set of alternatives; examples include threshold limits for chemical properties such as biodegradability, material selection priority lists based on recyclability, and environmental performance criteria for component suppliers.

2. *Assessment* methods are used to predict the expected environmental performance of designs; these methods may range from simple qualitative indices to sophisticated numerical simulation models. In concurrent engineering, the use of predictive tools for evaluating product performance is important to assure coverage of all product requirements and to identify design flaws or omissions. Design teams need to "close the loop" by receiving feedback about the benefits of design changes. Examples of assessment methods include:

   • Assessing anticipated waste streams and emission rates based on product composition data

   • Modeling product end-of-life costs based on available economic and operational data

   • Profiling life-cycle environmental and financial implications of design alternatives

3. *Trade-off* methods are used to compare the expected cost and performance of several alternative design approaches with respect to one or more attributes of interest. Such methods may include "what-if" simulations, quality-function matrices, or parametric analyses that draw upon the various assessment methods described above. Examples of key issues include:

   • Identifying interactions and trade-offs among eco-efficiency, cost, performance, manufacturability, maintainability, and other quality factors, so that candidate designs can be evaluated systematically

   • Assigning relative importance to various categories of environmental impacts (e.g., toxic material emissions may have widely varying impacts depending on their persistence, environmental fate, and surrounding ecosystems)

4. *Decision support* methods are used to help design teams select from among alternatives when the trade-offs are complex or when large uncertainties are present; examples include analytic hierarchy techniques, expert advisory systems, decision analysis and optimization methods. One of the most common types of decision support is computer-based DFE guidelines, which are important for today's fast-paced, high-performance, cross-disciplinary team approach. On-line guidance allows direct, real-time retrieval and maintenance of DFE knowledge by engineers and other professionals engaged in product development.

One of the pitfalls of computer-aided engineering is the tendency for different disciplines to develop their own specialized tools. To avoid this "islands of automation" syndrome, DFE tools should be implemented in a way that facilitates data exchange and interoperability with other CAE/CAD tools. The preferred solution is to adopt a "framework" architecture in which all tools share common data models and interface specifications, thus allowing incremental extension and maintenance of the tool kit with tools from multiple vendors as well as custom applications.

5.3.6 TOWARD A SUSTAINABLE ECONOMY

The greatest impacts of DFE often are achieved through initiatives that link the technologies of various companies involved in a production chain. By coordinating the development of environmentally conscious technologies with their upstream suppliers and their downstream customers, manufacturers can introduce significant improvements in both productivity and environmental performance.

It is well-known that sharing technical information, harmonizing specifications, and integrating business processes will improve communication, eliminate redundant or misdirected effort, increase overall efficiency, and lower costs. Moreover, strategic alliances among customers and suppliers can bestow a competitive advantage upon each party, allowing them to respond effectively to a dynamic business environment. This is especially true in the electronics industry, where short product lifetimes and rapidly changing technology make fast development cycle time a critical success factor.

Individual firms can only perform "local optimization" (i.e., efforts at source reduction, energy use reduction, and pollution prevention that are within their control). A more effective and powerful approach is to seek "global optimization" by identifying cross-cutting design changes that promise to decrease the life-cycle cost or increase the life-cycle efficiency of products in terms of resource usage, waste recovery, or other important metrics.

By lifting constraints and pooling their talents, cooperating companies can focus on innovative technologies that benefit the entire value-added chain. Such innovations might include short-term incremental improvements (such as standardizing the configuration of electronic components to sim-plify product disassembly at end of life) and long-term fundamental changes (such as introducing entirely new processes that eliminate the use of chemi-cal solvents). The overall societal benefits of this type of "vertically inte-grated" DFE can potentially be an order of magnitude greater than those achieved through conventional approaches.

5.4 Design for Environment

Design for environment (DFE) is defined as the systematic consideration of design performance of a product or service with respect to environmen-tal, health, and safety objectives over the full product life cycle. Thus, DFE is the design of safe and eco-efficient products and services (Fiksel, 1996). DFE responds to the necessity of moving toward a low-waste, low-risk approach to sustainability of the environment. In this approach, environ-mental considerations are incorporated into all aspects of product and process design, and technology plays a more active and positive role in achieving sustainable development.

An outline of the structure of DFE is shown in Figure 5.3. This architec-ture enables the designer to consider the issues that arise from the design of a new product or service.

Figure 5.3 **The design for environment structure.**

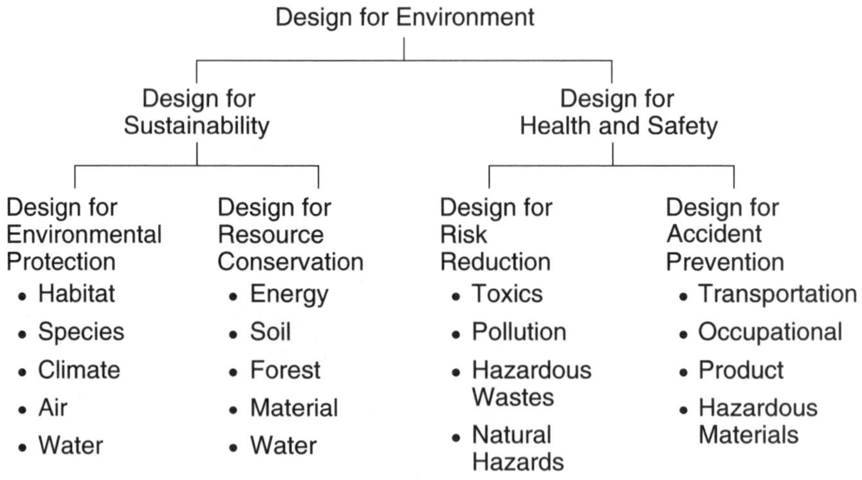

Design for Environmental Protection	**Design for Resource Conservation**	**Design for Risk Reduction**	**Design for Accident Prevention**
• Habitat	• Energy	• Toxics	• Transportation
• Species	• Soil	• Pollution	• Occupational
• Climate	• Forest	• Hazardous Wastes	• Product
• Air	• Material	• Natural Hazards	• Hazardous Materials
• Water	• Water		

Design for Environment → Design for Sustainability / Design for Health and Safety

SOURCE: Adapted from Fiksel, 1996.

DFE AT S.C. JOHNSON & SON, INC.

S.C. Johnson uses DFE for its new product development. The company looks for ways to deliver better product performance while simultaneously using fewer materials, resulting in less waste and risk. For products such as its Glade Plug-Ins air freshener, eco-efficiency translates into reduced costs and improved product competitiveness.

Glade Plug-Ins air freshener is a concentrated, refillable product. The concentrated product formulation resulted in the use of less packaging than in competitive products. The refill feature allows customers to replace only the air freshener cartridge, while keeping the base "plug-in" for up to 6 years.

Several product and packaging improvements since the introduction of the product have resulted in still greater eco-efficiency:

- *Glade Plug-Ins now last 50% longer, from 30 to 45 days, as a result of a product reformulation.*

- *Through redesign, 25% of the paperboard packaging has been eliminated since the introduction of the product.*

- *Packaging materials for both the refill cartridge and display carton are made entirely from 100% recycled materials.*

- *Marketing a "triple pack" of refills, in a package only slightly larger than the single refill, reduced material utilization per use by an additional 60%.*

Glade Plug-Ins deliver the same benefits as comparable products but with a significantly reduced VOC (volatile organic compound) content.

5.5 Designing Green Products

Green products are considered to be those products that meet one of more of the qualities of eco-efficiency. Typically, green products set themselves apart from their competitors by label claims or advertising claims. A recent study of green products used a definition of those who used terms such as *biodegradable, compostable, environmentally friendly,* or *organic* (*Green Business Letter,* Jan. 2000). Table 5.2 summarizes the attributes of green products. Some 983 products of 8,700 new products introduced in 1999 made such claims (11.3%). Most claims were for household products such as cleaning products and paper goods.

Table 5.2 **Attributes of Green Products**

Attribute of product	Methods
• Durability	Periodic maintenance, warranty
• Reusability	Interchangeable parts
• Reparability	Maintain spare parts
• Remanufactured	Volume products recycled
• Concentrated	Higher volume per purchase
• Less toxic	Avoid toxic ingredients
• Less packaging	Seek least materials package reusable containers

SOURCE: Adapted with data from Inform, Inc., *www.informinc.org.*

In the following section, the authors describe the design of green products and the resulting benefits.

5.6 Green Products[2]

BY NOELLETTE CONWAY-SCHEMPF[*] AND LESTER LAVE[†]
[*]*Automatika, Inc.*
[†]*Professor*
Carnegie Mellon University

A green product can be defined rather broadly as one that (1) uses less materials and energy in its production, use, and disposal than other products of similar function (particularly less nonrenewable resources) and (2) uses fewer toxic materials or results in lower discharges of hazardous materials than other products. These environmental improvements may be the result of dematerialization (less material usage), material substitution, processing changes, or increased recycling and reuse of materials, components, and products. Substituting green products for less environmentally conscious products is a step toward preserving the environment and giving future generations the same opportunities we enjoy: an ideal known as **sustainable economic development**.

5.6.1 GREEN PRODUCTS

A green product is not defined in any absolute sense, but only in comparison with other products of similar function. For example, a product could be entirely made of renewable materials, use renewable energy, and decay completely at the end of its life. However, this product would not be green if, for example, a substitute product uses less resources during production and use, or results in the release of fewer hazardous materials. Other things being

---

[2]This section has been reprinted with permission from *The Technology Management Handbook* (Richard Dorf, ed.), pp. 14-65–14-69, Boca Raton, FL: CRC Press, 1998. Copyright CRC Press, Boca Raton, Florida.

equal, a car that gets 50 miles per gallon is greener than one that gets 30 miles per gallon — unless the owner family cannot fit into the more fuel-efficient car, necessitating two trips. A fully loaded bus is greener than either car, but a bus with one passenger is not at all green. Still more important than green products is the concept of green systems. A 50-mile-per-gallon car whose components can be recycled easily is relatively green. However, as a means of getting from one place to another, it is much less green than a bicycle, even if the bicycle is made of materials that cannot be reused or recycled. A community where people can walk to work, school, and shopping is inherently greener than cities such as Los Angeles and Dallas where an automobile is required to get to work, shopping districts, schools, and recreation areas. Thus, while one product is greener than another, the more important attribute has to do with the system in which each product is used.

Rarely is one product greener in every dimension (resource and energy use, emissions, reyclability, etc.) than other products; there are usually trade-offs among characteristics. For example, making cars more fuel efficient requires making them lighter. This can be accomplished by substituting aluminum or plastic for steel. Both aluminum and plastic require more energy during production than does steel. How should we compare the energy required during production with the energy required for operation? One approach is to calculate how many miles the car must go to "pay back" the energy required during production. Another example is that some new materials and composites, such as carbon fibers, have many advantages but cannot be recycled. Which is more important, the ability to recycle or the lighter weight and strength? Finally, if petroleum is in extremely short supply, saving gasoline may be desirable even if the result is increasing total energy use, e.g., cars powered by methanol from coal.

In the sections that follow, we discuss the importance of focusing on the design phase when we desire green products, the principal tools for assessing the greenness of a product, comparing the environmental impacts of different materials, approaches to making products more recyclable, the importance of selecting the proper materials, and a tool for allowing companies and consumers to make green choices without having to become environmental experts. We end with a brief look at regulations designed to promote green products and the social benefits that flow from making the economy greener.

5.6.2 DESIGNING GREEN PRODUCTS

The largest potential for green products is to start with the design phase minimizing the environmental impact of a product, through all its life stages — from resource extraction to manufacturing and processing to disposal by designing out negative environmental attributes (Figure 5.4). Product design is an ideal point to address environmental problems (Hendrickson and McMichael, 1992). It is at the design stage that manufacturing decisions, resource requirements, toxic materials usage, energy use, waste disposal options, etc., are made. An oft-quoted NRC study (NRC 1991) estimates that over 70% of the environmental costs of product development, manufacture,

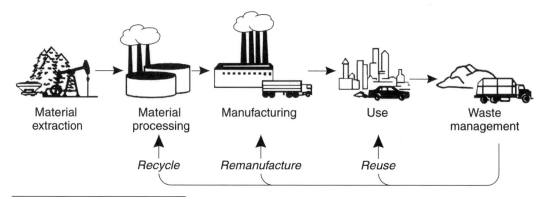

SOURCE: From "Green Products by Design: Choices for a Cleaner Environment," U.S. Congress, Office of Technology Assessment, OTA-E-541.

Figure 5.4 **Stages of a product life cycle.**

and use are determined during the design stage — this certainly holds true for environmental costs and impacts.

Design is a complicated messy process. Designers are faced with numerous competing requirements and constraints. For example, a personal computer must be fast and powerful and cheap. To be green it must also be energy efficient, and easily recycled. Designers have struggled to achieve the first set of attributes. Achieving environmental goals makes the task more difficult. Particularly as for most consumers, energy efficiency and recyclability are less important product attributes, which means that designers cannot compromise other product attributes in becoming green.

Designing and manufacturing green products requires appropriate knowledge, tools, production methods, and incentives. Aids for green design must be easy and quick to use and understand. Ideally, these design tools will help identify design changes that have lower costs while improving materials use and recyclability, for example, using snap fits rather than screws may have little additional cost burden at the design stage and may significantly increase recycling potential.

Green product design tools generally fit into the categories described in the following sections.

Life-Cycle Assessment

Life-cycle assessment (LCA) is a method of examining the environmental effects of a product or process throughout its life (EPA, 1993). A traditional LCA consists of (1) defining a system boundary, (2) carrying out an inventory of all the materials and energy used and all the environmental discharges resulting from the product's manufacture, use, and disposal within the defined boundary, (3) carrying out an assessment of the environmental implications resulting from the discharges and materials use identified in the inventory, and, finally, (4) carrying out an assessment of the opportunities for improvement. Unfortunately, the most commonly used tools for LCA have major flaws. They are generally expensive and time consuming and require data on environmental impacts that are not available. Most LCAs involve the use of boundaries that limit the analysis and generate controversy. A new approach solves many of these difficulties, although it has problems in carrying out detailed product or materials analysis (Lave *et al.*, 1995). The obvious need and demand for LCA data and tools, and the possibility of the development of international standards for LCA, should result in the availability of widely accepted tools within the next decade.

Environmental Impact or "Green" Indices

How can an analyst compare a pound of organic matter dumped into the environment with a pound of dioxin? Green indices or ranking systems attempt to summarize various environmental impacts into a simple scale. The designer or decision maker can then compare the green score of alternatives (materials, processes, etc.) and choose the one with the minimal environmental impact. This would contribute to products with reduced environmental impacts. Examples of green indices currently being used include Volvo's Environmental Priority Strategies system [EPS — this involves calculating "environmental load units" of alternatives (Graedel and Allenby, 1995)] and Carnegie Mellon University's CMU-ET toxicity weighting system [this uses U.S. EPA toxicity release inventory data and worker exposure safety data to compare alternatives (Horvath *et al.*, 1995)]. Although none of these tools is yet capable of incorporating many different types of environmental impact, all provide at least rudimentary guidance to the designer in choosing materials, components, or process alternatives that have reduced environmental impacts.

Design for Disassembly and Recycling Aids

Design for disassembly and recycling (DFD/R) means making products that can be taken apart easily for subsequent recycling and parts reuse. For example, Kodak's "disposable" cameras snap apart, allowing 87% of the parts (by weight) to be reused or recycled. Unfortunately, the economic costs associated with physically taking apart products to get at valuable components and materials often exceeds the value of the materials. Reducing the time (and thus cost) of disassembly might reverse this balance. Thus, DFD/R acts as a driver for recycling and reuse. DFD/R software tools generally calculate potential disassembly pathways, point out the fastest pathway, and reveal obstacles to disassembly that can be "designed out."

Material Selection and Label Advisors

Any of several materials can produce a particular quality component or product. However, they have different environmental implications. Material selection guidelines attempt to guide designers toward the environmentally preferred material. For example, Graedel and Allenby (1995) present the following guiding principles for materials selection:

- Choose abundant, nontoxic materials where possible.
- Choose materials familiar to nature (e.g., cellulose), rather than man-made materials (e.g., chlorinated aromatics).
- Minimize the number of materials used in a product or process.
- Try to use materials which have an existing recycling infrastructure.
- Use recycled materials where possible.

In addition to these generic guidelines, companies such as IBM and Chrysler have been developing and using specific materials selection guidelines for environmentally sound product development, which describe in detail the materials that should be used for specific applications.

Label advisors are generally marks on materials or products that reveal information about the material content relevant to materials handling and waste management. For example, the plastic bottles used in many consumer products usually have a plastics identification symbol, which can be used in plastics resorting and recycling efforts.

Full Cost Accounting Methodologies

Many corporations and consumers want to support green products and sustainability but do not know how to make greener decisions. Designers and plant managers are specialists who cannot be expected to be environmental experts capable of estimating the environmental and sustainability implications of their decisions. As a result, a company often incurs high costs from using a material or process that creates environmental problems when an environmentally benign material or process exists. Often consumers purchase products that create environmental problems because they do not know about green alternatives.

Companies need management information systems that reveal the cost to the company of decisions about materials, products, and manufacturing processes. This sort of system is called a "full cost accounting" system. For example, when an engineer is choosing between protecting a bolt from corrosion by plating it with cadmium vs. choosing a stainless steel bolt, a full cost accounting system could provide information about the purchase price of the two bolts *and* the additional costs to the company of choosing a toxic material such as cadmium.

In many cases, the choices that a designer or consumer makes also impose costs on society. For example, choosing a cadmium coating increases the possibility of human exposure to a toxic substance. The designer and consumer might be informed by showing them this social cost of the cadmium, i.e., the cost of preventing the exposure and the potential health risks of exposure. This information might be communicated by having a "social" cost listed on the price tag. A still stronger step would be to actually charge the designer and consumer for the social costs of environmentally damaging materials or products. Thus, the government might add an environmental tax or effluent fee that would account for the social damage.

A number of companies are beginning to launch preliminary full cost accounting efforts in order to spotlight products which result in relatively large environmental costs.

5.6.3 BENEFITS OF GREEN PRODUCTS

Developing and marketing green products is a concrete step toward sensible resource use and environmental protection and toward sustainable economic development. Green products imply more efficient resource use, reduced emissions, and reduced waste, lowering the social cost of pollution control and environmental protection.

Greener products promise greater profits to companies by reducing costs (reduced material requirements, reduced disposal fees, and reduced environmental cleanup fees) and raising revenues through greater sales and exports. Designing green products offers much to the current generation as well as providing future generations with a planet that will enable them to survive and prosper.

Governments, particularly in Europe, have been providing incentives for greener product development. As examples, Germany's **product takeback laws** have prompted European and U.S. industries to reduce packaging and start designing products with disassembly and recycling in mind. France and The Netherlands have special government agencies to foster clean technologies. In the United States, many government purchasing criteria specify the use of recycled materials; the federal government has ordered its employees to look for recycled products and has initiated a number of energy efficiency programs for products and buildings.

Significant progress has already been made as companies see that they can lower costs and increase revenues by making green products. Consumers have been slower to respond to green products. In the end, progress is limited by what consumers are willing to purchase.

Defining Terms

Product takeback laws: Laws in some European nations that transfer the responsibility of packaging and product disposal from the consumer to the producer. For example, in Germany, manufacturers are responsible for ensuring that packaging is recycled appropriately, and recent laws also require manufacturers of certain products to recover and recycle their products when the consumer no longer needs the product.

Sustainable economic development: Defined by WCED (1987) as "meeting the needs of the present without compromising the ability of future generations to meet their needs." For example, a vehicle powered by solar radiation would be sustainable at least in terms of energy for operation.

Further Information

http://www.ce.cmu.edu/greenDesign/.

WILL PEOPLE PAY FOR GREEN PRODUCTS?

A poll shows that some people will pay for green products. Respondents were asked if they have avoided or would consider avoiding a product for environmental reasons. Sixty percent said "yes" in the United States and 56% answered affirmatively in Mexico. When asked if they would pay a 10% premium for a greener cleaning product, two-thirds said "yes" in Venezuela and half of those in China agreed strongly. Of course, though consumers may tell pollsters they will pay a premium, getting them to do so at the shop may be more difficult.

Firms seeking global markets may be wise to include green benefits in their products.

SOURCE: "How Green Is Your Market." *The Economist,* Jan. 8, 2000, p. 66.

5.7 Green Manufacturing[3]

BY MARK ATLAS AND RICHARD FLORIDA, PROFESSOR
Carnegie Mellon University

There are many ways that industrial facilities can implement technologies and workplace practices to improve the environmental outcomes of their production processes (i.e., **green manufacturing**) and many motivations for doing so. Green manufacturing can lead to lower raw material costs (e.g., recycling wastes, rather than purchasing virgin materials), production efficiency gains (e.g., less energy and water usage), reduced environmental and occupational safety expenses (e.g., smaller regulatory compliance costs and potential liabilities), and an improved corporate image (e.g., decreasing perceived environmental impacts on the public) (Porter and van der Linde, 1995).

In general, green manufacturing involves production processes that use inputs with relatively low environmental impacts, that are highly efficient, and that generate little or no waste or pollution. Green manufacturing encompasses **source reduction** (also known as waste or pollution minimization or prevention), **recycling**, and green product design. Source reduction is broadly defined to include any actions reducing the waste initially generated. Recycling includes using or reusing wastes as ingredients in a process or as an effective substitute for a commercial product or returning the waste to the original process that generated it as a substitute for raw material feedstock. Green product design involves creating products whose design, composition, and usage minimizes their environmental impacts throughout their lifecycle.

Source reduction and recycling activities already have been widely adopted by industrial facilities. According to 1993 US. Environmental Pro-

---

[3]This section has been reprinted with permission from *The Technology Management Handbook* (Richard Dorf, ed.), pp. 13-85–13-89, Boca Raton, FL: CRC Press, 1998. Copyright CRC Press, Boca Raton, Florida.

tection Agency (EPA) Biennial Reporting System (BRS) data, which cover facilities that generate over 95% of the country's hazardous waste, 57% and 43% of these facilities had begun, expanded, or previously implemented source reduction and recycling, respectively. According to a 1995 survey of over 200 U.S. manufacturers, 90% of them cited source reduction and 86% cited recycling as main elements in their pollution prevention plans (Florida, 1996).

5.7.1 ORGANIZING FOR GREEN MANUFACTURING

Green manufacturing provides many opportunities for cost reduction, meeting environmental standards, and contributing to an improved corporate image. However, finding and exploiting these opportunities frequently involve more than solving technological issues. The ten most frequently cited hazardous waste minimization actions are listed in Table 5.3.

Table 5.3 **Most Frequently Cited Hazardous Waste Minimization Actions**

Percent of all actions	*Waste minimization action*
8.9	Improved maintenance schedule, recordkeeping, or procedures
8.0	Other changes in operating practices (not involving equipment changes)
7.1	Substituted raw materials
6.5	Unspecified source reduction activity
5.1	Stopped combining hazardous and nonhazardous wastes
4.8	Modified equipment, layout, or piping
4.6	Other process modifications
4.4	Instituted better controls on operating conditions
4.1	Ensured that materials not in inventory past shelf life
4.0	Changed to aqueous cleaners

$N = 81{,}547$ waste minimization actions.

SOURCE: Tabulations from 1989, 1991, 1993, and 1995 EPA BRS databases.

As the data show, only a small portion of these actions involves new or modified technology. Most involve improving operating practices or controls or fairly basic ideas — such as waste segregation or raw material changes — that production workers can suggest and implement. Thus, it is first necessary to organize production operations, management functions, and personnel for green manufacturing to facilitate the identification and development of both technical and common-sense waste minimization ideas (Dillon and Fischer, 1992).

There are several important prerequisites for this process. First, it is critical to have an accounting of inputs, wastes, and their associated costs at each point in the production process. According to 1994 EPA data, 31% of all reported source reduction actions were first identified through pollution prevention opportunity or materials balance audits (EPA, 1996). The normal financial incentives to reduce costs can be highly efficient within such an accounting system, but the actual efficiency greatly depends on the extent to which true costs are accounted for. The pinpointing of costs, particularly tracking them back to specific production processes, and the projection of future costs are challenging (Florida and Atlas, 1997; Todd, 1994). Second, the facility must thoroughly know the environmental laws with which it must comply now and in the foreseeable future. This includes environmental permits specifically applicable to it. The facility also must assess the legal implications of possible changes in its operations (e.g., the need for permits if certain changes are made or any restrictions on using particular chemicals).

Third, green manufacturing must be a central concern of the facility's top management (Florida and Atlas, 1997; Hunt and Auster, 1990). This is usually helped by outside pressure (from government or environmentalists) or by the convincing demonstration of its benefits (e.g., reduced production costs) (Lawrence and Morell, 1995). Fourth, it is typically very helpful to involve production workers in green manufacturing (Florida and Atlas, 1997; Makower, 1993). When they are involved in the environmental implications of their activities, they often make substantial contributions, especially improvements in industrial housekeeping, internal recycling, and limited changes in production processes. According to 1994 EPA data, 42% of all reported source reduction activities were first identified through management or employee recommendations (EPA, 1996).

Fifth, green manufacturing will greatly benefit from the easy availability of technical and environmental information about cleaner technology options. Both in-house technical and environmental experts and outside consultants can be useful. It also can be desirable to involve the facility's suppliers and customers in the effort (Georg *et al.,* 1992). Often they can provide solutions not easily perceived by the facility involved in the actual production. Finally, setting challenging objectives and monitoring the facility's progress toward achieving them can help in creating effective green manufacturing (Florida and Atlas, 1997). The targets may be financial (e.g., cost reduction), physical (e.g., input and/or discharge reduction), legal (e.g., lowering emissions to avoid the need for an environmental permit), and personnel (e.g., fewer injuries).

5.7.2 CHOOSING GREEN MANUFACTURING OPTIONS

Once the proper organizational approach is established, the first step in choosing options for green manufacturing is making an inventory by production operation of the inputs used (e.g., energy, raw materials, and water) and the wastes generated. These wastes include off-specification products, inputs returned to their suppliers, solid wastes, and other nonproduct outputs sent to treatment or disposal facilities or discharged into the environment. The second step is selecting the most important nonproduct outputs or waste streams to focus upon. Their relative importance could depend upon the costs involved, environmental and occupational safety impacts, legal requirements, public pressures, or a combination thereof.

The third step is generating options to reduce these nonproduct outputs at their origin. These options fall into five general categories: product changes, process changes, input changes, increased internal reuse of wastes, and better housekeeping. The fourth step is to pragmatically evaluate the options for their environmental advantage, technical feasibility, economic sufficiency, and employee acceptability. With respect to economic sufficiency, calculating the payback period is usually adequate.

This evaluation usually leads to a number of options, especially in better housekeeping and input changes, which are environmentally advantageous, easy to implement, and financially desirable. Thus, the fifth step is to rapidly implement such options. There typically are other options that take longer to evaluate but that also usually lead to a substantial number that are worth implementing.

5.7.3 POTENTIAL GREEN MANUFACTURING OPTIONS

As noted earlier, the options for green manufacturing can be divided into five major areas: product changes, production process changes, changes of inputs in the production process, internal reuse of wastes, and better housekeeping. The following discussion focuses on the physical nature of changes that can be implemented (excluding green product changes, which are discussed elsewhere).

Changes in Production Processes

Many major production process changes fall into the following categories: (1) changing dependence on human intervention, (2) use of a **continuous** instead of a **batch** process, (3) changing the nature of the steps in the production process, (4) eliminating steps in the production process, and (5) changing cleaning processes.

Production that is dependent on active human intervention has a significant failure rate. This may lead to various problems, ranging from off-specification products to major accidents. A strategy that can reduce the dependence of production processes on active human intervention is having machines take over parts of what humans used to do. Automated process control, robots used for welding purposes, and numerically controlled cutting tools all may reduce wastes.

With respect to using a continuous, rather than batch, process, the former consistently causes less environmental impact than the latter. This is due to the reduction of residuals in the production machinery and thus the decreased need for cleaning, and better opportunities for process control, allowing for improved resource and energy efficiency and reducing off-specification products. There are, however, opportunities for environmentally improved technology in batch processes. For chemical batch processes, for instance, the main waste prevention methods are (1) eliminate or minimize unwanted by-products, possibly by changing reactants, processes, or equipment, (2) recycle the solvents used in reactions and extractions, and (3) recycle excess reactants. Furthermore, careful design and well-planned use can also minimize residuals to be cleaned away when batch processes are involved.

Changing the nature of steps in a production process — whether physical, chemical, or biological — can considerably affect its environmental impact. Such changes may involve switching from one chemical process to another or from a chemical to a physical or biological process or vice versa. In general, using a more selective production route — such as through inorganic catalysts and enzymes — will be environmentally beneficial by reducing inputs and their associated wastes. Switching from a chemical to a physical production process also may be beneficial. For example, the banning of chlorofluorocarbons led to other ways of producing flexible polyurethane foams. One resulting process was based on the controlled use of variable pressure, where carbon dioxide and water blow the foam, with the size of the foam cells depending upon the pressure applied. An example of an environmentally beneficial change in the physical nature of a process is using electrodynamics in spraying. A major problem of spraying processes is that a significant amount of sprayed material misses its target. In such cases, waste may be greatly reduced by giving the target and the sprayed material opposite electrical charges.

Eliminating steps in the production process may prevent wastes because each step typically creates wastes. For example, facilities have developed processes that eliminated several painting steps. These cut costs and reduce the paint used and thus emissions and waste. In the chemical industry, there is a trend to eliminate neutralization steps that generate waste salts as by-products. This is mainly achieved by using a more selective type of synthesis.

Cleaning is the source of considerable environmental impacts from production processes. These impacts can be partly reduced by changing inputs in the cleaning process (e.g., using water-based cleaners rather than solvents). Also, production processes can be changed so that the need for cleaning is reduced or eliminated, such as in the microelectronics industry, where improved production techniques have sharply reduced the need for cleaning with organic solvents. Sometimes, by careful consideration of production sequences, the need for cleaning can be eliminated, such as in textile printing, where good planning of printing sequences may eliminate the need for cleaning away residual pigments. In other processes, reduced cleaning is achieved by minimizing carryover from one process step to the next. The switch from batch to continuous processes will also usually reduce the need for cleaning.

Changes of Inputs in the Production Process

Changes in inputs is an important tool in green manufacturing. Both major and minor product ingredients and inputs that contribute to production, without being incorporated in the end product, may be worth changing.

An example where changing a minor input in production may substantially reduce its environmental impact is the use of paints in the production of cars and airplanes. The introduction of powder-based and high solids paints substantially reduces the emission of volatile organic compounds. Also, substituting water-based for solvent-based coatings may lessen environmental impacts.

Internal Reuse

The potential for internal reuse is often substantial, with many possibilities for the reuse of water, energy, and some chemicals and metals. Washing, heating, and cooling in a countercurrent process will facilitate the internal reuse of energy and water. Closed-loop process water recycling that replaces single-pass systems is usually economically attractive, with both water and chemicals potentially being recycled. In some production processes there may be possibilities for **cascade-type reuse**, in which water used in one process step is used in another process step where quality requirements are less stringent. Similarly, energy may be used in a cascade-type way where waste heat from high-temperature processes is used to meet demand for lower-temperature heat.

Better Housekeeping

Good housekeeping refers to generally simple, routinized, nonresource-intensive measures that keep a facility in good working and environmental order. It includes segregating wastes, minimizing chemical and waste inventories, installing overflow alarms and automatic shutoff valves, eliminating leaks and drips, placing collecting devices where spills may occur, frequent inspections aimed at identifying environmental concerns and potential malfunctionings of the production process, instituting better controls on operating conditions (e.g., flow rate, temperature, and pressure), regular fine tuning of machinery, and optimizing maintenance schedules. These types of actions often offer relatively quick, easy, and inexpensive ways to reduce chemical use and wastes.

Defining Terms

Batch process: A process that is not in continuous or mass production and in which operations are carried out with discrete quantities of material or a limited number of items.

Cascade-type reuse: Input used in one process step is used in another process step where quality requirements are less stringent.

Continuous process: A process that operates on a continuous flow (e.g., materials or time) basis, in contrast to batch, intermittent, or sequential operations.

Green manufacturing: Production processes that use inputs with relatively low environmental impacts, that are highly efficient, and that generate little or no waste or pollution.

Recycling: Using or reusing wastes as ingredients in a process or as an effective substitute for a commercial product or returning the waste to the original process that generated it as a substitute for raw material feedstock.

Source reduction: Any actions reducing the waste initially generated.

Further Information

The Academy of Management has an Organizations and the Natural Environment section for members interested in the organizational management aspects of green manufacturing. For membership forms, contact The Academy of Management Business Office, Pace University, P.O. Box 3020, Briarcliff Manor, NY 10510-8020; phone (914) 923-2607. Also, many of the Web sites cited below lead to other organizations with particular interests in green manufacturing.

The quarterly *Journal of Industrial Ecology* provides research and case studies concerning green manufacturing. For subscription information, contact MIT Press Journals, 55 Hayward Street, Cambridge, MA 02142; phone (617) 253-2889.

There are numerous Web sites with green manufacturing-related information, including the following: *http://www.epa.gov/epaoswer/non-hw/ reduce/wstewise/index.htm; http://es.inel.gov/; http://www.epa.gov/greenlights. html; http://www.hazard.uiuc.edu/wmrc/greatl/clearinghouse.html; and http:// www.turi.org/P2GEMS/.*

5.8 Green Marketing

Marketing is the process of planning and executing the conception, pricing, promotion, and distribution of products or services offered by a firm. **Green marketing** is the marketing of environmentally sensitive products or services.

Ultimately, business responds to the needs of customers — their performance and cost requirements. Consumers need clear, accurate, and reliable information to make sound purchasing decisions and to provide the necessary pull from the marketplace. Advertising and marketing is one mechanism used to convey information and influence consumer choice.

The advertising and marketing of environmental issues is complex and the benefits of promoting environmentally improved products may be unclear. It is clear that environmental improvement alone will not sell products, but that when combined with value and performance it does sell.

Promoting the company as an environmentally responsible business (in other words building it into the company's corporate identity) may even provide a greater return than advertising and marketing the environmental attributes of products. It is important to note, however, that some companies, such as the producers of intermediate components, maintain minimal or no direct, visible link between the name of a company and its products.

For some companies, the environment is a focus. Fetzer Vineyards Corp., of Hopland, California, converted to all-organic farming in 1995 on the 1,000 acres it owns. Fetzer believes organic farming does not cost more than chemically based farming — although they pay suppliers a slight premium to grow grapes organically. Fetzer hopes to draw additional customers who value organic wine. Examples of other companies with green products and green marketing are Patagonia and Gap.

Seeking market advantage through "green product" claims and specific environmental features can be risky. Scientific knowledge can be uncertain, and the environment is not usually the main selling point. Environmental improvement alone will not sell products, but when combined with value and performance it may help. Therefore, the focus should be on the performance characteristics of products, not on their environmental features. It is better to show environmental merit through performance than words. It is valuable to offer clear, consistent, and accurate environmental messages if they can be backed up.

Environmental brands can be developed and furthered by developing positive relationships with all corporate environmental stakeholders. Environmental issues should be considered in all marketing steps to communicate a corporation's environmental dedication (Ottman, 1993).

5.9 Environmental Effects[4]

By Karen Blades[*] and Braden Allenby[†]
[*]*Founder and President*
December Technology
[†]*Environment, Health, and Safety Vice President*
AT&T

5.9.1 INTRODUCTION

The importance of electronics technology for consumers, and the electronics sector for the global economy, is already substantial and continues to grow rapidly. Such growth and innovation coupled with the global concerns for the environment and the need to better manage the resources of the earth pose many challenges for the electronics industry. While thought of as a "clean" industry, the technological advances made by the industry create a significant demand on the earth's resources. As an example, the amount of water required in the production of semiconductors, the engines that motor most of today's electronic gadgets, is enormous — about 2000 gallons to process a single silicon wafer. Building silicon chips requires the use of highly toxic materials, albeit in relatively low volumes. Similarly, printed wiring boards present in most electronic products and produced in

---

[4]This section has been reprinted with permission from *The Electrical Engineering Handbook* (Richard Dorf, ed.), 2nd ed., pp. 2505–2514, Boca Raton, FL: CRC Press, 1997.

high volume use large amounts of solvents or gases which are either health hazards, ozone depleting, or contribute to the greenhouse effect and contain lead solder. The challenge for the industry is to continue the innovation that delivers the products and services that people want yet find creative solutions to minimize the environmental impact, enhance competitiveness, and address regulatory issues without impacting quality, productivity, or cost; in other words, to become an industry that is more "eco-efficient." Eco-efficiency is reached by the delivery of competitively priced goods and services that satisfy human needs and support a high quality of life, while progressively reducing ecological impacts and resource intensity, to a level at least in line with the earth's estimated carrying capacity.

Like sustainable development, a concept popularized by the Brundtland Report, *Our Common Future,* the notion of eco-efficiency requires a fundamental shift in the way environment is considered in industrial activity. Sustainable development — "development that meets the needs of the present without compromising the ability of future generations to meet their own needs" (World Commission on Environment and Development, 1987) — contemplates the integration of environmental, economic, and technological considerations to achieve continued human and economic development within the biological and physical constraints of the planet. Both eco-efficiency and sustainable development provide a useful direction, yet they prove difficult to operationalize and cannot guide technology development. Thus, the theoretical foundations for integrating technology and environment throughout the global economy are being provided by a new, multidisciplinary field known as **industrial ecology**.

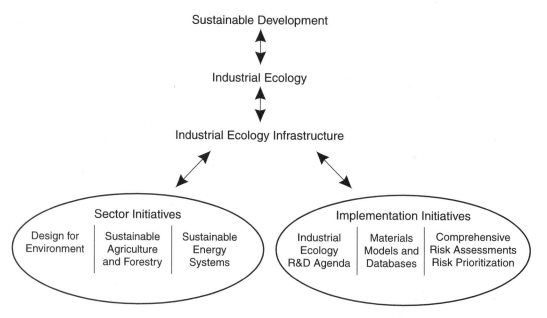

Figure 5.5 **Industrial ecology framework.**

The ideas of industrial ecology, which have begun to take root in the engineering community, have helped to established a framework within which the industry can move toward realizing sustainable development. The electrical, electronics, and telecommunications sectors are enablers of sustainability because they allow the provision of increasing quality-of-life using less material and energy, respectively, **dematerialization** and **decarbonization**. This chapter will provide an introduction into industrial ecology and its implications for the electronics industry. Current activities, initiatives, and opportunities will also be explored, illustrating that the concomitant achievement of greater economic and environmental efficiency is indeed feasible in many cases.

5.9.2 INDUSTRIAL ECOLOGY

Industrial ecology is an emerging field that views manufacturing and other industrial activity including forestry, agriculture, mining, and other extractive sectors, as an integral component of global natural systems. In doing so, it takes a systems view of design and manufacturing activities so as to reduce or, more desirably, eliminate the environmental impacts of materials, manufacturing processes, technologies, and products across their life cycles, including use and disposal. It incorporates, among other things, research involving energy supply and use, new materials, new technologies and technological systems, basic sciences, economics, law, management, and social sciences.

The study of industrial ecology will, in the long run, provide the means by which the human species can deliberately and rationally approach a desirable long-term global carrying capacity. Oversimplifying, it can be thought of as "the science of sustainability." The approach is "deliberate" and "rational," to differentiate it from other, unplanned paths that might result, for example, in global pandemics, or economic and cultural collapse. The endpoint is "desirable," to differentiate it from other conceivable states such as a Malthusian subsistence world, which could involve much lower population levels, or oscillating population levels that depend on death rates to maintain a balance between resources and population levels. Figure 5.5 illustrates how industrial ecology provides a framework for operationalizing the vision of sustainable development.

As the term implies, industrial ecology is concerned with the evolution of technology and economic systems such that human economic activity mimics a mature biological system from the standpoint of being self-contained in its material and resource use. In such a system, little if any virgin material input is required, and little if any waste that must be disposed of outside of the economic system is generated. Energetically, the system can be open, just as biological systems are, although it is likely that overall energy consumption and intensity will be limited.

Although it is still a nascent field, a few fundamental principles are already apparent. Most importantly, the evolution of environmentally appropriate technology is seen as critical to reaching and maintaining a sustainable state. Unlike earlier approaches to environmental issues, which tended to regard technology as neutral at best, industrial ecology focuses on development of economically and environmentally efficient technology as key to any desirable, sustainable global state.

Also, environmental considerations must be integrated into all aspects of economic behavior, especially product and process design, and the design of economic and social systems within which those products are used and disposed. Environmental concerns must be internalized into technological systems and economic factors. It is not sufficient to design an energy efficient computer, for example; it is also necessary to ensure that the product, its components, or its constituent materials can be refurbished or recycled after the customer is through with it — all of this in a highly competitive and rapidly evolving market. This consideration implies a comprehensive and systems-based approach that is far more fundamental than any we have yet developed.

Industrial ecology requires an approach that is truly multidisciplinary. It is important to emphasize that industrial ecology is an objective field of study based on existing scientific and technological disciplines, not a form of industrial policy. It is profoundly a systems-oriented and comprehensive approach which poses problems for most institutions — the government, riddled with fiefdoms; academia, with rigid departmental lines; and private firms, with job slots defined by occupation. Nonetheless, it is all too frequent that industrial ecology is seen as an economic program by economists, a legal program by lawyers, a technical program by engineers, and a scientific program by scientists. It is in part each of these; more importantly, it is all of these.

Industrial ecology has an important implication, however, of special interest to electronics and telecommunications engineers, and thus deserving of emphasis. The achievement of sustainability will, in part, require the substitution of intellectual and information capital for traditional physical capital, energy, and material inputs. Environmentally appropriate electronics, information management, and telecommunications technologies and services — and the manufacturing base that supports them — are therefore enabling technologies to achieve sustainable development. This offers unique opportunities for professional satisfaction, but also places a unique responsibility on the community of electrical and electronics engineers. We in particular cannot simply wait for the theory of industrial ecology to be fully developed before taking action.

5.9.3 DESIGN FOR ENVIRONMENT

Design for Environment (DFE) is the means by which the precepts of industrial ecology, as currently understood, can in fact begin to be implemented in the real world today. DFE requires that environmental objectives and constraints be driven into process and product design, and materials and technology choices.

The focus is on the design stage because, for many articles, that is where most, if not all, of their life cycle environmental impacts are explicitly or implicitly established. Traditionally, electronics design has been based on a correct-by-verification approach, in which the environmental ramifications of a product (from manufacturing through disposition), are not considered until the product design is completed. DFE, by contrast, takes place early in a product's design phase as part of the concurrent engineering process to ensure that the environmental consequences of a product's life cycle are understood before manufacturing decisions are committed.

It is estimated that some 80 to 90% of the environmental impacts generated by product manufacture, use, and disposal are "locked-in" by the initial design. Materials choices, for example, ripple backwards towards environmental impacts associated with the extractive, smelting, and chemical industries. The design of a product and component selection control many environmental impacts associated with manufacturing, enabling, for example, substitution of no-clean or aqueous cleaning of printed wiring boards for processes that release ozone depleting substances, air toxics, or volatile organic compounds that are precursors of photochemical smog. The design of products controls many aspects of environmental impacts during use — energy efficient design is one example. Product design also controls the ease with which a product may be refurbished, or disassembled for parts or materials reclamation, after consumer use. DFE tools and methodologies offer a means to address such concerns at the design stage.

Obviously, DFE is not a panacea. It cannot, for example, compensate for failures of the current price structure to account for external factors, such as the real (i.e., social) cost of energy. It cannot compensate for deficiencies in sectors outside electronics, such as a poorly coordinated, polluting, or even non-existent disposal and material recycling system in some areas of the world. Moreover, it is important to realize that DFE recognizes environmental considerations as on par with other objectives and constraints — such as economic, technological, and market structure — not as superseding or dominating them. Nonetheless, if properly implemented, DFE programs represent a quantum leap forward in the way private firms integrate environmental concerns into their operations and technology.

It is useful to think of DFE within the firm as encompassing two different groups of activities as shown in Figure 5.6. In all cases, DFE activities require inclusion of life-cycle considerations in the analytical process. The first, which might be styled "generic DFE," involves the implementation of broad programs that make the company's operations more environmentally preferable across the board. This might include, for example, development and implementation of "green accounting" practices, which ensure that relevant environmental costs are broken out by product line and process, so that they can be managed down. The "standard components" lists maintained by many companies can be reviewed to ensure that they direct the use of environmentally appropriate components and products wherever possible. Thus, for example, open relays might be deleted from such lists, on the grounds that

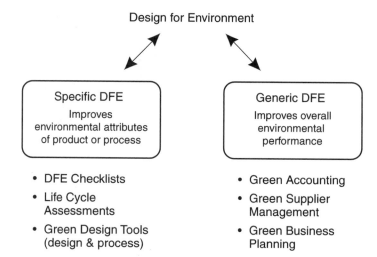

Figure 5.6 **Examples of DFE activities within the firm.**

they "can't swim," and thus might implicitly establish a need for chlorinated solvent, as opposed to aqueous, cleaning systems.

Contract provisions can be reviewed to ensure that suppliers are being directed to use environmentally preferable technologies and materials where possible. For example, are virgin materials being required where they are unnecessary? Do contracts, standards, and specifications clearly call for the use of recycled materials where they meet relevant performance requirements? Likewise, customer and internal standards and specifications can be reviewed with the same goal in mind.

The second group of DFE activities can be thought of as "specific DFE." Here, DFE is considered as a module of existing product realization processes, specifically the "Design for X," or DFX, systems used by many electronics manufacturers. The method involves creation of software tools and checklists, similar to those used in Design for Manufacturability, Design for Testability, or Design for Safety modules that ensure relevant environmental considerations are also included in the design process from the beginning. The challenge is to create modules which, in keeping with industrial ecology theory, are broad, comprehensive, and systems-based, yet can be defined well enough to be integrated into current design activities.

The successful application of DFE to the design of electronic systems requires the coordination of several design and data-based activities, such as environmental impact metrics; data and data management; design optimization, including cost assessments; and others. Failure to address any of these aspects can limit the effectiveness and usefulness of DFE efforts. Data and methodological deficiencies abound, and the challenge is great, yet experience at world class companies such as AT&T, Digital, IBM, Motorola, Siemens Nixdorf, Volvo, and Xerox indicate that it can be done. AT&T, for example, is testing a draft DFE practice; baselining the environmental attributes of a telephone at different life cycle stages to determine where meaningful environmental improvements in design can be achieved; and developing software tools that can inform environmentally preferable design decisions (Seifert, 1995). In Sweden, the government and Volvo have developed a relatively simple Environment Priority Strategies for Environmental Design, or EPS, system which uses Environmental Load Units, or ELUs, to inform materials choices during the design process. In Germany, Siemens Nixdorf has developed an "Eco-balance" system to help it make design choices that reflect both environmental and economic requirements. Xerox is a world leader in designing their products for refurbishment using a product life extension approach.

More broadly, the American Electronics Association (AEA) Design for Environment Task Force has created a series of White Papers discussing various aspects of Design for Environment and its implementation. The Microelectronics and Computer Technology Corporation (MCC) has published a comprehensive study (Lipp *et al.,* 1993) of the environmental impacts of a computer workstation, which is valuable not only for its technical findings, but for the substantial data and methodological gaps the study process identified. The Society of Environmental Toxicology and Chemistry (SETAC) and others, especially in Europe, are working on a number of comprehensive life-cycle assessment (LCA) methodologies designed to identify and prioritize environmental impacts of substances throughout their life cycle. The International Organization for Standards (ISO) is in the process of creating an international LCA standard. The IEEE Environment, Health and Safety Committee, formed in July, 1992, to support the integration of environmental, health, and safety considerations into electronics products and processes from design and manufacturing, to use, to recycling, refurbishing, or disposal has held a series of annual symposium on electronics and the environment. The proceedings from these symposia are valuable resources to the practitioners of DFE.

5.9.4 ENVIRONMENTAL IMPLICATIONS FOR THE ELECTRONICS INDUSTRY

Global concerns and regulations associated with environmental issues are increasingly affecting the manufacturing and design of electronic products, their technology development, and marketing strategies. No point illustrates this better than the German Blue Angel **eco-labeling** scheme for personal computers (the Blue Angel is a quasi-governmental, multi-attribute eco-labeling program). The Blue Angel requirements are numerous and span the complete life-cycle of the computers. Examples of some the requirements include: modular design of the entire system, customer-replaceable subassemblies and modules, use of non-halogenated flame retardants, and take back by manufacturers at the end of the product life. Market requirements such as these, focused on products and integrating as they do environmental and technology considerations, cannot possibly be met by continuing to treat environmental impact as an unavoidable result of industrial activity, i.e., as overhead. These requirements make environmental concerns truly strategic for the firm.

Perhaps the most familiar example of "a new generation of environmental management" requirements which will have enormous effects on electronics design is **product take back**. These programs, such as the one mentioned in the Blue Angel labeling scheme, are being introduced in Germany and other countries for electronics manufacturers. They generally require that the firm take its products back once the consumer is through with them, recycle or refurbish the product, and assume responsibility for any remaining waste generated by the product. Other members of the European Union and Japan are among others considering such "take back" requirements. Similarly, the emergence of the international standard, **ISO 14000**, which includes requirements for environmental management systems, methodologies for **life cycle assessment** and environmental product specifications will have vast implications for the electronics industry. Though technically voluntary, in practice these standards in fact become requirements for firms wishing to engage in global commerce. These examples represent a global trend towards proactive management of business and products in the name of the environment.

5.9.5 EMERGING TECHNOLOGY

New tools and technologies are emerging which will influence the environmental performance of electronic products and help the industry respond to the regulatory "push" and the market "pull" for environmentally responsible products. In the electronics industry, technology developments are important not only for the end-products, but for components, recycling, and materials technology as well. Below is a brief summary of technology developments and their associated environment impacts as well as tools to address the many environmental concerns facing the industry.

The electronics industry has taken active steps toward environmental stewardship, evidenced by the formulation of the IEEE Environment, Safety and Health Committee, the 1996 Electronics Industry Environmental Roadmap published by MCC, and chapters focused on environment in The National Technology Roadmap for Semiconductors. Moves such as this, taken together with the technical sophistication of control systems used in manufacturing processes, have allowed the electronics industry to maintain low emission levels relative to some other industries. Despite the industry's environmental actions, the projected growth in electronics over the next 10 to 20 years is dramatic and continued technological innovation will be required to maintain historically low environmental impacts. Moreover, the rapid pace of technological change generates concomitantly high rates of product obsolescence and disposal, a factor that has led countries such as Germany and the Netherlands to focus on electronics products for environmental management.

Environmental considerations are not, of course, the only forces driving the technological evolution in the electronics industry. Major driving forces, as always, also include price, cost, performance, and market/regulatory requirements. However, to the extent that the trend is toward smaller devices, fewer processing steps, increased automation, and higher performance per device, such evolution will likely have a positive environmental impact at the unit production level, i.e., less materials, less chemicals, less waste related to each unit produced. Technology advances that have environmental implications at the upstream processing stage may well have significant benefits in the later stages of systems development and production. For example, material substitution in early production stages may decrease waste implications throughout the entire process. Since both semiconductors and printed wiring boards are produced in high volume and are present in virtually all electronic products ranging from electronic appliances, to computers, automotive, aerospace, and military applications, we will briefly examine the impact of these two areas of the electronic industry.

Integrated Circuits

The complex process of manufacturing semiconductor integrated circuits (IC) often consists of over a hundred steps during which many copies of an individual IC are formed on a single wafer. Each of the major process steps used in IC manufacturing involves some combination of energy use, material consumption, and material waste. Water usage is high due to the many cleaning and rinsing process steps. Absent process innovation, this trend will continue as wafer sizes increase, driving up the cost of water and waste water fees, and increasing mandated water conservation.

Environmental issues that also require attention include the constituent materials for encapsulants, the metals used for connection and attachment, the energy consumed in high-temperature processes, and the chemicals and solvents used in the packaging process. Here, emerging packaging technologies will have the effect of reducing the quantity of materials used in the packaging process by shrinking IC package sizes. Increasing predominance of plastic packaging will reduce energy consumption associated with hermetic ceramic packaging.

Printed Wiring Boards

Printed wiring boards represent the dominant interconnect technology on which chips will be attached and represents another key opportunity for making significant environmental advances. PWB manufacturing is a complicated process and uses large amounts of materials and energy (e.g., 1 MegaW of heat and 220 kW of energy is consumed during fabrication of prepeg for PWBs). On average, the waste streams constitutes 92% — and the final product just 8% — of the total weight of the materials used in PWB production process. Approximately 80% of the waste produced is hazardous and most of the waste is aqueous, including a range of hazardous chemicals. Printed wiring boards are not recycled because the removal of soldered subassemblies is costly and advanced chip designs require new printed wiring boards to be competitive. As a result, the boards are incinerated and the residual ash buried in hazardous waste landfills due to the lead content (from lead solder).

5.9.6 TOOLS AND STRATEGIES FOR ENVIRONMENTAL DESIGN

The key to reducing the environmental impact of electronic products will be the application of DFE tools and methodologies. Development of CAD/CAM tools based on environmental impact metrics, materials selection data, cost, and product data management are examples of available or clearly foreseeable tools to assist firms in adopting DFE practices.

These tools will need to be based on life cycle assessment, the objective process used to evaluate the environmental impacts associated with a product and identify opportunities for improvement. Life cycle assessment seeks to minimize the environmental impact of the manufacture, use, and eventual disposal of products without compromising essential product functions. Figure 5.7 shows the life stages that would be considered for electronic products (i.e., the life cycle considered has been bounded by product design activities). The ability of the electronics industry to operate in a more environmentally and economically efficient mode, use less chemicals and materials, and reduce energy consumption will require support tools that can be used to

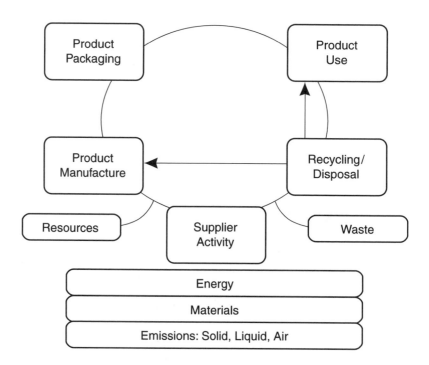

Figure 5.7 **Design for environment: systems-based, life cycle approach.**

evaluate both product and process designs. To date, many firms are making immediate gains by incorporating basic tools like DFE checklists, design standards and internal databases on chemicals and materials, while other firms are developing sophisticated software tools that give products environmental scores based on the product's compliance with a set of predetermined environmental attributes. These software tools rely heavily on environmental metrics (typically internal to the firm) to assess the environmental impact and then assign a score to the associated impact.

Other types of tools that will be necessary to implement DFE will include tools to characterize environmental risk, define and build flexible processes to reduce waste, and support dematerialization of processes and products. The following sections provide a brief review of design tools or strategies that can be employed.

Design Tools

Environmental design tools vary widely in the evaluation procedures offered in terms of the type of data used, method of analysis, and the results provided to the electronic designer. The tool strategies range in scope from assessment of the entire product life cycle to the evaluation of a single aspect of its fabrication, use, or disposal. Today's DFE tools can be generally characterized as either life cycle analysis, recyclability analysis, manufacturing analysis, or process flow analysis tools.

The effectiveness of these design tools is based both on the tool's functionality as well as its corresponding support data. One of the biggest challenges designers face with regards to DFE is a lack of reliable data on materials, parts, and components needed to adequately convey the impact and trade-offs of their design decisions. To account for these data deficiencies, a number of environmental design tools attempt to use innovative, analytical methods to estimate environmental impacts: while necessary, this indicates they must be used with care and an understanding of their assumptions.

Although DFE provides a systems-based, life cycle approach, its true value to the system designer is lost unless the impact of DFE decisions on other relevant economic and performance measures (i.e., cost, electrical performance, reliability, etc.) can be quickly and accurately assessed. Trade-off analysis tools that have DFE embedded can perform process flow-based environmental analysis (energy/mass balance, waste stream analysis, etc.) concurrently with non-environmental cost and performance analysis so that system designers can accurately evaluate the impact of critical design decisions early.

Design Strategies

Design strategies such as lead minimization through component selection, and the reduction of waste resulting from rapid technological evolution through modular design, help to minimize the environmental impact of electronic products. Although at this time no suitable lead-free alternatives exist for electronic interconnections, designers can still minimize the lead content of electronic designs. Surface mount technology requires less solder than through-hole technology. New interconnection technologies, such as microball grid array and direct chip attachment, also require less solder. The environmental benefits increase with the use of these advanced interconnection technologies.

The rapid advancement of the electronics industry has created a time when many products become obsolete in less than five years' time. Electronic products must be built to last, but only until it is time to take them apart for rebuilding or for reuse of material. This means employing modular design strategies to facilitate disassembly for recycle or upgrade of the product rather than replacement. Designers must extend their views to consider the full utilization of materials and the environmental impact of the material life cycle as well as the product life cycle.

Conclusion

The diverse product variety of the electronics industry offers numerous opportunities to curtail the environmental impact of the industry. These opportunities are multidimensional. Services made possible through telecommunications technology enable people to work from home reducing emissions that would be generated by traveling to work. Smaller, faster computers and the Internet require less material usage, reducing the energy demand during processing and waste generated during fabrication. All these represent examples of how the electronics industry provides enablers of sustainability.

Global concerns and regulations associated with environmental issues are increasingly affecting the manufacturing and design of the electronics industry. Environmental management standards, "take back" programs, ISO 14000 standards development activity, and eco-label requirements represent a sample of the initiatives driving the industries move to more environmentally efficient practices. While the industry has initiated some activities to address environmental concerns, the future competitiveness of the industry will depend on improvements in environmental technology in manufacturing, accurate assessment of the environmental impact of products and process, and design products that employ design for environment, reuse, and recycleability. Industrial ecology offers a framework for analyzing the environmental effects of the electronics industry which is complicated by the rapid pace of change.

Acknowledgments

This work was performed under the auspices of the U.S. Department of Energy by Lawrence Livermore National Laboratory under Contract W-7405-Eng-48.

Disclaimer

This document was prepared as an account of work sponsored by an agency of the United States Government. Neither the United States Government nor the University of California nor any of their employees, makes any warranty, express or implied, or assumes any legal liability or responsibility for the accuracy, completeness, or usefulness of any information, apparatus, product, or process disclosed, or represents that its use would not infringe privately owned rights. Reference herein to any specific commercial product, process, or service by trade name, trademark, manufacturer, or otherwise, does not necessarily constitute or imply its endorsement, recommendation, or favoring by the United States Government or the University of California. The views and opinions of authors expressed herein do not necessarily state or reflect those of the United States Government or the University of California, and shall not be used for advertising or product endorsement purposes.

Defining Terms

Decarbonization: The reduction, over time, of carbon content per unit energy produced. Natural gas, for example, produces more energy per unit carbon than coal; equivalently, more CO_2 is produced from coal than from natural gas per unit energy produced.

Dematerialization: The decline, over time, in weight of materials used in industrial end products, or in the embedded energy of the products. Dematerialization is an extremely important concept for the environment because the use of less material translates into smaller quantities of waste generated in both production and consumption.

Design for environment (DFE): The systematic consideration of design performance with respect to environment over the full product and process life cycle from design through manufacturing, packaging, distribution, installation, use, and end of life. It is proactive to reduce environmental impact by addressing environmental concerns in the product or process design stage.

Eco-label: Label or certificate awarded to a product that has met specific environmental performance requirements. Some of the most widely known eco-labels include Germany's Blue Angel, Nordic White Swan, and U.S. Green Seal.

Industrial ecology: The objective, multidisciplinary study of industrial and economic systems and their linkages with fundamental natural systems.

ISO 14000: Series of international standards fashioned from the ISO 9000 standard which includes requirements for environmental management systems, environmental auditing and labeling guidelines, life cycle analysis guidelines, and environmental product standards.

Life cycle assessment (LCA): The method for systematically assessing the material use, energy use, waste emissions, services, processes, and technologies associated with a product.

Product take back: Program in which manufacture agrees to take back product at the end-of-life (typically at no cost to the consumer) and disposal of product in an environmentally responsible matter.

Further Information

The IEEE Environment, Health and Safety Committee annually sponsors and publishes the proceeding of the *International Symposium on Electronics and the Environment*. These proceedings are a valuable resource for practitioners of DFE.

The National Technology Roadmap for Semiconductors, published by the Semiconductor Industry Association contains information on the environmental impacts of semiconductor fabrication as well as initiatives begun to address these concerns.

The *AT&T Technical Journal* has a dedicated issue on Industrial Ecology and DFE entitled AT&T Technology and the Environment, Volume 74, no. 6, November/December 1995.

Other suggested reading:

American Electronics Association, "The hows and whys of design for the environment," 1993.

B.R. Allenby and D.J. Richards, Eds., *The Greening of Industrial Ecosystems,* Washington, D.C.: National Academy Press, 1994.

P. Eisenberger, Ed., *Basic Research Needs for Environmentally Responsive Technologies of the Future,* Princeton, NJ: Princeton Materials Institute, 1996.

T.E. Graedel and B.R. Allenby, *Design for Environment,* Englewood Cliffs, NJ: Prentice-Hall, 1996.

5.10 Information Technology in Sustainable Development

By James R. Sheats
Program Manager
Hewlett Packard Laboratories

5.10.1 INTRODUCTION

The discussion and practice of sustainability has come a long way since the Brundtland commission enunciated its now-famous definition in the 1987 report "Our Common Future." The Rio "Earth Summit" of 1992 gave extensive publicity to the grim future faced by a world that did not accept the consequences of its overuse of physical resources. Today, the companies who are actively engaged in a discussion of how to achieve sustainability represent more than 10% of the world's gross domestic product, and many environmental NGOs emphasize working with businesses to find profit-oriented solutions to environmental problems.

On the other hand, the ecological condition of the planet continues to worsen; a useful summary with additional references can be found in Lubchenco (1998). One of the most ominous of recent observations is a quantitative analysis of the extent to which humans dominate earth's ecosystems: one-third to one-half of the land surface has been transformed by human activity (Vitousek *et al.,* 1997), with both land and water supplies nearing fundamental limits. These statistics have received somewhat less attention in the popular press than the rather well-publicized global warming issue.

At a National Academy of Sciences conference, Cassman (1999) summarized a substantial body of evidence suggesting that continued major gains in food grain output will not be obtained from breeding or other genetic engineering, yet a 50% increase is needed to meet the needs of the rising human population during the next 25 years. Accomplishing this cannot be achieved by blind application of more fertilizer (which is a major source of aquatic eutrophication) (Tilman, 1999), nor by building dams to provide more water (Postel *et al.,* 1996). The United Nations estimates that the fraction of the world's population living in countries experiencing moderate to high water stress (where consumption is more than 20% of available supply) is about one-third, and may rise to two-thirds by 2025 (World Resources, 1999). At least one billion people are without safe drinking water today, and two billion without adequate sanitation (World Business Council, 1997).

It is against this backdrop that sustainability must be understood at every company that is positioned to be a source of new product innovation. Such companies face a different landscape than those whose business involves supplying a certain type of product that can be enhanced or extended by innovation, but not fundamentally altered or eliminated. Interface Corp. is an example of this type. One can change the way carpet is manufactured and delivered to the customer, but as long as people want to walk on something soft, the product itself cannot be eliminated or drastically altered. A company that is currently in the business of producing oil, on the other hand, could find that the sustainability mandate would direct it to get into a completely different business (solar energy, for example). A company that manufactures cars could find that there are other ways to deliver the service of transportation, or even the human interaction that is often the goal of transport. These are not unfamiliar themes in the sustainability discussion. The primary goal of this section is to highlight the connection with the discussion of resources and ecosystem stress, identify the critical resource issues, and see how they relate to the business of information technology.

5.10.2 THE FUNDAMENTAL PROBLEM OF SUSTAINABILITY

Sustainability is fundamentally a matter of living within the framework of earth's systems, both thermodynamically and kinetically. Change is a normal characteristic of nature; but it occurs on definite timescales, and large, sudden perturbations may be catastrophic for many ecosystem members. The desired goal is a social system in which we satisfy our basic needs, and as many of our desires as possible, without introducing perturbations into nature that rapidly shift ecosystems irreversibly away from the steady-state conditions in which we found them. The timescale of any changes that we cause should be comparable to the timescale with which the system can adjust continuously.

An analysis of environmental impact into factors of population and consumption is discussed elsewhere in this book. Considerations of population stabilization imply that standard of living for most of the world's people must rise by around a factor of 5 (even with a doubling of population), yet this must not be translated into a comparable increase in consumption if environmental breakdown is to be avoided. People's desires can be met with many different technologies representing different resource intensities. In some cases the desire for a technology (e.g., mobility) is really a desire for something else (e.g., communication), which can be satisfied with no physical motion at all (e.g., telecommuting).

Lifestyle changes will very likely be necessary to achieve this goal; some of the overconsumption of the developed nations could easily be eliminated simply by examining our motivations and sense of what constitutes fulfillment. Businesses of the future will have a different set of customer demands to fulfill. However, there is ample evidence to suggest that reducing environmental impact to an acceptable level is not only well within our capability using the tools of modern technology and business practice, but that the business community is the portion of society best positioned to do so. This viewpoint does not ignore the importance of regulatory systems, tax codes, trade agreements, or social institutions. However, business can use the conventional profit motive, coupled with a clear view of the long-term market trends, to start immediately on the path to sustainability. As it does so, the public elements will be encouraged to follow and make the path easier.

5.10.3 THE IMPLICATIONS OF CURRENT ECOSYSTEM STRESS

As mentioned earlier, the planet is currently facing an increase of roughly a factor of 10 in ecosystem stress, arising from population increase and rising standard of living. This is a prediction that is essentially demographic in origin, and therefore has an inherent momentum that cannot be substantially altered by any simple policy change. Since the global environment is already under substantial stress, this factor of 10 increase represents the makings of a major catastrophe, and we can expect that powerful forces will arise to counteract it. These forces might be coercive in nature, and painful human social upheavals would be the result. With proper anticipation, however, we can see that providing the technology to reduce this incipient impact is a major business opportunity. The market driving force is clearly there; the challenge is to formulate a strategy that will meet this long-range market demand in a way that satisfies short-term business requirements as well.

Perhaps the most important principle to heed is that recycling and even reuse are not sufficient solutions. *If every aspect of U.S., European and Japanese industry were instantaneously transformed into a 100% cradle-to-cradle system in which every material were completely reused for the same purpose, the planet would still be headed for disaster.* This is because the volume of material required to raise the affluence of the rest of the world up to our own (using existing technology), combined with unavoidable population increase, would be sufficient to overwhelm many aspects of the global ecosystem. Those materials are today still sequestered in nature in various forms, and it is their extraction and processing that causes the important impact. The fact that they would thereafter be indefinitely reused is of scant consolation. Thus, the fundamental challenge of sustainability is to obviate the need for these materials ever to be extracted, and this can happen only with dramatic technological and business innovation in the way that we provide the services that people demand.

The developing world is thus of central importance. Because, although material waste is highest in the highly developed world, it is the inexorable motion of the remainder toward a similar condition (along with a larger population) that constitutes the bulk of the predicted impact. Without doubt the people living in these countries have a right to expect a rising standard of living. It is our duty, for sheer survival as well as from a sense of social justice, to see that they do so without incurring the same magnitude of material flows. It is also likely to be a successful business strategy: In many cases major technology innovations can be more easily introduced into the developing world where there is less established infrastructure to displace.

This analysis provides a meaningful quantitative goal. Avoiding the 10-fold increase in impact projected above, coupled with reducing current impact by about a factor of 2, would go a long way toward eliminating the major global stresses and providing breathing room to work toward full sustainability. Prioritization is vital. All economic progress requires investment, and investment resources are always limited. Thus it is important not to devote these valuable resources to achievements that have little clear environmental value. For example, the amount of space devoted to landfills is not a major ecosystem-disturbing problem at the present time (Brower and Leon, 1999). We should be wasting less material, but the impact in most cases is much more severe in the extraction phase than in the disposal.

The following set of topics covers the critical issues as suggested by the current literature. They are not intended to be ranked within this list, which would require data and analytical tools not available at present.

1. Carbon dioxide and other global warming gases
2. Agriculture (soil loss, water source depletion and pollution, wildlife habitat loss)
3. Water (effects of diversions)
4. Overfishing
5. Deforestation.

Other common environmental themes are either subsumed under these (for example, the severe air pollution that plagues many megacities would disappear with CO_2-minimizing energy solutions) or are less important to the global ecosystem (localized air, water, and soil pollution may be locally devastating, but typically do not threaten the extinction of entire species or the health of the planet that sustains all of us).

5.10.4 SOME APPLICATIONS OF INFORMATION TECHNOLOGY

The words "information technology," often abbreviated IT, are commonly understood to refer to computers and peripherals, such as displays, printers, mass storage, and networking technology. Here we expand the definition to include also the means by which the information is acquired and used (basically sensing and actuating technology). IT so defined has some rather obvious applications to sustainability. Hewlett-Packard has begun to address a few of these, building on its general competence base in IT as broadly defined here. Partnerships must certainly be a key element of these solutions, and HP's projects involve important partnerships with other organizations with a similar interest in sustainability. Some of our current accomplishments, while only a beginning, illustrate the exciting potential of these concepts. Many other companies are engaging in these arenas as well.

The following discussion, while by no means exhausting the possibilities, attempts to provide some comment on the majority of ways in which currently known information technologies can dramatically affect resource usage on a large scale, although one may be confident that more such inventions will arise. I have attempted to give as much quantitative justification of these as possible; more detailed impact analysis would be of great value for the prioritization process.

Control of Energy Use

While transformation to a totally renewable energy economy is an essential step, increasing energy efficiency will go a long ways toward reduction of CO_2 emissions as well as saving money. Electronic control systems with sophisticated sensing and computational capabilities can help in almost every use of energy. A simple example is in the area of lighting, which uses about one-third of the world's electricity. I estimate that the amount of energy wasted by unused lighting is at least 90%. While the ideal design is yet to appear for a cost-effective system that would turn lights on only as they are needed (and only as intense as needed) in a manner that is effortless, unobtrusive, and natural for the user, the payoff makes it well worthwhile. Coupled with greater efficiency in light generation (for example, by the use of solid-state LEDs), a factor of 100 savings in material flows for lighting could be achieved.

Additional opportunities can be found in the optimization of highly dispersed, small-scale energy production (to achieve maximum efficiency of the aggregate generating capacity for a given load) and in distributed process control.

These systems may be greatly enhanced by the capability of aggregating information at a central processing station and distributing control signals accordingly, rather than attempting to have each local control station contain the entire intelligence required for the optimum decisions. Hewlett-Packard (in what is now its spin-off, Agilent Technologies) has developed an

Internet-based distributed measurement and control system (originally targeted at the decentralizing utility industry) that makes it possible to easily connect virtually any sensor or actuator to the Internet, and interact with it remotely by standard Web protocols. Included in this system is the ability to accurately control the time of measurements.

Solid-State Lighting

Hewlett-Packard and Philips have formed a joint venture (LumiLeds) to produce inorganic LEDs for illumination, initially for traffic and street lights, but eventually for office lighting. While some wavelengths still require improvement to match the best discharge lamps, they are projected to eventually exceed the latter by around a factor of 2. The large cost of the light bulb (which is recovered from saved energy over many years) is likely to make the home market very difficult to penetrate with this technology, but organic LEDs (Sheats *et al.,* 1996), as they mature in performance and manufacturing cost, will probably play a role analogous to inorganics in the office; eventually organic solid-state lighting (manufactured on recyclable polyester substrates) may be the dominant technology. Organic LEDs are already more efficient than incandescents in about half of the visible spectrum, and improvements in both luminescence efficiency and light extraction efficiency can be reasonably expected to yield home lighting panels with efficiency equal to compact fluorescent bulbs at far lower cost and with longer lifetimes. In addition, these lights will allow color balancing to achieve whatever effect is desired by the user.

Transportation

The transportation industry is another major user of energy [over 25% of U.S. energy use, and two-thirds of total oil; this energy comprises around a third of the total material flow by weight (World Resources Institute, 1999). in the highly industrialized countries]. I have already alluded to two obvious contributions of IT to this realm: telecommuting and teleconferencing. The former is becoming a significant practice for many companies today; this and some other initiatives are discussed in a recent World Resources Institute (WRI) report (Horrigan *et al.,* 1998). The fact that videoteleconferencing is far from being as ubiquitous as it could and should be is primarily a technological issue. While there will always be an essential role for personal contact in business and scientific life, a great many business interactions that now require air travel could certainly be replaced by teleconferencing, even without the psychologically important enhancement of 3-D display. But along with adequate bandwidth and display resolution, it must be readily available and convenient to use. It will probably not have a major impact until it is as common as the speaker phone is today.

There are many other ways, potentially much more powerful, in which IT can reduce transportation impact. Carsharing is a private, for-profit business in Portland, Oregon, with several dozen cars that are distributed around the city for members to use whenever they are needed. The owners have calculated that anyone driving less than 10,000 miles per year will benefit from using the system. Such systems could almost certainly be made much more effective, and extended to a wider customer base, by more sophisticated IT. The phrase "station car" is a term describing cars that use a smart card to enable operation, and which can therefore be rented essentially automatically; they are picked up at and returned to specific, but widely distributed, locations. The advantage to the environment is that they can encourage the use of mass transit for commuting, by making it as easy to get to and from the transit center as if by one's own car, and by providing cars at work for short-term use at any time of day. Another advantage is that the optimum size car can be used for these trips (a single passenger can use a much lower consumption vehicle than the family vacation car or SUV). In the fairly near future, today's buses, poorly utilized due to their inconvenient routes and connections as well as rider discomfort, could be replaced by dynamically reschedulable vans, whose drivers would be in wireless communication with computers that would compute optimum routes for multiple passengers as needed. The bus could become nearly as convenient as a taxi, at a small fraction of the cost (and fuel usage). Extremely effective, user-friendly wireless communications will be an essential component of such a system, along with all the other elements of e-services.

Electronic commerce has the potential to substantially reduce intracity errand traffic. While some shopping will always require human contact, much of it could be (and increasingly is) done on the internet. Merchandise delivery by a van can be optimized to use less fuel than individuals traveling between store and home. It is likely that the amount of space (and hence energy and building materials) devoted to the storage and display of merchandise will diminish with the increase of consumer e-commerce.

Remote Sensing

As discussed earlier, food and water constitute two of the most important sustainability concerns. The problem has multiple, interrelated components: the increasing demand for more diverse foods by people with rising incomes, the need for basic food and water for a growing population, and the impact that agriculture [which consumes around 70% of the water and much of the land (Postel *et al.*, 1996; World Resources, 1999)] has on ecosystem health in general. All of these themes lead to a common conclusion that new technological solutions are needed for growing more food on less land using less water.

A specific and dramatic example of how IT can enable such solutions is provided by Cassman (1999), who describes studies of rice paddies in Indonesia showing a variation in yield of a factor of 3 for unfertilized plots (no exogenous nitrogen) in 42 contiguous fields. Other data from several countries show that this is not an atypical case. Since the attainment of an adequate basic food supply will depend primarily on moving closer to the theoretical yield of the plant, rather than higher yielding plants (Cassman,

1999), very detailed optimization of soil and crop treatment will be essential. The technology basis for sensing the required information (basic soil chemistry) and transmitting it wirelessly to a major database server, which analyzes it and returns an action recommendation for the farmer, exists today, and can be done at far less cost than the added value. A business structure to implement such a system is described later.

Precision monitoring and control of irrigation water, pest control solutions, and pollution problems will be enabled by the same techniques. The health of ecosystems can be monitored much as we monitor the health of our bodies by the diagnostic techniques of preventive medicine. As understanding of the economic cost (distributed through the entire society) of pollution becomes more evident, the commercial value of such activities will increase further.

One type of remote sensing technology is the electronic ID tag, which allows remote identification of tagged objects. Consider just one example of what these could do if they were cheap enough (as they soon will be). Suppose every package going into a refrigerator could have such a tag on it. On first applying the tag, one simultaneously tells a computer (which has recorded the identity signal of the tag) what it corresponds to. Thereafter, the computer knows not only if the object is in the refrigerator, but how long it has been there and approximately where it is. Refrigerators could be designed to allow much higher packing density with greater convenience of access than today, and the result would be lower energy consumption (by perhaps a factor of 4 or more) along with less spoilage.

Information Display

Paper usage increased in the United States by more than 50% during the 1980s and 1990s (Horrigan *et al.,* 1998), to about 100 million tons, and constitutes 10% of total wood fiber use worldwide; some of that is due to general economic growth and the greater fraction of the workforce devoted to services and information-related activity, and some to the increased use of computers. This is not surprising: the concept of a "paperless office" can only be related to a technology that replaces paper. The computer, by itself, only transforms (and proliferates) information, and the associated storage devices do not make it visible. Information becomes useful only when it is observed, and today's CRT display technology is little changed from 50 years ago. Without major innovations in display quality, people will continue to find that paper is the best digital information display medium that we have. LCDs and other flat-panel displays suffer from various combinations of high cost and effectively the same inconvenient physical form as the CRT.

The requirements for a paper replacement are daunting. Nevertheless, there are technologies, both reflective (Comiskey *et al.,* 1998) and emissive (Sheats *et al.,* 1996), that can meet this challenge. Although the commercialization of these systems is still in early stages, the scientific foundations are relatively well understood; primarily business-related barriers separate us from having media, at reasonable cost, with the physical form factor, resolution, and contrast of paper yet electronically refreshable and long lived (and recyclable upon wearout). While there will always be a role for paper, these developments should displace more than 90% of our current paper usage.

HP Labs carried out fundamental research in the area of organic LEDs (Sheats *et al.*, 1996), and has helped facilitate a commercialization path for flexible displays using this approach. Another project has resulted in a full-color, high-resolution wearable (eyeglass) display. While probably not suitable for office use, this type of display could very well be usable at home, for viewing newspapers, catalogs, and similar information.

Manufacturing

The process known as "micromachining" has been used for around two decades now to make miniature devices useful as sensors: acceleration detectors used to trigger airbags are a common example. Today this type of technology is on the verge of a major advance in complexity and power. Several companies have demonstrated complete microchip DNA analyzers (including synthetic and purification steps) for the pharmaceutical industry. Scientists at DuPont have taken this a step further and fabricated a complete chemical plant using just three silicon wafers taken from a modern IC process (Service, 1998). This "microfactory" is capable of synthesizing 18,000 kg/yr of the toxic industrial chemical methylisocyanate.

The purpose of this project was to provide a way to generate toxic chemical feedstocks on-site, on demand, from benign precursors, thus eliminating both the transportation and storage of such materials, with the attendant risks of environmental damage. The implications, however, are far broader. Small chemical reactors, made with planar processing technology, could provide a new paradigm in manufacturing for many materials, allowing them to be produced at the point of use from biologically derived feedstocks. Transportation is eliminated, and fewer materials are required to construct the reactor (the small scale reduces structural demands compared with large industrial plants).

Digital Technology for the Developing World

In December 1998 Jose Maria Figueres, the most recent president of Costa Rica and head of the Foundation for Sustainable Development of Costa Rica, approached the Massachusetts Institute of Technology (MIT) Media Lab with a bold vision: a modern, Internet-based communications network that would help developing countries leapfrog from their legacy systems to state-of-the-art global connectivity, and the economic and social benefits that would be thus enabled. The heart of the system that was devised is a standard ISO shipping container (8 × 8 × 20 ft), into which can be built a computer room, with about six computers and a printer and scanner, a communication room with a cell phone base station, telephone, and printer/fax/copier, and a medical diagnostic bay with another computer.

Other technologies planned include a smart card reader, analytical equipment, and large-area screen on the outside of the unit. Connectivity is provided by a VSAT (very small aperture terminal) antenna and transceiver, connected to a geostationary satellite Internet node capable of offering up to 4 Mbps downlink data rates. Prototypes, using HP IT equipment, were demonstrated at MIT and in Costa Rica in the spring of 1999. The project was named "LINCOS," for Little Intelligent Communities.

The purpose envisioned for these units is to provide a multiuse information center for isolated regions, with high-speed Internet access and integrated local wireless communications, at an affordable cost for developing nations. Specific applications include telemedicine and simple medical diagnostics, analytical capabilities, distance education, computer services and training, electronic commerce, and entertainment and teleconferencing.

It is readily apparent how such a unit might provide the infrastructure required for the agricultural information service described earlier, as well as other kinds of remote interactions. The degree to which this simple unit, with a total capital and installation cost of around $50,000 to $100,000 (depending on what is actually put into it), can provide the most sophisticated information availability to people with no other connectivity outside their locality is evident. Coupled with an effective business development process to foster local entrepreneurs and Internet service providers, these centers can also achieve a rapid and consequential effect on the world's current ecological crisis. Rural villages can achieve a standard of living that discourages migration into overcrowded and polluted megacities; people can earn a good living by doing things that preserve the land, rather than exploiting it through resource extraction and heavy industrial construction. Finally, as such populations develop a level of affluence that supports the kind of mobility and consumer goods that we have come to take for granted, they will already be highly sophisticated in the use of IT, and will be well positioned to use this knowledge to obtain the highest possible efficiency in the systems they develop.

Pioneering this framework makes good business sense for HP, which has already defined its current identity around e-services. Addressing the need just identified requires us to conceptualize how to extend these capabilities to a market where e-services are as yet unknown, and much fundamental development is needed. In the developing world, it will be necessary to first work with customers to understand what problems could be solved by these techniques, and then assist them in setting up the necessary business structures. The presence in countries like Costa Rica (for example at the Institute of Technology) of a substantial base of expertise in entrepreneurial training and business incubation provides a springboard that can be used to achieve this development, and evolve replicable, scalable models for other countries. There have already been many successful telecenters in nearly every part of the world, which have made substantial positive differences in the lives of area inhabitants. Nevertheless, these activities have not proliferated rapidly. It is our hope that partnership with transnational companies like HP will enable the global networking and integration that can facilitate such proliferation.

Global solutions, however, do not imply global dominance. The businesses that we enable should be locally owned and controlled, and this caveat is not in conflict with the presence of multinationals such as HP as service providers. As we embrace new business models in the pursuit of sustainability, it is important that the sociological implications be carefully thought through and the input of all stakeholders be sought.

5.10.5 CONCLUSIONS

The discussion of what technologies are needed to solve the unsustainability of the present industrial system, and who will implement them, has barely begun. Even at this stage, however, we can be sure that this is a rich area in which to prospect for new business, for any company with a culture of innovation and a willingness to invest in the future. I would like to close with a few observations concerning the business and social aspects of the process.

A key theme of all of the scenarios I have described here is dematerialization: accomplishing desired goals with less material. From the business perspective, this means making service the product of the company. We live in a material world, but it is the *interactions* of the materials that are perceived by our minds as having meaning, utility, and value. Thus, our goal should be: "What you market is what you sell" (Bertolucci, 1998). What we *market* is always the function of the product: What can it do for the customer? Yet what we *sell* (i.e., have a price tag attached to) is usually the material object that provides the function. *In a sustainable economy market forces will efficiently minimize material flows because the price tag will be attached to the function.* Incorporation of this concept into the foundation of every organization (corporate, NGO, and government) is a necessary (and possibly sufficient) condition for achieving sustainability for the 10 or more billion people expected to inhabit the earth in the near future.

A second vital issue is prioritization. How do our proposed actions relate to the list of ecological problems listed earlier? This list constitutes a guide that will maintain focus on the high-profit, high-contribution formula of being positioned to meet long-term market demand and maximizing material productivity. It also prevents "greenwashing," and similarly avoids being distracted into expending excessive effort on relatively minor problems. Further scientific study of these priorities and their scientific bases will be very useful for keeping industry on track.

The issue of social justice plays several roles in this discussion. We have emphasized the necessity of profit and the existing free enterprise global economic system. Yet both in the developing world and in our own society, local empowerment and control is essential for social stability and progress (Serageldin, 1999). Mechanisms must be found that make investment attractive for financial sources yet avoids their domination. Novel revenue-sharing models, many of them made more feasible by the Internet, will be essential. Clearly these are important issues that have to be given serious attention. It is very much in our self-interest to do so, because failure will harm us all.

Concerns are sometimes raised that by assisting development, we will be doing the opposite of what we want, because the higher standard of living will inevitably lead to heavy resource usage just as it has in the United States. It is important to realize that the rise in affluence of the developing world is going to occur, independent of what any of us thinks or wants. The people in the economies where consumer wealth is just emerging are (understandably) not concentrating on the global environmental implications of their actions, any more than the majority of executives of multinational corporations are thinking of these implications. As one examines the list of specific impacts given earlier, it is clear that each one of them can be lowered by a 10- to 20-fold amount with respect to current developed-world levels. Along the way, the discussion of how we will all have to live in the more distant future can be nurtured and expanded, with everyone participating in an informed and constructive manner.

My final comment relates to this issue of motivation. I have emphasized the profit-motivated, free enterprise system must be the basis of achieving sustainability, yet this same system has in the past been a key contributor to creating the crisis we find ourselves in today. Many corporations still today seem more or less unconcerned about any obligations they might have to stakeholders other than stockholders interested in short-term high returns. The fundamental *social* challenge before us is to achieve an integration of our behavior outside of the corporations in which most of us work with that practiced inside. Business is fundamentally a social activity that addresses community needs, and it should not be considered separate from the community in which it is practiced. With at best several billion people on earth, we have no alternative to business to provide for our physical needs. We also therefore have no option but to understand how to seamlessly integrate this process with the individual and collective goals of humanity.

Acknowledgments

Many of the concepts in this paper originate from conversations with Stuart Hart, to whom I am greatly indebted. I would also like to acknowledge Joe Laur and Sara Schley for starting me on the journey of discovery in sustainability. An earlier draft appeared in the proceedings of the Greening of Industry Network Conferance, November 1999, Durham Hill, NC.

The largest privately funded industrial photovoltaic system in the United States covers one-half acre and supplies about 200,000 kilowatt-hours of power to Interface's Bentley Mills in California. Interface, an international floor covering service, is committed to "take nothing harmful or nonrewable from the Earth's crust and to emit nothing harmful into the biosphere as a result of its operations," according to Ray C. Anderson, Chairman and CEO. Photograph courtesy of Interface, Inc.

CHAPTER 6

Engineering Design and Innovation

6.1 Engineering

In many ways the story of the development of new technologies to aid man is the story of civilization itself. Persons who are professionally involved in the development of new technologies are usually called engineers. Engineers are constantly seeking to understand the laws of nature so as to modify the world beneficially.

One view of engineering is to describe it as the activity of problem solving under constraints. An engineer defines a problem, asks questions that determine its true nature, and ascertains whether the problem is overconstrained, so that no solution exists, or underconstrained, so that several exist. The engineer relaxes constraints selectively, seeking better solutions. Also, she studies the range of possible solutions, seeking the one that is optimal in as many ways as possible. Usually he studies the solutions using an iterative process which examines a cycle of steps toward a suitable design.

Engineering represents a variety of activities associated with technology, but we shall use a somewhat expanded definition as follows: **Engineering** is the profession in which a knowledge of mathematical, natural, and social sciences gained by study, experience, and practice is applied with judgment to develop ways to utilize economically the materials and forces of nature for the use of mankind. Thus, for example, the design and construction of a bridge such as the Golden Gate Bridge in San Francisco requires the knowledge of the sciences that an engineer would gain in college and in practice. This knowledge is coupled with good judgment to choose from alternative designs for a bridge and to examine the economics of various approaches. The engineer can then utilize materials such as steel and cement to construct a bridge for the benefit of society.

The ultimate objective of engineering work is the design and production of specific items. Engineers may design and build a ship, a radio station, a digital computer, or a software program. Each item is designed to meet a different set of requirements or objectives. The idea of design is to bring into existence something not already existing. Engineering design is thus distinguished from a reproduction of identical items, which is the manufacturing process.

The essential features of the engineering process are as follows:

1. Identification of a feasible and worthwhile technical objective and definition of this objective in quantitative terms
2. Conception of a design which, in principle, meets the objective (This step involves the synthesis of knowledge and experience.)
3. Quantitative analysis of the design concept to fix the necessary characteristics of each part or component and to identify the unresolved problems — usually problems of modules, of component performance, and of the interrelationship of components and modules
4. Exploratory research and component tests to find solutions to unresolved problems
5. Concepts for the design of those components and modules that are not already developed and available
6. Reanalysis of the design concept to compare the predicted characteristics with those specified
7. Detailed instructions for fabrication, assembly, and test
8. Production or construction
9. Operational use, maintenance, and field service engineering.

If possible, it is desirable to satisfy the technical objective of step 1 with a best or optimal design. The design objective of a bridge might call for a certain number of traffic lanes and a specified loaded strength at a minimal cost. The best design, then, is the one that meets all nine steps in the process and results in satisfying the objective with minimal cost. Of course, the engineer must recognize that minimal construction cost may not yield a minimal maintenance cost, and both must be accounted for in the design objective. The engineering solution is the optimum solution, the most desirable end result taking into account many factors. It may be the cheapest for a given performance, the most reliable for a given weight, the simplest for a given level of safety, or the most efficient for a given cost.

For the engineer, one important quality is efficiency, which is defined as the ratio of the output to input and written as

$$\eta = \frac{\text{Output}}{\text{Input}}.$$

The ratio may be expressed in terms of energy, materials, money, time, or work-hours. Commonly, energy or money is used in this indicator. A highly efficient device yields a large output for a given input. An example of a high-efficiency device is an electric generator in a power plant. An example of a low-efficiency device is a passenger automobile, which ordinarily has an efficiency of approximately 0.15 (or 15%). (An auto provides only 15% of the energy stored in the gasoline as energy output to the tires of the car.)

The engineer is called on to conceive an original solution to a problem or need, and to predict its performance, cost, and efficiency. Then, she is required to construct the device. The engineer uses the sciences and materials to build desirable products. The engineer is concerned with design and construction, whereas the scientist is concerned with developing the fundamental knowledge of nature. The engineer must know science, but he is not necessarily a scientist, although many engineers are involved in pure science as well as their engineering work.

Engineering is built on the use of quantitative data. Thus, measurement of important data is often carried out as part of an engineering task. The engineer often utilizes empirical approaches to the solution of a problem.

Engineering is concerned with the proper use of money, a capital resource, as well as all other resources. Engineering requires the economical use of energy and materials so that there is a minimum of waste and a maximum of efficiency. It is also the safe application of the forces and materials of nature. Safety is an important quality and must be incorporated in all objectives.

Engineering is intimately involved with the process of technological change, whereby new methods, new machines, new organizational techniques, and new products are (1) discovered and developed to the market stage; (2) introduced in the fields of both production and consumption; and (3) improved, refined, and replaced as they become obsolete.

6.2 Engineering Design Methods

Engineering design is the systematic, intelligent generation and evaluation of products whose form and function achieve stated objectives and satisfy specified constraints. Most design methods are based on the following sequence:

1. Determine the customers' needs and translate them into design goals and product attributes.
2. Define the essential problems that must be solved to satisfy the needs.
3. Conceptualize the solution through synthesis, which involves the task of satisfying several different functional requirements using a set of inputs such as product design parameters within given constraints.

4. Analyze the proposed solution to establish its optimum conditions and parameter settings.

5. Check the resulting design solution to see if it meets the original customer needs.

Design proceeds from abstract and qualitative ideas to quantitative descriptions. It is an iterative process by nature: New information is generated with each step, and it is necessary to evaluate the results in terms of the preceding step. Thus, design involves a continuous interplay between the requirements the designer wants to achieve and how the designer wants to achieve these requirements.

Designers often find that a clear description of the design requirements is a difficult task. Therefore, some designers deliberately leave them implicit rather than explicit. Then, they spend a great deal of time trying to improve and iterate the design. To be efficient and generate the design that meets the customer needs, the designer must specifically state the users' requirements before the synthesis of solution concepts can begin.

Solution alternatives are generated after the requirements are established. Many problems in engineering can be solved by applying practical knowledge of engineering, manufacturing, and economics. Other problems require far more imaginative ideas and inventions for their solution.

Engineering effort is best described by the words **designing** and **inventing**. The difference between them is quantitative rather than qualitative. The terms imply "putting parts together so that they can work together" — or, more precisely, so that natural forces can work through these parts to satisfy human needs.

Specifications will normally include environmental factors such as waste, recyclability, and resource use. A sample of design goals that might be considered is provided in Table 6.1.

Table 6.1 **Design Goals for a Product**

- Performance
- Maintainability
- Aesthetics
- Cost
- Availability of parts and materials
- Ease of use
- Versatility
- Portability
- Recyclability of parts
- Readily manufactured
- Safety

6.3 The Design Process

The design process can be illustrated by Table 6.2. The process always begins with defining and documenting the customer's needs. A useful tool for doing this is quality function deployment (QFD), which is discussed in the next section. After the customer's needs are determined, the design goes through successive generations as the design cycle is repeated. The requirements are set and a model or prototype is created. Each validation of a model or test of a prototype provides key information for refining the requirements.

Table 6.2 **The Design Process**

1. Identify the customer's needs.
2. Analyze the requirements provided by the customer and the manufacturer.
3. Create alternative system design concepts that might satisfy these requirements.
4. Build, validate, and simulate a model of each product design concept.
5. Select the best concept by doing a trade-off analysis.
6. Update the customer requirements based on experience with the models.
7. Build and test a prototype of the product.
8. Update the customer requirements based on experience with the prototype.
9. Build and test a preproduction version of product and validate the manufacturing processes.
10. Update the customer requirements based on experience with the preproduction analysis.
11. Build and test a production version of the product.
12. Deliver and support the product.

For example, when designing and producing a new electric car, the initial task is to develop a model of the expected performance. Using this model, the engineers make initial estimates for the partition and allocation of system requirements. The next step is to build a demonstration unit of the most critical functions. This unit does not conform to the form and fit requirements but is used to show that the system requirements are valid and that an actual car can be produced. Requirements are again updated and modified, and the final partitioning is completed.

The next version is called the proof-of-design car. This is a fully functioning prototype. Its purpose is to demonstrate that the design is within specifications and meets all form, fit, and function requirements of the final product. This prototype is custom made and costs much more than the final production unit will cost. This unit is often built partly in the production factory and partly in the laboratory. Manufacturing capability is an issue and needs to be addressed before the design is complete. More changes are made and the manufacturing processes for full production are readied.

The next version is the proof of manufacturing or the preproduction unit. This device will be a fully functioning automobile. The goal is to prove the capability of the factory for full-rate (speed) production and to ensure that the manufacturing processes are efficient. If the factory cannot meet the quality or rate production requirements, more design changes are made before the drawings are released for full-rate production. Not only the designers but the entire design and production team must take responsibility for the design of the product and processes so that the customer's requirements are optimized.

Most designs require a model on which analysis can be done. The analysis should include a measure of all the characteristics the customer wants in the finished product. The concept selection will be based on the measurements done on the models. The models are created by first partitioning each conceptual design into functions. The decomposition often occurs at the same time that major physical components are selected. For example, when designing the electric vehicle mentioned above, we need to examine the electric motor, the controls, the battery, and other subsystems.

An example of design process is seen in the sterile bandage, which was designed by Earle E. Dickson. He recognized the need for such a bandage when his wife burned her fingers. He decided to design a bandage that would be easily applied to the wound, retain its sterility, and remain in place after its application.

After studying the problem, Dickson placed strips of surgical tape on top of a table and affixed gauze to the center of each strip's exposed adhesive surface, taking care that some of the adhesive remained uncovered by the gauze. Finally, he covered the gauze with crinoline to complete his design. Whenever an injury occurred, a portion of these bandage strips could be cut with scissors to the desired length, the crinoline removed, and the bandage applied directly to the wound.

Dickson's original design managed to achieve each of his three specific goals: the use of crinoline ensured that the bandage would retain its sterility, the gauze and adhesive strips allowed the user to apply the preformed bandage with relative ease, and the adhesive of the surgical tape ensured that the bandage would remain in place after its application to the wound. Formulating the problem in terms of the specific goals that must be achieved was a critical step toward the solution.

Dickson placed his invention with Johnson & Johnson for commercialization and the product became the now-familiar Band-Aid.

6.4 The House of Quality[1]

By Derby A. Swanson
Former Principal
Applied Marketing Science

6.4.1 WHAT IS THE HOUSE OF QUALITY?

The **House of Quality** is a management design tool that first experienced widespread use in Japan in the early 1970s. It originated in Mitsubishi's Kobe Shipyard in 1972, helped Toyota make tremendous improvements in their quality in the mid-1970s, and spread to the United States (via the auto industry) in the mid-1980s. A significant step in popularizing its use in the United States was the 1988 publication, in the *Harvard Business Review,* of "The House of Quality," by John Hauser and Don Clausing. Hauser and Clausing (1988) stated that "the foundation of the house of quality is the belief that products should be designed to reflect customers' desires and tastes — so marketing people, design engineers, and manufacturing staff must work closely together from the time a product is first conceived."

Now, the House of Quality and similar total quality management-related quality tools have moved well beyond their initial application in the U.S. auto industry. These techniques are used in all types of settings for a wide variety of product and service improvement/development projects.

The House of Quality[2] refers to the first set of matrices used in **quality function deployment (QFD)**. It acts as a building block for product development by focusing efforts on accurately defining customer needs and determining how the company will attempt to respond to them.

Building a House of Quality matrix starts by defining a project around product and/or service development/improvement and putting together a cross-functional team with responsibility for making it happen. The team begins by working to fully understand the customers' wants and needs for the particular product, service, or combination — known as the **voice of the customer (VOC)** or customer attributes (CAs). A determination of how important the needs are to customers and how well they are currently being met is used to help assess their criticality in the development effort.

The team then identifies the possible measures or product characteristics that could impact those customer needs. In product development customer needs are related to physical characteristics, commonly called engineering characteristics (ECs), of the product. For example, a customer's need for an *easy-to-carry* container will be impacted by the physical characteristic of its *weight*. For issues having to do with service development — for example,

---

[1]This section has been reprinted with permission from *The Technology Management Handbook* (Richard Dorf, ed.), pp. 12-27–12-33, Boca Raton, FL: CRC Press, 1998. Copyright CRC Press, Boca Raton, Florida.

[2]The name "House of Quality" comes from the shape of the matrix components.

how to set up an order desk — the customer needs (e.g., *quick and accurate from order to delivery*) would be impacted by performance measures (PMs) such as *time it takes to complete an order.*

The next step is to correlate the two — to assess the relationship (or impact) of the measures on the needs. This step fills in the matrix that makes up the center of the House (Figure 6.1).

The strength of those relationships (example in Figure 6.2) will be an important factor in telling the team where they should focus. After the links to customer needs are identified it will be possible to prioritize product characteristics or measures based on which have the biggest net impact on customers. The more impact a measure/specification has on important customer needs, the more attention it should get.

In addition to defining the customer needs, it is important to understand that all customer needs may not be created equal! Some will be more important to customers than others (e.g., *safety* vs. *easy to buy* for a medical product). Also, there may be areas that the team wants to target for improvement — where the company's performance is lacking vs. competition or where there are particular quality problems that need to be addressed. Both the importance to customers and any relative weight on the needs will also be addressed in the House of Quality matrix.

Since both the importance/weighting of the customer needs and the relationships at the center of the House are expressed as numbers, the result

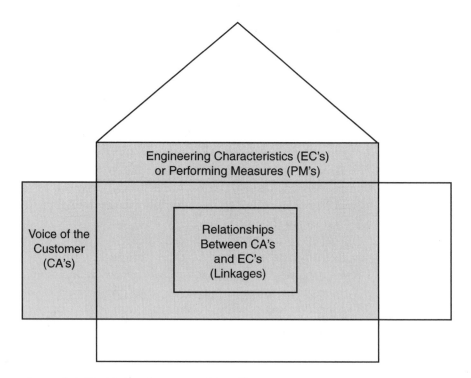

Figure 6.1 **Diagram of House of Quality components.**

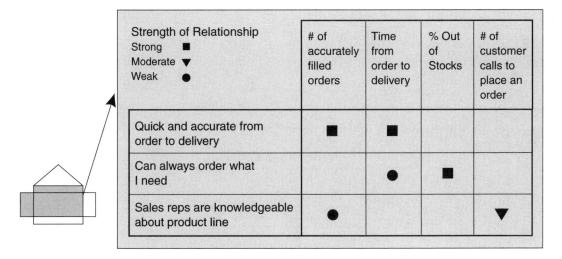

Strength of Relationship Strong ■ Moderate ▼ Weak ●	# of accurately filled orders	Time from order to delivery	% Out of Stocks	# of customer calls to place an order
Quick and accurate from order to delivery	■	■		
Can always order what I need		●	■	
Sales reps are knowledgeable about product line	●			▼

Figure 6.2 **Example of CAs, PMs/ECs, and linkages.**

of "crunching the numbers" in the matrix is a prioritization of the engineering characteristics/performance measures that tells the team where it should put its energy and resources.[3]

Finally, the roof of the House contains information regarding how the ECs and PMs interact with each other. They may be independent or there may be positive or negative correlations, for example, attempts to reduce a copier product's footprint will correlate positively with making it lighter in *weight* but there may be a negative effect on *the amount of paper it can hold* (Figure 6.3).

6.4.2 WHY USE THE HOUSE OF QUALITY?

The discipline of building a House of Quality leads to a design that is truly driven by the voice of the customer rather than the "voice of the loudest vice-president." Development time is invested at the beginning rather than in redesigning a product/service that doesn't work. A cross-functional team of people working together, focused on the customer instead of on their own departments, becomes a cohesive unit. Finally, a decrease in cycle time and increase in responsiveness from understanding the marketplace can lead to the kind of results that helped Japanese automakers to launch their successful foray into Detroit's marketplace.

Businesses succeed by offering products and services that meet customer requirements profitably. Businesses retain their customers by continuing to please them, which sounds very basic — but is, in fact, difficult. One of the causes has to do with the ways in which companies attempt to "hear" their customers. There are many stakeholders to be satisfied and customers speak

---

[3]Commercially available software such as QFD Capture helps manage this process.

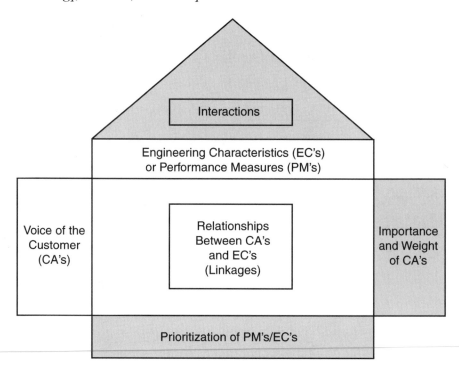

Figure 6.3 **Diagram of House of Quality components.**

different languages from one another and, most especially, from a company. Greising (1994) gives an example of what companies face as they pay homage to quality without understanding what it takes to be truly customer driven:

> A major computer manufacturer spent years using mean time between failures (MTBF) as their measure of reliability. However, along with their industry-leading MTBF went a customer reputation for unreliable systems. It turned out that to customers reliability had to do with availability — how much of the time did they have access to their equipment. The manufacturer needed to include diagnostic time as well as technician travel time, etc. to really understand how they were doing.

Customers have different concerns and speak a different language than companies, and it is that language and thought process that must be understood. Commonly used "quality-oriented" market research words such as "reliable," "convenient," or "easy to use" will likely have very specific meaning to individuals on a project team, as in the MTBF example. However, they may have very different meanings for customers. Those differences, and similarities, must be understood in detail before a company can begin successful use of the House of Quality.

Another problem companies face is trying to aggregate data too soon, to look for the one number, the one answer on how to design a product. A rating of "85% satisfaction" tells you very little about the cause and effect of what the company is doing. Scoring on typical market research survey attributes doesn't give engineers and designers the level of detail they need to design. Plus, this type of data gives you information after the fact; it measures results based on history. The House of Quality is proactive — it enables you to build a model that relates the actions taken (or the things that you measure) directly to the impact that they will have on customers.

6.4.3 BEGINNING A HOUSE OF QUALITY

There are many tools available for helping a team interested in using QFD and the House of Quality — including firms and individuals that can act as facilitators and books (e.g., Lou Cohen's 1995 *Quality Function Deployment — How to Make QFD Work for You*) that can act as manuals. The discussion that makes up the remainder of this section talks about some of the basic issues to be addressed while putting together a team and beginning a House of Quality. The emphasis on customer needs is intentional — if the VOC that begins the exercise is flawed, so is everything that follows.

Getting Started

As with any successful team effort the key here is assembling the team and defining the problem/issues. It is important the team be cross-functional — with representation from all interested and affected parties. This means marketing plus engineering plus manufacturing plus information technology, etc. It can be extremely helpful to have an experienced facilitator from outside. This is someone who has been through the process before and whose responsibility is taking the team through the exercise — not trying to promote his own agenda. An outsider can be objective, help resolve conflicts, and keep the team on track.

It is important to spend time up front assessing what the project is about and who the "customers" really are. First, there is likely more than one buyer for a product or service, e.g., medical instruments (medical technicians, purchasing agents, and doctors) or office furniture systems (facilities managers, architects, designers, MIS professionals, and end users). A university must consider students, faculty and industry recruiters. Utilities deal with everything from residential to industrial to agricultural customers — and in literally millions of locations.

There are also internal customers who may feel very differently from external ones. Employees are also customers and may well have significant impact on how external customers evaluate the service they receive.

Collect the Voice of the Customer

A focused and rigorous process for gathering, structuring and prioritizing the VOC will ensure that you have the customers' true voice and that the data are in a form that will be useable for a House of Quality (as well as other types of initiatives, e.g., developing a customer satisfaction survey).

Customers will not be able to tell you the solution to their problems, but they will be able to articulate how they interact with a product or service and the problems they encounter, e.g., wanting to get printer output that is "crisp and clear with no little jaggy lines." By focusing too early on a solution, e.g., resolution of *600 dots per inch,* you lose sight of other, even-better solutions. The goal is products that are "inspired by customers — not designed by them."

Beginning with one-on-one interviews (or focus groups) to gather the customer needs and using more quantitative customer input to structure and prioritize customer needs enables the customer's own words to drive the development process forward. It maintains the purity of the customer data while allowing the developers to focus on the company voice. As Hauser and Clausing (1988) state,

> CAs are generally reproduced in the customers' own words. Experienced users of the house of quality try to preserve customers' phrases and even clichés — knowing that they will be translated simultaneously by product planners, design engineers, manufacturing engineers, and salespeople. Of course, this raises the problem of interpretation: what does a customer really mean by "quiet" or "easy"?

The issue of interpretation is addressed by asking customers themselves to define these words. Twenty to 30 one-on-one interviews from a relatively homogeneous group of customers will ensure that 90% or more of the needs have been identified (Griffin and Hauser, 1993). However, the topic and type of customer being interviewed should also be considered — more complex topics may require more interviews.

What you talk about in a VOC interview is critical to its success. Customers tend to talk in targets, features, and solutions — *"Why don't they just give me one rep?"* The goal of the interview is to understand the needs behind the solutions by asking "why." The key is to question everything — *"Why is that important?"; "Why is A better (worse) than B?"* That way you hear things such as *"If I had a dedicated rep then I'd know that that person could **get my questions taken care of right away.***" By understanding underlying needs, the team can think about a number of different ways to meet them.

Team members can gain significant insight into customers by listening to them discuss their issues and concerns. However, this does not mean that team members should necessarily conduct the interviews. The keys are training and objectivity, which outside interviewers are more likely to have. For example, in one case an engineer spent an hour interviewing a customer; the engineer started out by being defensive, then tried to solve the customer's problem, and ended up learning nothing about customer's needs. However, with the right focus and training, company personnel have found it very useful to conduct interviews themselves.

A formal process (such as VOCALYST®4) will help efficiently reduce

---

[4]VOCALYST® stands for Voice of the Customer Analyst and is used by Applied Marketing Science, Inc.

the thousands of words and phrases collected from the interviews to the underlying set of needs (e.g., 100 to 200) that can then be categorized into groups that provide a "primary," "secondary," and "tertiary" level of detail (Figure 6.4).

The best way to do the necessary data reduction is to have customers develop their own hierarchy of needs (e.g., through a card-sorting process). Just as the language that customers use is different from a company's, so is the way in which they aggregate ideas. Members of a company team often organize ideas based on functional areas (things that are marketing's role or part of the information services department). Companies think about how they deliver or make a product, and customers think about a product based on how they use it or interact with it. No customer ever separated his needs into hardware and software — but many computer companies have organized themselves that way. For example, Griffin and Hauser (1993) show two hierarchies — team consensus and customer based — to a number of professionals (including the team!) and find that the "customer-based (hierarchy) provided a clearer, more believable, easier-to-work-with representation of customer perceptions than the (team) consensus charts."

Develop Engineering Characteristics (ECs) or Performance Measures (PMs)

Next the team will develop lists of ECs/PMs, assess the relationship to customer needs, and set priorities. Each customer need should have at least one EC or PM that the team feels will strongly impact the customer need. One

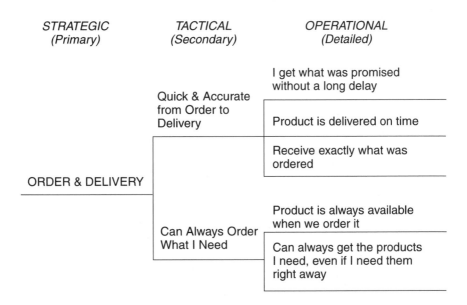

Figure 6.4 **Section of attribute hierarchy.**

way to get a list is to brainstorm a list of possible contenders for each need and then have the team decide which ones are most likely to be predictive of meeting the need. As with the customer needs, it is important that the "company voice" fit the criteria for a "good" measure:

- Strongly linked to need
- Measurable
- Controllable
- Predictive of satisfaction
- Independent of implementation
- Known direction of improvement or desired target value
- Impact of hitting the target value is known

Assign Relationships between ECs or PMs and Needs

The team votes on every cell in the matrix to determine the strength of relationship the measure will have with each need. For the most part, these relationships are positive (since that is the point of the House!) but there may be customer needs that end up negatively impacted by an EC/PM and that should be accounted for. There are many scales and labels, but the most common process is to assign the following types of values:

- Strong relationship "9"
- Moderate relationship "3"
- Weak relationship "1"
- No relationship

Calculate Priorities for the ECs/PMs

The "criticality" of each EC or PM comes from the strength of its relationship with the customer needs, weighted by both the importance and any weight assigned to the need. The importance comes directly from customers (either when they create the hierarchy of needs or via a survey). A weight can also be assigned to each need based on a target for improved performance (from customer data), a desire for improved quality in that area, etc. In the example here, the weight is expressed as an index. In the case of the example in Figure 6.5, the PM with the highest priority is *number of accurately filled orders* — calculated by summing down the column for each linkage with that PM:

- (Importance of the need · the strength of the relationship) · weight of the need
- $[(100 \cdot 9) \cdot 1.0] + [(75 \cdot 1) \cdot 2.0] = 1050$

The ECs with the highest priority are the ones that the team absolutely needs to focus on to ensure that they are delivering on customer needs.

Finally

As companies make decisions about where to allocate resources for programs and development, they should be focused on the things that are important to customers but also play to their own strengths and core competencies. In "The House of Quality," Hauser and Clausing (1988) said "strategic quality management means more than avoiding repairs. It means that companies learn from customer experience and reconcile what they want with what engineers can reasonably build." Using the House of Quality can help make that so.

Defining Terms

House of Quality: Basic design tool of management approach known as quality function deployment.

Quality function deployment (QFD): A series of planning matrices that allow planners to continually integrate customer needs with the technical responses to meeting those needs when designing a product or service

Voice of the customer (VOC): Customer wants and needs for a particular product or service — developed and structured in such a way as to be used in a House of Quality or other initiative.

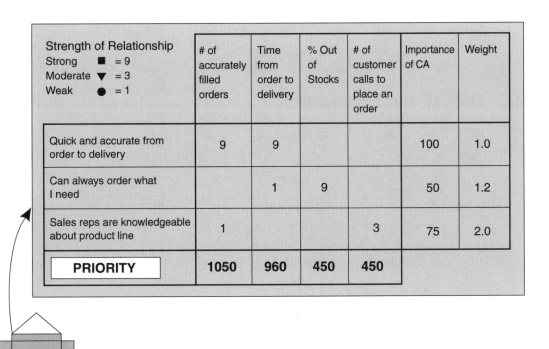

Strength of Relationship Strong ■ = 9 Moderate ▼ = 3 Weak ● = 1	# of accurately filled orders	Time from order to delivery	% Out of Stocks	# of customer calls to place an order	Importance of CA	Weight
Quick and accurate from order to delivery	9	9			100	1.0
Can always order what I need		1	9		50	1.2
Sales reps are knowledgeable about product line	1			3	75	2.0
PRIORITY	**1050**	**960**	**450**	**450**		

Figure 6.5 **Example of CAs, PMs/ECs, and linkages plus weighting and prioritization.**

Further Information

Anyone interested in the basics should read the *Harvard Business Review* article "The House of Quality" by John Hauser and Don Clausing. A good overall introduction to the whole topic of new product design is the 2nd edition of Glen Urban and John Hauser's book, *Design and Marketing of New Products* (Prentice Hall, 1993).

One example of a "primer" for conducting a QFD/House of Quality program would be Lou Cohen's book *Quality Function Deployment — How to Make QFD Work for You,* published by Addison-Wesley (1995) in their Engineering Process Improvement Series.

INTERFACE, INC.: A NEW APPROACH TO LEASING "COMFORT"

Interface is the worldwide leader in the modular carpet sector for the commercial and institutional markets. With sales of $1.3 billion in 2000, the firm is actively designing and manufacturing green products. Interface has launched an innovative program aimed at selling "functionality" of comfort. Working in collaboration with fiber producers, a new product line has been created by "remanufacturing" products, thereby converting the "old" product into new carpeting or floor tile. The customer "leases" the product, or the "comfort" provided by the carpet. Once the customer's carpet reaches the end of its useful life, a new floor covering is supplied to replace the old, and the "spent" product is then reintroduced to the marketplace after refurbishment, remanufacture, or a fashion facelift.

6.5 Discontinuous Innovation[5]

By Christopher M. McDermott
Associate Professor
Lally School of Management
Rensselaer Polytechnic Institute

When managing projects in an innovative environment, it is critical to be aware that projects differ in the extent to which they depart from existing knowledge. From a managerial perspective, these differences in the uncertainty of projects affect the way it needs to be directed and managed. As will be discussed in this section, the approaches used in managing an incremental project are often quite inappropriate when dealing with a project at the more discontinuous end of the spectrum. However, the relative infrequency of these discontinuous innovations (leading to the lack of experience most

---

[5]This section has been reprinted with permission from *The Technology Management Handbook* (Richard Dorf, ed.), pp. 3-11–3-16, Boca Raton, FL: CRC Press, 1998. Copyright CRC Press, Boca Raton, Florida.

managers have dealing with these projects), coupled with the critical importance of such "leaps" to a firm's long-term viability, makes this a very important topic. This section provides an overview of research and practice in the management of projects at this more discontinuous end of the spectrum. It begins with a discussion of the measurement of innovativeness and the importance of discontinuous innovation within a firm's portfolio of projects. The section then moves on to discuss the impact of discontinuous innovation at a (macro) industry level and (micro) firm level, including a discussion of managerial approaches that appear to be appropriate in more discontinuous environments.

6.5.1 FUNDAMENTALS OF DISCONTINUOUS INNOVATION

Traditionally, the bulk of academic research and practitioner attention in the management of technological innovation has focused on the development and implementation of practices that are appropriate and effective in directing incremental projects. This makes sense for a number of reasons:

- The majority of projects that are ongoing at any point in time in a firm *are* incremental.
- The short-term return on investment (ROI) mindset prevalent in many firms discourages discontinuous "leaps."
- Recent focus on lean production and reengineering absorbs many of the prerequisites for considering such projects.

However, the argument has been made (Morone, 1993; Tushman and Anderson, 1997) that only through a balanced portfolio approach to innovation (with firms pursuing both incremental and discontinuous projects) can a firm continue to prosper in the long term. While incremental innovation can maintain industry leadership in the short term, it is the projects at the discontinuous end of the spectrum that put firms in the lead in the first place.

There is no easy answer to the question: What makes a project discontinuous? Certainly, one common theme tied to **discontinuous innovations** is the high degree of risk and uncertainty throughout the process. The technologies involved are often unproven, and the projects are typically longer in duration and require a significantly greater financial (and managerial) commitment than their incremental counterparts. In addition, the end products themselves are often targeted at markets that are currently nonexistent or at least have the potential of being so significantly "shaken-up" by the introduction of the final product that existing, traditional market research is of questionable value. This discussion suggests one common way to conceptualize such projects which is shown in Figure 6.6.

In this figure, the vertical axis relates to the technological uncertainty associated with the project. At the low end of this axis, a firm is dealing with technology that is essentially an extension of current capabilities. The question here is more of *when* as opposed to *if* the technological advance embodied in the project will work. At the high end of this axis, however, firms manage projects where the functionality of the technology itself is in

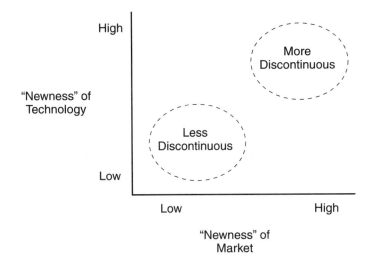

Figure 6.6 A diagram illustrating the dimensions of discontinuous innovations

doubt or at least yet unproven. This may mean that a technology is being transferred to an entirely new set of applications (and production environment) where it has yet to be proven, or it might be a technology that emerged from research and development (R&D) as a promising opportunity but has yet to have all the kinks worked out.

The horizontal axis describes market uncertainty, again going from low to high. At the low end, the project is targeted at a market that is existing, well defined, and familiar to the firm. At the high end, the firm is moving into uncharted territory — a market that may not yet perceive the need for the product or understand the value it will add. The critical infrastructure required to make the innovation a reality (e.g., charging stations for the electric car) may also not exist.

Looking more closely within the matrix, projects positioned in the upper right face high degrees of uncertainty in both dimensions. However, if successful, such projects have the potential to be home runs (to create whole new markets even industries) while at the same time positioning the developers to be technological leaders in the future. Projects in the upper left quadrant of this scheme (low market uncertainty, high technological uncertainty) are going after existing (perhaps their own) markets with a new way of delivering the product. A completely new, unproven technology that offers a major (30 to 40%?) decrease in product costs or enables significant improvement in product features to an existing market would fit here (e.g., Gillette's development of the Sensor line of razors). Projects in the lower right quadrant would be discontinuous in terms of the market while being more incremental in terms of the technology in the product. Projects here might be providing a new combination of existing technologies to a nonexistent market (e.g., Sony's Walkman).

Recent research by Green *et al.* (1995) divides characteristics of innovation into categories, indicating that it can be thought of as being comprised of at least four distinct, independent items:

- *Technological uncertainty* — Will the technology itself work as planned?
- *Technological inexperience* — Does the *firm* know enough about the innovation?
- *Business inexperience* — Is the firm familiar with the market and industry?
- *Technology cost* — What is the overall cost of the project?

6.5.2 EFFECTS ON INDUSTRY STRUCTURE

Tushman and Anderson (1997) and Ettlie *et al.* (1984) cite the difficulties of large firms in creating an environment conducive to innovation. To summarize, a central problem in the management of more discontinuous innovation is the phenomenon that large, complex, decentralized, formal organizations tend to create a form of inertia that inhibits them from pushing toward the more discontinuous end of the spectrum. Their size and past successes push them toward patterns that strove to further enhance existing, dominant industry designs and practices. As a result, there is often a tendency for smaller firms to be innovation leaders. The larger firm's existing positions of leadership act to limit their desire to change current market paradigms, and their large, formal, structures make it difficult to do so even if they chose to. Companies develop core competencies (Prahalad and Hamel, 1990) and work to extend them incrementally.

Tushman and Anderson (1997) suggest that a form of technology cycles exists and "are made up of alternating periods of technological variation, competitive selection, and retention (or convergence)." These cycles are initiated by discontinuous innovations, which either enhance existing competencies (i.e., build off current expertise) or destroy existing competencies (i.e., render existing technological capabilities noncompetitive). Thus, when a discontinuous innovation occurs, there is first a period of rapid change and multiple, competing designs vying for industry leadership. Finally, there emerges a winner — the dominant design — after which there is a prolonged era of incremental changes and elaborations on this industry accepted model. This period will continue until the next discontinuous innovation occurs. This cycle has been shown to exist across industries — from chemicals to computers.

6.5.3 MANAGING DISCONTINUOUS INNOVATIONS AT THE FIRM LEVEL
Challenges to the New Product Development Process

As was noted earlier, many of the approaches to the management of incremental innovation may not be appropriate as a given innovation becomes more discontinuous in nature. For example, the combination of speed to market pressures and the need for product innovation in the 1980s forced many firms to experiment with new ways of bringing new products to market. It was recognized early on that Japanese auto makers were consistently able to design and build a new automobile in less than 30 months. Until very recently, the Big Three automotive manufacturers required from 48 to 60 months to accomplish the same set of tasks. One common strategy that emerged during this time was to view the development of innovative new products as team based, rather than a sequential process, stressing the importance of getting the functional areas together early and frequently in bringing the product to market.

However, recent research on more discontinuous innovation (McDermott and Handfield, 1997) indicates that this parallel approach may be less than effective in discontinuous projects. Findings of this study show that much of the richness of this interaction is lost when the technological leap associated with the project is large. Often, team members find that the uncertainty levels are so high that there is little benefit to the frequent interactions encouraged by the more concurrent approach. Quipped one senior designer, "All we could prove to a typical manufacturing engineer is that we don't know enough to be wasting his time." This suggests that a return to some of the elements of the more sequential approach to innovative product development might be more appropriate in the face of such high uncertainty.

Evaluating Progress

Another interesting challenge associated with the management of discontinuous innovation is how a manager gauges success — both at the personal and project level. At the personal level, it quickly becomes clear that traditional evaluation methods based on meeting a set of articulated and predetermined goals fall apart in environments of discontinuous innovation. Unforeseen obstacles are the norm rather than the exception in this realm. If discontinuous project managers are measured by traditional, short-term methods such as payback, meeting technical milestones, etc., one will quickly find that it is difficult to find a person who will be willing to fill this leadership position. Further, technical team members need reassurance that their involvement in such projects, which typically have "failure" rates approaching 90%, will not limit their career. If a firm wants their best and brightest working on their leading-edge discontinuous projects, it becomes absolutely necessary to measure them on substantial progress toward technical goals rather than on performance against some external yardstick. The focus becomes "What have we learned?" and "Can we now do things better?" as opposed to "Did we meet our goals?"

At the project level, it is important to remember that these discontinuous projects are typically longer in duration and more costly than most of the other projects in a firm's portfolio. If a firm's traditional new product development metrics are used on discontinuous projects, they would most certainly be flushed out of the system. They are simply too high a risk for conventional management metrics. This is why many discontinuous projects in large firms are nurtured in corporate R&D, as opposed to the division's R&D. The short-term pressures on divisions (such as ROI performance) are frequently at odds with the very nature of discontinuous innovation.

Creating an Environment that Supports Change

The discussion above regarding short-term pressures is a key reason why upper-management support is critical for these projects. Clearly, without the protection and support of individuals in power, these more discontinuous projects could not continue to push on year after year. Dougherty and Hardy (1996) and Nadler and Tushman (1990) stress the importance of both upper-management and *institutionalized* organizational support for discontinuous change. Leonard-Barton (1990) argues that core competencies and capabilities too often become core "rigidities" in firms—limiting future moves and creating blinders that deny access to and interest in new ways of doing things. Dougherty and Hardy (1996) concur, arguing that it is the very nature of the organization systems in large firms that serves to limit their ability to innovate effectively. Their study of innovation in 15 large firms found that those searching for improvements need a new way of thinking, including

- More strategic discussions about innovation between senior and middle management.
- An overall increase in the relevance and meaning of innovation throughout the organization, through such means as reconciling conflicts between cost cutting and innovation performance metrics within business units.

Thus, in large organizations, it is often not enough to have one senior manager or champion behind a project. This is a necessary but not sufficient requirement. The whole context of the firm and the way in which it views discontinuous activities is a critical part of building an environment that supports such projects. If the importance of such innovation is communicated and understood at all levels of the firm and throughout the management system, there is a greater chance of the survival. In short, senior management is very helpful, but a corporate environment that understands the importance of discontinuous innovation and nurtures promising projects is better.

One way in which a firm can nurture discontinuous projects is by providing **slack** resources. The creation of slack resources for discontinuous projects shows recognition that they *will* run into unforeseen difficulties and roadblocks along the way. There are simply more hurdles to go over and unknown entities in these projects than in other, more traditional ones. To fail to give them slack resources (in terms of extra time, people, or financial support) would essentially kill them off. Nohria and Gulati (1996) found that there is an inverse U-shaped relationship between slack and innovation. Firms that focus too much on lean operation may be hurting their long-term opportunities for innovation. Too much slack, on the other hand, was often seen to lead to a lack of focus or direction in innovating firms. The challenge for managers in environments of discontinuous innovation is then to provide enough slack so as not to stifle good projects, while at the same time provide enough direction and guidance to keep projects from floundering.

Core Competence and Discontinuous Innovation

The idea of building from core competencies and capabilities is also relevant to the management of discontinuous innovation. Prahalad and Hamel (1990) define core competencies as the "collective learning in the organization, especially how to coordinate diverse production skills and integrate multiple streams of technologies within an organization." It makes a disproportionate contribution to the perceived customer value or to the efficiency with which that value is delivered. A core competence is an organization's hidden capability of coordination and learning that competitors cannot easily imitate, often providing a basis for access to new markets and dominance in existing ones. Prahalad and Hamel (1990) assert that it is necessary to seek competitive advantage from a capability that lies behind the products that serve the market. Clearly, working on projects that have some relevance to the firm's existing strengths increases the probability of success.

For example, Morone (1993) documented the effectiveness of Corning in their development of industry-leading, *discontinuous* process innovation. Beginning in the late 1970s with their development of an industry-leading "inside chemical vapor deposition process," Corning produced generation after generation of new-to-the-world, leading-edge manufacturing processes — each of which yielded both *substantially* improved fiber optics (with better attenuation, bandwidth, etc.) *and* lower costs. Often, these successive innovations were making obsolete the very processes that Corning had developed (and patented) just a few years earlier. Just as the competition began to benchmark their industry-leading practice, Corning would change the rules. As their capacity doubled and redoubled many times during the 1980s, fiber optics grew to be a central part of Corning's business. Due in no small part to these industry-leading advances in their manufacturing processes, Corning continues to be the dominant figure in fiber optics in the 1990s, with annual sales at an estimated $600 million.

Marketing and Discontinuous Innovation

One difficulty in working on discontinuous projects is in the development of and learning about new markets for products. Attempts to be customer driven are often missing the big picture — the customer may not yet realize the need for the product. Imagine trying to map the future of today's PC market 20 years ago. A recent study of this phenomenon (Lynn *et al.*, 1996) reached the conclusion that conventional marketing techniques proved to be of limited value at best, and were often wrong in these environments. What proved to be more useful, they found, was what they termed the **"probe and learn"** process. In this process, the companies "'developed their products by probing potential markets with early versions of the products, learning from the probes, and probing again. In effect, they ran a series of market experiments — introducing prototypes into a variety of market segments." This process, as described, understands the nature of emerging markets and uncertain technology. Probes are simply "feelers" to get a better sense of the market. From the outset, it is viewed as an iterative process, with the goal of the first several attempts to learn as much as possible about the market so that later attempts might be more on target.

6.5.4 CONCLUDING COMMENTS

Discontinuous innovation provides both unique challenges and opportunities to the firm. When working on a project that is a significant leap from existing technologies or markets, it is important to bear in mind that the techniques required to effectively manage the project are significantly different too. This section provides a brief overview of some of the critical issues relating to this topic at both the micro and macro level.

Defining Terms

Discontinuous innovation: Product development characterized by technical and/or market change ("newness") on a grand scale. Often creates new markets or destroys existing ones. Typically long term, with high risk and high potential return.

Slack: Excess resources (time and money) within a business unit. Often deliberately created to enhance discontinuous innovation.

Probe and learn: Running market "experiments" with new products, with the goal of gaining a greater understanding of end user needs and wants. Typically not viewed as product launch but rather just a means of further refining the product.

Technicians at Xerox analyze returned copiers as part of a remanufacturing effort that saves on raw materials and reduces waste. PHOTOGRAPH COURTESY OF XEROX CORPORATION.

CHAPTER 7

Business and the Corporation

7.1 Green Companies

A static view of business is unrealistic. Companies operate in a world of dynamic competition. They are constantly finding innovative solutions to pressure of all sorts — from competitors, customers, and regulators. Properly designed environmental standards can trigger innovations that lower the total cost of a product or improve its value. Such innovations allow companies to use a range of inputs more productively, thus offsetting the costs of improving environmental impact. Ultimately, this enhanced resource productivity makes companies more competitive, not less.

All businesses are called to respond to environmental issues. Many believe their response is: "Does it pay to be green?" The responsible answer is not simply yes or no, but rather what is the right strategy given the customer's values and the business circumstances? The use of recycling and remanufacturing methods should reduce costs and thus increase profits. Environmental problems are correctly viewed as business issues. Companies seek to create products or processes that increase value and reduce costs while using all applicable means to respond to social and environmental needs.

Willingness to pay for environmentally differentiated products arises in industrial markets if, and only if, the products lower the overall costs to the customer. This is a straightforward requirement and one that applies very broadly to differentiation in industrial markets. However, environmental differentiation is possible in industrial markets even if the costs the customer is avoiding are uncertain. Also, environmental differentiation is possible even if all of the benefits to the customer are difficult to quantify.

Not all companies will be able to capture new profits through environmental product differentiation, but many will do so by joining with other companies within an industry to set private standards or by convincing government to create regulations that favor their product.

For many businesses, environmental management means risk management. They need to avoid accidents or an environmental lawsuit. Petroleum and forest product companies are excellent examples of companies that need to avoid accidents and lower environmental risks.

An example of an innovative initiative is the Xerox program of taking back its product at the end of its life. Machines are disassembled, remanufactured to incorporate new technology, and resold as new machines. This practice enables Xerox to reduce its overall costs and also to differentiate them from their competitors who lack similar capabilities.

Companies that increase their resource efficiencies will decrease their costs. Thus, innovations can improve quality while lowering costs. Like defects, pollution often reveals flaws in a product design or production process. Efforts to eliminate pollution can therefore follow the same basic principles widely used in quality programs: Use resource inputs more efficiently and eliminate the need for hazardous, hard-to-handle materials. Process changes that reduce emissions and use resources more productively often result in higher yields. Recognizing environmental improvements is often an economic and competitive opportunity for product differentiation. Companies can encourage their engineers to seek innovation-based, productivity-enhancing solutions. They can trace their own and their customers' discharges, scrap, emissions, and disposal activities back into company activities to gain insight about beneficial product design, packaging, raw material, or process changes. A summary of environmental business opportunities is provided in Table 7.1.

Table 7.1 Environmental Opportunities for Companies

- Reduce costs through recycling and remanufacturing.
- Increase value through better methods that impact the environment less.
- Set private environmental standards with cooperating partners.
- Convince government to create regulations that favor your product.
- Reduce risk of accidents and environmental impacts.

REMANUFACTURING AT XEROX

Xerox has been remanufacturing products for more than 20 years. Xerox remanufactures reprographic machines and has remanufactured cartridges for copiers while maintaining its quality standards. Allowing for the costs associated with recovering cartridges, the unit manufacturing cost over the lifetime of the cartridge and its components is lower than single-use cartridges. For this reason, the company is providing customers with a cash incentive to return their cartridges.

Refurbishing demonstration equipment for resale is more profitable than remanufacturing it, but remanufacturing in turn yields greater value than equipment conversion or parts reuse, with material recycling resulting in the lowest return. The remanufacturing strategy also benefits the environment because it saves on the use of virgin raw materials, and means that fewer parts are manufactured and less waste is generated.

IBM AND JOHNSON & JOHNSON

IBM and Johnson & Johnson (J&J) have agreed to develop a program to make verifiable reductions in greenhouse gas emissions. The World Wildlife Fund provides the firms with options for reducing energy usage and cutting their emissions of carbon dioxide, methane, and other gases.

For J&J the program will include cutting emissions at 150 facilities in 50 countries. For IBM the program will cover 30 plants in 14 nations. They plan to accomplish this through the use of more energy-efficient equipment, eco-generation, and renewable energy sources.

Both companies expect that these strategies will not hinder their business growth.

Source: *Wall Street Journal,* Mar. 1, 2000, p. A4.

7.2 Building a Business Model

The primary sources of a successful business are its core competencies and its business model, which builds on its core competencies. **Core competencies** result from the collective learning in a firm, especially the ability to coordinate diverse skills and streams of technologies within the organization. These competencies are difficult for a competitor to imitate. For example, Intel possesses a unique set of skills and technologies that is used to design and manufacture integrated circuits. The core competencies and a company's know-how, routines, processes, and culture add up to its organizational capital.

The five dimensions of a business model are provided in Table 7.2 (Slywotzky and Morrison, 1997). These five dimensions are linked as an integrated design. The best business model is one that is designed for customer relevance and high profitability. We start with a customer-centered model and attempt to choose the right customer and fully understand him or her. Businesses learn how their customers use their products and help them with their business needs. As an example let us consider the business model of Dell Computer. Their business model is summarized in Table 7.3. Dell is highly profitable because it sells direct to its business customers, who are willing to pay for a customized product.

The business model of a successful environmentally sustainable firm, Patagonia, is given in Table 7.4. Patagonia, a private corporation, is successful in making and marketing high-quality sportswear. Most of the fabric is made of recycled (PET) bottles and organic cotton. Their prices often exceed their competitors' prices by 40%. Patagonia's market research prior to the switch to organic cotton convinced company executives that the most significant reason for purchasing from Patagonia was quality while environmental concerns, either in terms of the company's performance or characteristics of the product, were less important to customers. In response to this research, Patagonia's marketing strategy linked the environment and product quality.

Any business model that will lead to successful profitability must be durable and meet the test of competitive copying.

Table 7.2 **The Five Dimensions of a Business Model**

Dimension	Key questions
1. Customer selection	Which customers do I want to serve? To which customers can I add value? Which customers will allow us to profit?
2. Value capture	How do we make a profit? How do we capture, as profit, a portion of the customer value?
3. Differentiation and strategic control	How do we protect our profit? What differentiates my value proposition? How do we retain strategic control to counterbalance the customer and competitors?
4. Scope	What activities do we perform? What is the range of our products? What activities do we perform in-house versus subcontracted activities?
5. Channels	What market channels do I choose? Are these profitable?

Table 7.3 **The Dell Computer Business Model**

Dimension	
1. Customer selection	• Business, companies
2. Value capture	• Product, service, solution
3. Differentiation and strategic control	• Brand
	• Direct marketing
	• Fast shipping
	• Make to order
4. Scope	• Products for business
	• Assemble on order
5. Channels	• Catalog and phone — direct
	• Internet — direct

Table 7.4 **The Business Model of Patagonia**

Dimension	
1. Customer selection	Higher income individuals who will pay more for desirable sportswear and outdoor clothing
2. Value capture	Attract higher income customers who will pay for environmentally friendly products with good design
3. Differentiation and strategic control	Fabrics made from recycled polyethylene (PET) bottles and organic cotton
4. Scope	High-quality recreational clothing
5. Channels	• Catalogs
	• Own retail stores
	• Outdoor wear retailers

7.3 An Entrepreneurial Future

Wealth is created when entrepreneurs see opportunities to work and invest in situations where large economic and social disequilibriums exist. New businesses grow rapidly when they take advantage of technological or social disequilibriums. Currently most businesses have the opportunity to respond in an entrepreneurial manner to environmental business opportunities that they may not have considered in the past. There are no institutional substitutes for individual entrepreneurial change agents. The entrepreneur winners of the marketplace become wealthy and powerful, but without entrepreneurs, economies become poor and weak.

Entrepreneurs recognize that economic and environmental progress can coexist and may be strongly linked. Natural and environmental resources undergird the foundations of our wealth. We use them to survive. We build our civilizations on them. Our civilization is also built on the natural environment. Man-made technologies for reducing usage and increasing economically reachable supplies have vastly expanded the natural resources that can be used effectively.

The dynamic pattern of new innovators creating new firms that displace established firms is called **creative destruction**, which recognizes that disequilibrium is the driving force of change. Coal-age technologies give way to oil-age technologies, which are displaced by information-age technologies. One current disequilibrium is the rise of sustainable technology and business. During disequilibrium, the established companies engage in incremental change while the new firms seek radical changes. Table 7.5 shows the strategies employed by existing firms and new firms (Hart, 1999). An example of a potentially radical change is the fuel cell, which may displace the internal combustion engine.

Table 7.5 **Strategies for Change During a Disequilibrium**

Existing firms	*New entrepreneurs*
Continuous incremental improvement	Radical innovation
	Creative destruction
Focus on existing:	Focus on emerging:
• Products	• Technologies
• Processes	• Markets
• Customers	• Customers

7.4 Measures of Corporate Sustainability Efforts

One of the greatest obstacles to the expansion of sustainable business is the lack of clarity about exactly what it is, the difficulty of designing measures for it, and the resulting difficulty in rewarding behavior that contributes to it. Measurements are important for three reasons. The first is to monitor environmental and social progress: Are the environmental impacts declining? How are we viewed by our communities and our employees?

Second, developing new measurements that are consistent with traditional business and investment ones is essential in order to engage senior management. Since business is usually measured by shareholder return or return on assets, the new measurements need to communicate how environmental and social investments enhance those goals.

Finally, measurement is critical for reporting. Communicating clear and consistent numbers of performance to internal and external stakeholders is essential for building trust and connections to communities.

The four categories for measuring environmental performance are (1) materials use, (2) energy consumption, (3) waste output, and (4) pollutant releases. A corporation provides value to its customers and needs new measures of customer value related to materials, energy, waste, and pollutant output. Several examples of such measures are provided in Table 7.6. The key to resource productivity lies in making creative use of knowledge to drive resource use down and the value to a customer up. Knowledge can increase the efficiency of an operation, and also the value of a product.

Table 7.6 **Measures of Customer Value for Resource Use**

- Value per material used
- Value per energy used
- Material per customer served
- Recovered material and energy per customer
- Service activity per resources used

As we saw in Section 7.2, competitive success depends on differentiation of a product or service as evidenced by added value that the competitors cannot match. Competitive strategy is about being different. It means deliberately choosing a different set of activities to deliver a unique mix of value. An example of a simple value added to an otherwise standard commodity, the personal computer, is offered by Intel. Intel's value-added feature allows a PC to enter a standby mode using less than 5 watts (compared to the normal 180 watts) that reduces energy use and yet returns to normal operation very quickly.

7.5 Managing the Value Chain

The value chain begins with the corporation's assets, core competencies, and organizational capital as shown in Figure 7.1. The company then builds a business model so that it can access inputs such as natural resources to build a product or service offering, which is ultimately provided through the chosen channels to the customer.

A value chain is most efficient when there are fewer transactions between producers and their final customers. Companies that are serious about their environmental legacy will strive to manage their value chain from natural resource to final customer. This will facilitate the exchange of materials

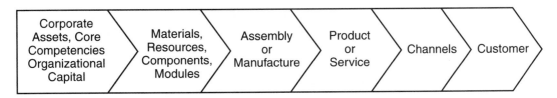

Figure 7.1 **The value chain.**

from waste streams to raw material. It will reduce transaction costs among suppliers, and it will add customer value by making customer needs known throughout the entire chain.

Managing the value chain may also be a way to accelerate revenue growth. Many corporations have moved downstream and captured the customer interface, acting as the main intermediary and delivery channel for products and services to the customer. An example is Dell Computer, which sells directly to its customers.

A strategy that places more importance on the delivery of value and away from the production of products requires a new set of skills in forming alliances and partnerships. Interconnecting supplier relationships become the key competitive advantage. Instead of a linear chain, the relationships among suppliers shift toward a web, with ample connections for new entrants and innovation.

The ecological sustainability of the entire system is enhanced by a tighter value chain. Information about customer needs is immediately available to producers, who can design greater efficiency into the system. The company that owns the customer interface can offer recovery of product, maintenance, and redistribution. The company that controls its supplier relationships can mandate the use of fewer, more responsible inputs. In this tightened supply chain, material flow can be reduced, while profits and returns to capital improve.

The goal is to get all participants on the value chain aligned together to find environmentally sound substitute materials, production techniques, and storage and inventory management systems. The corporation can seek environmentally sound components and materials. If these need to be developed, the corporation can share the risk as well as the development effort with their suppliers.

7.6 Four Visions of the Century Ahead: Will It Be Star Trek, Ecotopia, Big Government, or Mad Max?[1]

By Robert Costanza
Director, Institute for Ecological Economics
University of Maryland

Probably the most challenging task facing humanity today is the creation of a shared vision of a sustainable and desirable society, one that can provide permanent prosperity within the biophysical constraints of the real world in a way that is fair and equitable to all of humanity, to other species, and to future generations. This vision does not now exist, although the seeds are there. We all have our own private visions of the world we really want, and we need to

---

[1]This section has been reprinted with permission from *The Futurist,* Vol. 33, No. 2, Feb. 1999, pp. 23–29.

overcome our fears and skepticism and begin to share these visions and build on them, until we have built a vision of the world we want.

The most effective ingredient to move change in any particular direction is having a clear vision of the desired goal that is also truly shared by the members affected by it, whether an organization, a community, or a nation.

Social observer Daniel Yankelovich has described the need for governance to move from public opinion to public judgment. Public opinion is notoriously fickle and inconsistent on those issues for which people have not confronted the broader implications of their opinions. For example, many people are highly in favor of more effort to protect the environment, but at the same time they are opposed to any diversion of tax revenues to do so. Coming to public judgment is the process of resolving these conflicts.

To start the dialogue and move quickly to public judgment, we may consider issues in the form of "visions" or scenarios. This section lays out four such visions, each presented as a "future history" written from the vantage point of the year 2100. These visions include both positive and negative scenarios — hopes and fears — allowing us to fully explore what the future may hold and thus to make informed choices among complex alternatives with a range of implications.

While there are an infinite number of possible future visions, I believe these four visions embody the basic patterns within which much of this variation occurs. Each of the visions is based on some critical assumptions about the way the world works, which may or may not turn out to be true. This format allows one to clearly identify these assumptions, assess how critical they are to the relevant vision, and recognize the consequences of them being wrong.

7.6.1 FOUR VISIONS OF THE FUTURE

The four visions derive from two basic world views that reflect one's faith in technological progress. The "technological optimist" world view is one of continued expansion of humans and their dominion over nature. This is the "default" vision in current Western society and represents the continuation of current trends into the indefinite future.

There are two versions of this vision, however: one in which the underlying assumptions are actually true in the real world and one in which those assumptions are false. The positive version of the "technological optimist" vision I'll call "Star Trek," named for the popular TV series that is its most articulate and vividly fleshed-out manifestation. The negative version of the "technological optimist" vision I'll call "Mad Max" after the popular postapocalyptic Australian movie of 1979 that embodies many aspects of this vision gone bad.

The "technological skeptic" vision focuses much less on technological change and more on social and community development. The version of this vision that corresponds to the skeptics being right about the nature of the world I'll call "Ecotopia" after a book of the late 1970s. If the technological skeptics turn out to be wrong, and the optimists right, about the real state of the world, we see the version I'll call "Big Government" come to pass — a scenario of protective government policies overriding the free market.

Each of these future visions is described below from the perspective of the year 2100. The visions are described as narratives with specific names and events, rather than as vague general conditions, in order to make them more real and vivid. They are, of course, only caricatures, but I hope they capture the essence of the visions they represent.

Star Trek: The Default Technological Optimist Vision

The turning point came in 2012, when population pressure was mounting and natural resources were being strained. The greenhouse effect caused by burning fossil fuel was beginning to cause some major disruptions. But the development of practical fusion energy allowed a rapid reduction of global fossil-fuel burning to practically zero by the year 2050, eventually reversing the greenhouse effect. Fusion energy was infinitely better and cheaper than any alternative, and it was inexhaustible.

Air pollution was essentially eliminated between 2015 and 2050, as cars were converted to clean-burning hydrogen produced with energy from fusion reactors. Electricity for homes, factories, and other uses came increasingly from fusion, so the old, risky nuclear fission reactors were gradually decommissioned; even some hydropower stations were eliminated to return some great rivers to their wild state. In particular, the dams along the Columbia River in Oregon were completely eliminated by 2050, allowing the wild salmon runs and spawning grounds to be reestablished.

While clean, unlimited energy significantly lowered the impact of humans on the environment, the world was still getting pretty crowded. The solution, of course, was space colonies, built with materials taken from the moon and asteroids and energy from the new fusion reactors. The initial space colonies were on Earth's moon, the moons of Jupiter, and in free space in the inner solar system. From there it was a relatively short step to launch some of the smaller space colonies off toward the closer stars.

By 2050, about one-tenth of the total population of 20 billion was living in space colonies. Currently (A.D. 2100), the total human population of 40 billion is split almost equally between Earth and extraterrestrial populations. The population of Earth is not expected to rise above about 20 billion, with almost all future growth coming in space-based populations.

Since food production and manufacturing are mainly automated and powered by cheap fusion energy, only about one-tenth of the population actually needs to work for a living. Most are free to pursue whatever interests them. Often the biggest technological and social breakthroughs have come from this huge population of "leisure thinkers." People also have plenty of time to spend with family and friends, and the four-child family is the norm.

Mad Max: The Technological Skeptic's Nightmare

The turning point came in 2012, when the world's oil production finally peaked and the long slide down started. The easy-to-get oil was simply exhausted, and the price started to rise rapidly. All the predictions about the rapidly rising price of oil causing new, cheaper alternatives to emerge just never came to pass. There were no cheaper alternatives — only more expensive ones. Oil was so important in the economy that the price of everything else was tied to it, and the alternatives just kept getting more expensive at the same rate. Solar energy continues to be the planet's major power source — through agriculture, fisheries, and forestry — but direct conversion using photovoltaics never achieved the price/performance ratios to allow it to compete, even with coal.

Of course, it didn't really matter anyway, because the greenhouse effect was kicking in, and the earth's climate and ecological systems were in a complete shambles. Rising sea levels inundated most of the Netherlands, as well as big chunks of Bangladesh, Florida, Louisiana, and other low-lying coastal areas, by about 2050.

Once the financial markets figured out what was happening, the bubble really burst. During the stock market crash of 2016, the Dow Jones average dropped 87% in a little over three days in December. Although there was a brief partial recovery, it has been downhill ever since.

Both the physical infrastructure and the social infrastructure have been gradually deteriorating, along with the natural environment. The human population has been on a long downward spiral since the global airbola (airborne Ebola) virus epidemic killed almost a quarter of the human population in 2025–2026. The population was already weakened by regional famines and wars over water and other natural resources, but the epidemic came as quite a shock. The world population peaked in 2020 at almost 10 billion. More than 2 billion died in the epidemic in the course of a little over a year and a half. Since then, death rates have exceeded birthrates almost everywhere, and the current population of 4 billion is still decreasing by about 2% per year.

National governments have weakened, becoming mere symbolic relics. The world has been run for some time by transnational corporations intent on cutthroat competition for the dwindling resources. The distribution of wealth has become more and more skewed. The dwindling few with marketable skills work for global corporations at good wages and lead comfortable and protected lives in highly fortified enclaves. These people devote their lives completely to their work, often working 90- or 100-hour weeks and taking no vacation at all.

The rest of the population survives in abandoned buildings or makeshift shelters built from scraps. There is no school, little food, and a constant struggle just to survive. The majority of the world's population lives in conditions that would make the favellas of twentieth-century Rio seem luxurious. The almost constant social upheavals and revolutions are put down with brutal efficiency by the corporate security forces (governments are too broke to maintain armies anymore).

Big Government: Public Interest Trumps Private Enterprise

The turning point came in 2012, when the corporate charter of General Motors was revoked by the U.S. federal government for failing to pursue the public interest. Even though GM had perfected the electric car, it had failed to make its breakthrough battery technology available to other car makers, even on a licensing basis. It preferred, instead, to retain a monopoly on electric cars, to produce them exclusively in China with cheap labor, and to gouge the public with high prices for them. After a series of negotiations broke down, government lawyers decided to invoke their almost forgotten power to revoke a corporation's charter and made the technology public property. This caused such a panic throughout corporate America that a complete rethinking of the corporate/public relationship took place, which left the government and the public with much more control over corporate behavior.

Strict government regulations had kept the development of fusion energy slow while safety issues were being fully explored. No one wanted a repeat of fission energy's problems: The Three Mile Island and Chernobyl accidents were nothing compared to the meltdown of one of France's fission breeder reactors in 2005, which left almost one-quarter of the French countryside uninhabitable, killing over 100,000 people directly and causing untold premature cancer deaths throughout Europe.

Fusion energy therefore got a very long and careful look. Government regulators also required the new fusion power plants to bear the full financial liability, causing a much more careful (albeit slightly slower) development of the industry.

High taxes on fossil energy counteracted the greenhouse effect and stimulated renewable energy technologies. Global carbon-dioxide emissions were brought down to 1990 levels by 2005 and kept there through 2030 with concerted government effort and high taxes. Later, the new fusion reactors — along with new, cheaper photovoltaics — gradually eliminated the need for fossil fuels, and the worst of the predicted climate-change effects was thus averted.

Government population policies that emphasized female education, universal access to contraception, and family planning managed to stabilize the global human population at around 8 billion, where it remained (give or take a few hundred million) for almost the entire twenty-first century.

A stable population allowed many recalcitrant distributional issues to finally be resolved, and income distribution has become much more equitable worldwide. While in 1992, the richest fifth of the world's population received about 83% of the world's income and the poorest fifth received only a little more than 1%, by 2092, the richest fifth received 30%, and the poorest, 10%. The income distribution "champagne glass" had become a much more stable and equitable "tumbler." Some libertarians have decried this situation, arguing that it does not provide enough incentive for risk-taking entrepreneurs to stimulate growth. But governments have explicitly advocated slow or no-growth policies, preferring to concentrate instead on assuring ecological sustainability and more-equitable distribution of wealth.

Stable human population also took much of the pressure off other species. The total number of species on earth declined during the twentieth century from about 3 million to a low of about 2.2 million in 2010. But that number has stabilized and even recovered somewhat in the twenty-first century, as some species previously thought to be extinct were rediscovered and some natural speciation of fast-growing organisms has occurred. The current estimate of the number of species on earth is about 2.5 million, and there are strict regulations in effect worldwide not only to prevent any further loss, but also to encourage natural speciation.

Ecotopia: The Low-Consumption Sustainable Vision

The turning point came in 2012, when ecological tax reform was enacted almost simultaneously in the United States, the European Union, Japan, and Australia after long global discussions and debates, mostly over the Internet. In the same year, Herman Daly won the Nobel Prize for Human Stewardship (formerly the prize for economics) for his work on sustainable development.

A broadly participatory global dialogue had allowed an alternative vision of a sustainable world to emerge and gain very wide popular support. People finally realized that governments had to take the initiative back from transnational corporations and redefine the basic rules of the game if their carefully constructed vision was ever going to come to pass.

The public had formed a powerful judgment against the consumer lifestyle and for a sustainable lifestyle. The slogan for the new revolution became the now famous "sustainability, equity, efficiency."

All depletion of natural capital was taxed at the best estimate of the full social cost of that depletion, and taxes on labor and income were reduced for middle-income and lower-income people. A "negative income tax," or basic life support, was provided for those below the poverty level. Countries without ecotaxes were punished with ecological tariffs on goods they produced.

The QLI (Quality of Life Index) came to replace the GNP as the primary measure of national performance. The reforms were introduced gradually over the period from roughly 2012 to 2022 in the United States, European Union, Japan, and Australia, giving businesses ample time to adjust. The rest of the world followed soon thereafter, with almost all countries completing the reforms by 2050. They had very far-reaching effects.

Fossil fuels became much more expensive, both limiting travel and transport of goods and encouraging the use of renewable alternative energies. Mass transit, bicycles, and sharing the occasional need for a car became the norm. Human habitation came to be structured around small villages of roughly 200 people, whether these were in the village which provided most of the necessities of life, including schools, clinics, and shopping, all within easy walking distance. It also allowed for a real sense of "community" missing from late-twentieth-century urban life. Such changes drastically reduced the GNP of most countries, but drastically increased the QLI.

Because of the reduction in consumption and waste, there was only moderate need for paid labor and money income. By 2050, the workweek had shortened in most countries to 20 hours or fewer, and most full-time jobs became shared by two or three workers. People could devote much more of their time to leisure, but rather than consumption-oriented vacations taken far from home, they began to pursue more community activities (like participatory music and sports) and public service (like caring for children and the elderly).

Unemployment became an almost obsolete term, as did the distinction between work and leisure. People were able to do things they really liked much more of the time, and their quality of life soared (even as their money income plummeted). The distribution of income became an almost unnecessary statistic, since income was not equated with welfare or power and the quality of almost everyone's life was relatively high.

While physical travel decreased, people began to communicate electronically over a much wider web. The truly global community could be maintained without the use of resource-consuming physical travel.

7.6.2 JUDGING THE FOUR VISIONS

How should society decide among these four visions? A two-step process starts with forming and expressing values with the goal of finding a rational policy for managing human activities. Social discourse and consensus is built around the broad goals and visions of the future and the nature of the world in which we live. When a consensus is formed, institutions and analytical methods are marshaled to help achieve the vision.

Three of the four visions are sustainable in the sense that they represent continuation of the current society (only "Mad Max" is not), but we need to take a closer look at their underlying world views, critical assumptions, and the potential costs of those assumptions being wrong. I have already set up the four visions with this in mind.

The world view (and attendant policies) of the "Star Trek" vision is technological optimism and free competition, and its essential underlying assumption is unlimited resources, particularly cheap energy. If that assumption is wrong, the cost of pursuing this world view and its policies is something like the "Mad Max" vision.

Likewise, the world view (and attendant policies) of the "Ecotopia" vision is technological skepticism and communitarianism (the community comes first), and its essential underlying assumptions are that resources are limited and that cooperation pays. If the assumption that resources are limited is wrong, the cost of pursuing this world view and its policies is the "Big Government" vision, where a "community first" policy slows down growth relative to the free market "Star Trek" vision.

The next step toward coming to public judgment is to discuss the four visions with a broad range of participants and then have them evaluate each vision in terms of its overall desirability. Most of those I have already surveyed found the "Ecotopia" vision "very positive"; very few expressed a negative reaction to such a world. "Star Trek" was the next most-positive vision.

7.6.3 QUESTIONING TECHNOLOGY

After discussing and evaluating these scenarios, we can choose between the two world views (technological optimism or skepticism) and their attendant policies, but we face pure and irreducible uncertainty. Who knows whether or not practical fusion or something equivalent will be invented? Should we choose the "Star Trek" vision (and the optimist policies) merely because it is the most popular or because it is the direction things seem to be heading in already?

From the perspective of game theory, this problem has a fairly clear answer: The game can only be played once, and the relative probabilities of each outcome are completely unknown. In addition, we can assume that society as a whole should be risk averse in this situation.

For the optimistic policy set, "Mad Max" would be considered the worst case. For the skeptical policy set, "Big Government" would be the worst case. If "Big Government" is viewed as more positive (or less negative) than "Mad Max," then it would make sense to choose the skeptic's policy set at least until more information is available.

In fact, the way I have set up the game, "Mad Max" is the one really negative outcome and the one really unsustainable outcome. We should develop policies that assure us of not ending up in "Mad Max," no matter what happens.

One could also argue that the probabilities of each state of the world in the scenario matrix are not completely unknown. If the prospects for cheap, unlimited, nonpolluting energy were, in fact, known to be very good, then the choices would have to be weighted with those probabilities.

But the complete dependence of the "Star Trek" vision on discovering a cheap, unlimited, nonpolluting energy source argues for discounting the probability of its occurrence. By adopting the skeptic's policies, the possibility of this invention is preserved, but we don't have to be so utterly dependent on it.

It's like leaping off the World Trade Center and hoping to invent a parachute before you hit the ground. It's better to wait until you have the parachute (and have tested it extensively) before you jump.

7.7 System Design for a Post-Corporate World

By David C. Korten
President
PCD Forum

During the past 20 years we have seen a dramatic shift in power from national governments to global corporations. Now we are waking up to the fact that our largest corporations command internal economies larger than those of the majority of countries, are reshaping our culture through saturation advertising and their control of the mass media, and dominating our political processes through their use of the media to shape national dialogue and political contributions to buy the loyalty of politicians. For all practical purposes, most of the world is now ruled by publicly traded corporations and the global financial markets to which they owe their allegiance. This poses a barrier to sustainability comparable to the barrier once posed to human freedom by the institutions of monarchy.

7.7.1 THE GLOBAL CAPITALIST ECONOMY

Consider the characteristics of the global corporate capitalist economy under which most of the world now lives. Money is its defining value, the maximization of returns to financial capital, its defining goal. Any contribution to the creation of new wealth is as an incidental by-product of a primary concern with increasing shareholder value.

Though living capital — human, social, institutional, or natural — is the ultimate source of all real wealth, corporate capitalism assigns it no value and makes no accounting for its depletion. People and nature are valued only for their contribution to money making.

People are reduced to their purely economic roles as consumers and workers. Competition, individualism, and materialism are nurtured as favored cultural norms. Stock prices and gross domestic product (GDP) are the accepted measures of progress and well-being. Inflation of land and stock values is encouraged, while wages are held constant or depressed — resulting in a massive upward redistribution of wealth.

Capitalism's favored institution is the publicly traded, limited liability corporation — a direct descendant of the British crown corporations, such as the British East India and Hudson's Bay corporations, created to exploit the markets and resources of colonial territories. One of the least democratic of human institutions, it concentrates power in the hands of a chief corporate executive accountable only to absentee owners who themselves are shielded from public accountability for the decisions made on their behalf. With a legal fiduciary responsibility to maximize short-term returns to its shareholders, management is pressed to use the corporation's power to privatize the gains of economic activity to the exclusive benefit of its shareholders, while forcing the larger community to absorb the related costs. The legal structure of the corporation virtually compels it to mimic a cancer — pursuing its own unlimited growth without regard to consequences for either itself or its host.

Under corporate capitalism the ultimate power over both governments and corporations resides with global financial markets in which speculators gamble with hundreds of billions of dollars in borrowed money. Corporate ownership of media and politicians renders democracy meaningless as the institutions of money rewrite laws to free themselves from the restraints on their power imposed by public regulation and national borders. Global institutions like the World Bank, International Monetary Fund (IMF), and World Trade Organization (WTO) are used to override surviving democratic processes at national and local levels to mold all countries into a seamless laissez–faire capitalist economy in which property rights take priority over human rights.

From a systems design perspective the global capitalist economy is well suited to the task of rapidly exploiting natural and human resources for the purpose of inflating the financial assets of a small elite. It is ill suited to the task of creating a just, sustainable, and compassionate society that works for all on a finite living planet.

7.7.2 AN ALTERNATIVE: A PLANETARY SOCIETY

Fortunately, there are alternatives more consistent with the values and preferences of the vast majority of the world's people. For example, instead of a global economy that reduces us all to our purely economic roles of consumers and laborers, we might create a planetary society in which we are encouraged to function as whole persons. It would have an ethical culture in which life is the defining value and democracy and markets would place decision making in the hands of free and aware people.

In short, the planetary society is nearly the mirror opposite of the existing global capitalist economy. Imagine the possibilities. Its defining goal is to ensure the happiness, well-being, and creative expression of every person — each recognized as a whole person encouraged to participate fully in the social, political, cultural, and economic life of the community. Indices of the vitality, diversity, and productive potential of the whole of society's living capital — its human, social, institutional, and natural capital — are the key indicators of well-being and progress.

Human rights and political sovereignty reside in real persons on the basis of one person, one vote. Civic associations facilitate the practice of direct democracy. Public funding of elections and free access for political candidates to media minimize the role of money in elections.

Economic life centers on well-regulated, self-organizing markets that function within a strong ethical culture of cooperation and mutual responsibility. I call them *mindful markets*, because the underlying culture encourages their participants to act with a mindfulness of the needs of both self and community. Firms are human scale and owned by real human stakeholders — their workers, customers, suppliers, and community members. There are many forms of enterprise, including proprietorships, cooperatives, partnerships, and stakeholder-owned corporations — but the once common publicly traded, limited-liability corporation no longer exists.

The right of each person to a means of livelihood is considered to be the most basic of human rights — a right secured in part through each individual owning a share in the assets on which their livelihood depends. Concerns for equity and public accountability are hallmarks of economic life.

Money is society's servant, not its master, and is used solely to facilitate productive investment and beneficial exchange. Its creation is a public function. Financial speculation is strongly discouraged by regulation and tax policy. Local currencies are common, as are independent community banks and credit unions. Thus, finance is predominantly local, as are most enterprises and most production. Countries trade their surplus production based on their comparative natural endowments.

Cultural diversity is highly valued, as is economic diversity and experimentation. Individual local and national economies vary substantially in their mix of individual and cooperative ownership and in the extent of their participation in a planetary trading system — depending on their circumstances and preferences. Experience, culture, information, and technology are freely shared among people, communities, and nations through individual travel and electronic communication, thus facilitating rapid social learning toward constant improvements in real living standards and the quality of life of all.

Each community or nation has the right to determine what and how much it will trade, with whom, and under what circumstances. Similarly it has the right to decide on the terms, if any, under which it will invite others to participate in its economy through investment. These rights, as well as other appropriate standards for an equitable and beneficial planetary trading system, are secured by international agreements implemented under the supervision of the United Nations.

From a system design perspective, the central task in transforming a global capitalist economy into a healthy planetary society is to bring market incentives into a natural alignment with the public interest through the full implementation of political and economic democracy. The more decision-making power that resides equitably in living persons with a natural interest in their community and the living earth that sustains it, the greater the possibility that humanity as a whole will make the choices necessary to live in a loving and mutually productive relationship with the planet and one another. There is no place in such a system design for either financial speculation or for institutional forms such as the limited liability, publicly traded corporation that concentrate power without accountability. Supporting public policy would seek:

- *Electoral and legislative integrity.* Place strict limits on political contributions and spending. Provide public financing of elections, free air time for candidates on media using the public airwaves, and criminal penalties for anyone using corporate resources to influence elections or legislation.
- *Corporate accountability to the public interest.* Eliminate the legal fiction of corporate personhood and the related Bill of Rights protections, remove limited liability protections of corporations, and prohibit one corporation from owning another.
- *Full cost internalization.* Eliminate corporate subsidies and tax breaks, establish high regulatory standards, and charge cost recovery fees in any instance in which a corporation is found to be externalizing costs onto the public.

- *Economic democracy.* Use regulatory and fiscal incentives to achieve broadly based participation in the ownership of enterprises and productive assets, with the goal of eliminating absentee ownership and ensuring ownership participation by each person in the assets on which their livelihood depends.
- *Human-scale enterprises.* Implement fiscal incentives such as a graduated corporate income and assets tax to encourage the breakup of existing corporate concentrations into human-scale enterprises. Strengthen antitrust provisions and pursue their rigorous enforcement. Approve mergers and acquisitions only where a *compelling* public interest is demonstrated.
- *The integrity of money.* Reduce reliance on debt-based money by imposing a 100% reserve requirement on demand deposits, prohibit lending for financial speculation (such as buying stocks on margin and lending to hedge funds), place a confiscatory tax (both personal and corporate) on short-term financial gains from financial assets, and provide federal deposit insurance only for unitary community banks.
- *The integrity of national economies.* Eliminate the foreign financial debts of Third World countries and the mechanisms by which they are created — including the World Bank. Close the WTO and assign responsibility for matters relating to the international regulation of trade, finance, and corporations to a strengthened United Nations. Reorganize and restaff the IMF and place it under the direct authority of the United Nations with responsibility for monitoring and facilitating the balancing of international accounts between nations. Restore national ownership and control of productive assets — with special attention to low-income countries. Eliminate international financial speculation by allowing the exchange of national currencies only for the purchase and exchange of real goods and services.

Several generations ago our ancestors decided that the institutions of monarchy no longer served the human interest and invented the institutions of political democracy to replace them. Contemporary rule by financial speculators and global corporations not only negates the democracy achieved when humanity turned away from monarchies, it is inconsistent with humanity's present need to achieve an equitable and sustainable relationship to nature. The time has come to recognize that it is our individual and collective right and responsibility to transform or replace the unjust and destructive institutions of the global capitalism economy with the life-nurturing, people-affirming institutions of a planetary society.

The Boeing paint hangar in Everett, Washington, is designed to protect the environment and worker health in a cost-effective way. It shows that environmental concerns such as air quality and waste reduction can be met while achieving worker safety and morale goals within the constraints of economic factors such as capital costs and return on investment. PHOTOGRAPH COURTESY OF BOEING.

CHAPTER 8

Financial and Cost Issues in Sustainable Business and Government

8.1 Financing Sustainable Business

It is a reasonable assumption that a society that prioritizes the production of essential goods, produces these with ever-cleaner technology, and operates to increasingly high standards of energy and resource-use efficiency will make a greater contribution to global sustainability than one that fails on any or all of these grounds. A key problem is the development of a strategy for a transition to sustainable business. Significant amounts of capital may be required. A transition to renewable energy sources, for example, would require an expensive restructuring of the generating and supply industries, because renewable energy sources have different characteristics than nonrenewable ones. Thus the transition would be expensive, although the more efficient use of energy flows and resources might well reduce costs in the long run.

Thus, if a change to sustainable business is to be achieved, the world's financial markets will need to support and enable this change (Schmidheiny and Zorraquin, 1998). Financial markets are, at base, tools that reflect society's values and thus should reward firms that rationally manage environmental resources. Otherwise, governmental intervention will be seen as inevitable. Investments in energy, transport, agricultural, water, and sewer systems will be required worldwide. The necessary capital must come from the financial markets: the equity markets, bond markets, and the banks. It may be possible in some instances to finance much of the necessary investment from profits.

Some of the issues surrounding the matter of financing sustainable business are summarized in Table 8.1, in a list that may seem discouraging. However, many companies are viewing becoming eco–efficient as an opportunity to reap strategic advantages. But first the companies must find financing for what will be inherently long-term investments, such as the reconstruction of polluting industries, the development of energy–efficient transport systems, and the cleaning up of historical pollution.

Table 8.1 **Issues of Financing the Transition to Sustainable Business**

- Sustainable business often requires long-term payback.
- Eco-efficiency often provides returns in the future, beyond 5 to 10 years.
- Financing is difficult when resource prices are relatively low.
- Environmental costs are often external to the firm and thus are not included in the calculation of profits.
- Sustainable development requires investments in emerging countries where the risk of loss may be higher and thus the market demands a higher risk premium.
- Accounting practices do not always convey potential environmental or hazard risks.
- Sustainable development investment is often burdened by high discount rates due to risk premiums and higher interest rates and inflation.

This suggests that there would have to be a number of changes in the function and structure of the market itself. These might include the use of financial incentives and penalties to encourage long-term investment moves to link the remuneration of managers to factors other than share price alone, and a redefinition of the legal responsibilities of directors to encompass wider concerns than the maximization of profits. Another approach would be to use the banking system. The concept of internalizing environmental costs is worth considering. This concept holds that the price of a good or service should reflect all the costs associated with it. For example, the cost of electricity from a coal-fired power station rarely reflects the costs of the damage done by the acid rain it causes or the health problems related to its pollution. We consider this concept in the next section.

8.2 Internalizing Environmental Costs

Traditionally, government's main tool for achieving environmental goals has been command-and-control regulations; these often tell a company precisely what technology to use and precisely what can be emitted and in what quantities. There will always be a need for such restrictions in situations where major risks and uncertainties exist. However, a regulation requires a company to reach a certain standard and then do no more.

An alternative approach is the use of economic instruments such as taxes, charges, and tradable permits. A tax or charge on pollution or resource use encourages a company to become ever more eco-efficient by producing a steady effect on that company's profit and loss figures. There is a growing consensus that the use of economic instruments is increasing and that — if the instruments are well constructed and combined well with other approaches — such instruments can help meet four needs: (1) provide incentives for continuous improvements and continuous rewards, (2) use markets more effectively in achieving environmental objectives, (3) find more cost-effective ways for both government and industry to achieve these same objectives, and (4) move from pollution control to pollution prevention.

An example of a tax is a landfill tax. A landfill tax on waste can raise significant sums, but it generally falls on the waste collection company and not the individuals who make the waste. Another waste tax would be a virgin-materials tax that should make new materials more expensive, thus encouraging recycling. But how should this tax be imposed, on volume or value of the material?

Possibly useful methods for governments to use to impose the full cost of using a material or energy or of creating pollution are summarized in Table 8.2.

Table 8.2 **Means for Internalizing the Costs of Materials and Energy**

- Eliminate subsidies for fossil fuel exploration and exploitation.
- Eliminate regulatory barriers to competition in the energy industry.
- Incorporate standards for fuel or materials use with penalties for exceeding the standard.
- Create tradable permits so that companies who become eco-efficient can trade their permits to "create pollution" to other companies that need it. Tradable permits are discussed in detail in Section 8.3.

8.3 Tradable Permits

Tradable permits for the use of fuels or materials or the creation of pollution are utilized in some cases. These allow trades between groups when there is an agreement on a goal of allowable total pollution levels or resource usage such as fishery catches or air pollution. For example, fisherman can buy the right to a share of the allowed catch and then sell that right to others.

Tradable permits assign "rights" to use a resource or emit certain levels of pollution. These rights can be bought or sold on a market. In practice, governments issue the permits or assign a quota for emissions of a specified pollutant or consumption of a resource over a given period. In the case of pollution releases, governments typically cap aggregate emissions from sources within a particular geographic region at a level consistent with environmental goals, and issue the corresponding number of permits.

Once an overall level of pollution or resource use has been set, permits or quotas can be bought and sold among the industries and individuals involved in the targeted activity, allowing the marketplace to determine how the scarce resources or rights to pollute are allocated. This strategy can be extremely cost effective: When tradable air pollution permits were introduced to control sulfur emissions (precursors of acid rain) in the United States, emission reductions were achieved at about half the cost that analysts had predicted before the system was implemented.

The Kyoto Protocol, an international treaty aimed at reducing global warming, seeks restrictions on carbon-dioxide emissions. It also calls for trade in emissions. A company that is more efficient than the standard will be able to sell its excess credits to less efficient companies. Thus, an incentive is provided to all companies to install means to reduce their emissions.

In the case of electric power plants, with a credit program, energy companies with "dirty" power plants can buy the right to exceed pollution standards or install new equipment to cut emissions. Pollution has declined because the total allowances available add up to substantially lower emissions than before the program started, while each company gets to map its own strategy for meeting its limits.

8.4 Innovation and Environmental Sustainability

By Ronald Mascitelli
President
Technology Perspectives

Since the industrial revolution, economic development has been based on two rather convenient assumptions: (1) that the resources of the earth are available in unlimited supply as inputs to production and (2) that the environmental damage caused by firms, industries, and nations in their pursuit of wealth has no economic significance. Clearly, in these enlightened times the falsity of such assumptions can no longer be ignored. In this section I propose that the self-serving assumptions of the past can be supplanted by pro-growth/pro-environment strategies enabled by the power of human creativity and innovation.

The World Commission on Environment and Development has defined **sustainable development** as economic growth that "meets the needs of the present without compromising the ability of future generations to meet their own needs" (Esty, 1994). From a less human-centered perspective, sustainable development implies that the rates at which we consume resources and create pollutants must be controlled in such a way that economic growth can continue indefinitely without causing irreversible environmental damage. (For further discussion, see OECD, 1997b.) Fortunately, such a balance is achievable: The natural environment has a powerful capacity to cleanse itself. Although the ecosystem of Prince William Sound, for example, has been irrevocably altered by the *Exxon Valdez* disaster, life has returned, and its former beauty has largely been restored.

Economically speaking, environmental damage from industrial activity is a negative externality that, in cases such as global warming, can create undesirable spillovers to every corner of the planet. These negative spillovers are perpetuated by the fact that environmental costs are not accounted for in business transactions. This market failure causes producers to have no financial incentive to reduce waste or eliminate emissions. In the absence of either effective government regulation or countervailing market forces, shared resources such as lakes and rivers, the atmosphere, rain forests, and fossil fuels will be subjected to unrestrained use by profit-seeking individuals. If left unchecked, the inevitable result will be environmental tragedy. Fortunately, however, the environmental debate progressed during the 1990s from the anti-industry, antiprofit, and antigrowth rhetoric of radical environmentalism to the realization that business must play a central role in achieving the goal of sustainable development (Elkington, 1994, p. 91).

There are two general approaches to engaging industry in the battle to save our environment: either through government regulation or market-driven technological innovation. Unfortunately, the costs of government-mandated protection measures such as the "polluter pays" principle can be high, and it is hard to find a politician who would willingly impose such a steep bill on domestic industries. Yet without some reasonable safeguards, there is no guarantee that increased wealth will make people better off in the future. Unrestrained development can make society richer, but at what point does the degradation of our surroundings and the risks to our health render growth in per-capita income sadly irrelevant?

In an ideal world, one might imagine that altruism alone would drive companies to innovate new ways to conserve resources and reduce harmful emissions. Given the competitive realities of global commerce, however, it seems unlikely that solutions will be put forward without a suitable economic incentive being provided by government. In fact, I suggest that the most optimistic scenarios for sustainable development involve synergistic innovations in both regulatory policy *and* environmentally friendly technologies. Governments can provide creative economic incentives that will induce industries to innovate, while the actual methods and technologies used to capture these incentives would be left to the discretion of industry. Such a partnership between the public and private sectors would reduce the potential economic hardship that could result from targeted environmental policy, while catalyzing industrial innovation through the allure of profit maximization.

From a policy perspective, the most problematic environmental concerns are those that impact common resources, with global warming being the most prominent among these issues. (For a broad overview of the issue of global warming, see Read, 1994.) Both the atmosphere and the greenhouse gases being emitted into it are common to everyone on the planet, yet it is in no country's best interest to take unilateral action. Indeed, such action would be fruitless: Even if the United States reduced its substantial greenhouse gas emissions to near zero, the rate of atmospheric degradation over the long term would hardly be affected. It is only through multilateral action by both advanced and developing nations that further destabilization of the earth's climate can be avoided (Schelling, 1997).

From the viewpoint of industry, the range of attitudes toward environmental protection is striking, with energy-intensive sectors typically giving the cold shoulder to the implementation of multilateral environmental agreements. This position has been taken to an extreme by some self-interested industry leaders, whose antiregulation rhetoric sounds disconcertingly similar to the "scientific" propaganda of major cigarette manufacturers during the 1970s and 1980s. Capitalizing on the indeterminate results of scientific investigations into global warming, influential leaders such as Exxon Corporation's chairman Lee Raymond have campaigned both domestically and in developing nations for a 20-year moratorium on greenhouse gas controls, ostensibly to give science sufficient time to "thoroughly understand the problem." A sensible counterpoint to this cynical perspective has been offered by John Browne, the chief executive of British Petroleum: "The time to consider policy dimensions of climate change is not when the link between greenhouse gases and climate change is conclusively proven, but when the possibility cannot be discounted and is taken seriously by the society of which we are a part" ("Exxon Urges . . . ," 1997).

In December 1997, a global-warming treaty was forged in Kyoto, Japan, by negotiators representing 159 nations. According to the so-called Kyoto Protocol, the United States is required to reduce its emissions of greenhouse gases by the year 2010 to a level that is 7% below domestic emissions in 1990. European nations are assigned a target of 8% below their emissions in that same year, while Japan somewhat reluctantly agreed to a target reduction of 6%. To meet the goals of the Kyoto Protocol, the United States will have to reduce emissions of carbon dioxide, carbon monoxide, methane, and other carbon-based gases by roughly one-third of the current projections for U.S. output of these gases in 2010. (For additional discussion of the Kyoto Protocol, see Coppock, 1998.)

One of the most promising policy innovations resulting from the Kyoto meeting was a recommendation that an international market be established in **emissions credits**. The concept is simple: Nations whose firms reduce their emissions below a specified level can sell their excess reductions to firms in nations that are over their limits. In this way, market forces can be engaged in the battle to control global warming. In a speech to Congress, Senator Robert Byrd stated that "reducing projected emissions by a national figure of one-third does not seem plausible without a robust emissions-trading and joint-implementation framework" (Swift, 1998, p. 75).

The international trading of emissions credits allows firms and nations alike to reduce their output of carbon-based gases in the most economically efficient way. Under such a regime, firms would have the flexibility to select the most efficient methods for achieving their emissions targets, either through immediate action or by purchasing credits from other firms or nations until an optimal improvement plan could be implemented. Command-and-control policy measures that force firms to adopt "quick-fix" solutions can be suboptimal in terms of the cost to firms and, more important, in terms of overall effectiveness. If firms are given time to implement the most efficient and cost-effective innovations, the economic impact of emissions reductions can be dramatically reduced ("Global Warming . . . ," 1998).

The trading of emissions credits between nations can result in far greater economic efficiency, since the cost of reducing the emissions of advanced manufacturing processes in the North can be as much as 10 times higher than the cost of the same improvement in the South. In a credit-trading system, firms in the United States that buy emissions credits from China, for example, would essentially be subsidizing the modernization of China's energy-related industries. Since all emissions are equivalent from a global perspective, this could be a far more cost-effective solution than attempting to make incremental improvements at home. An international "commodity market" in emissions credits would allow the market price for these credits to approach the marginal cost of emissions reductions worldwide. Assuming that the transaction costs are kept low, credit trading would provide an efficient economic incentive for firms to innovate environmental solutions.

There is an excellent recent example of the synergy between the incentive provided by credit trading and the power of technology to reduce environmental damage. In 1990, the U.S. Acid Rain Program was created with the goal of halving the emissions of sulfur dioxide by domestic utilities. According to a study by the Government Accounting Office, this credit-trading system has decreased the cost of pollution reduction to half of what was expected under the previous rate-based measures, and well below industry and government estimates. Moreover, by 1995, one-third of all utilities that complied with this measure did so at a net profit, due to innovations that yielded unforeseen savings upon changeover to low-sulfur coal (Swift, 1998, p. 77).[1]

Fortunately, government policy is not the only source of economic incentives; the marketplace offers a price premium for many environmentally safe products, potentially offsetting much of the recurring costs associated with these socially responsible strategies. There has been a surprising shift in the preferences of consumers in recent years toward environmentally friendly products, a movement that is often referred to as the "greening of the marketplace." Ultimately, such ethical consumers may have the final word in global environmental protection, by insisting on high standards of corporate citizenship. Programs such as eco-labeling provide an opportunity for firms to advertise their social responsibility, while imparting a "green tinge" to their corporate brands. The development of so-called **green products** is an

---

[1]The U.S. Acid Rain Program is a notable model of a broader multilateral credit-trading scheme for another reason. The use of high-quality monitoring, the implementation of a public allowance tracking system, and the imposition of steep penalties have led to 100% compliance among U.S. utilities. Similar tough enforcement regimes will be needed to avoid rampant cheating on any multilateral global warming agreement.

example of how a pull from the marketplace can spur environmentally friendly innovations. The two complementary goals of green design are (1) to prevent waste, by reducing the weight, toxicity, and energy consumption of products, and by extending their service life; and (2) to improve the management of energy and materials, through techniques such as remanufacturing, recycling, composting, and energy recovery (OTA, 1992).

Many companies have already adopted corporate-level environmental strategies. Techniques such as voluntary environmental audits and product life-cycle impact analyses are the first steps toward integrating environmental costs into the core strategies of firms. Guidance for enterprise-wide management of environmental issues is provided by global standards such as the International Standards Organization's ISO-14000 series. Those firms requiring outside expertise will find that there is no shortage of consultants available, spanning all aspects of environmental protection.

Restoration of the earth's environment will offer tremendous innovation opportunities well into the 21st century. The market for environmental products and services is expected to reach $300 billion by 2000 (Elkington, 1994, p. 67). According to one estimate, 40% of global economic output in the first half of the 21st century will be derived from environmental or energy-linked products and technologies (OECD, 1997a). U.S. firms currently hold an edge in important nonpolluting energy technologies, including reliable solar power, gasoline alternatives based on agriculture, zero-emissions fuel cells, and so on. (For more information on environmentally critical technologies, see *World Resources Institute Annual Yearbook,* 1992.)

Once market forces begin to act on problems such as global warming, the profit incentive will fuel the creative fires of entrepreneurs and innovators, potentially yielding faster-than-expected emission reductions. A decade ago, an international negotiating team met in Montreal to establish a protocol for eliminating the use of chlorofluorocarbons (CFCs), the compounds linked to depletion of the earth's ozone layer. At that time, both government and industry predicted an economic catastrophe. Instead, CFC emissions have declined so rapidly that replenishment of the ozone layer now is expected to occur early in the 21st century. Much of this tremendous progress was the result of a skillful realignment by the manufacturers of air conditioners and refrigerators to non-ozone-depleting refrigerants, a transition that went virtually unnoticed by consumers ("Hot Air Treaty," 1997).

The ethical use of technology, coupled with responsible action on the part of industries and governments, can simultaneously raise both the quality of life *and* the quality of the environment throughout the world. If we can all learn to cooperate toward such a goal, the effects could be significant within our lifetimes. Thus, the human capacity to invent, adapt, and solve intractable problems represents our greatest hope in the battle to save the natural world.

8.5 Life-Cycle Costs

The **life–cycle approach** to products simply means analyzing each step of the product life cycle for ways of reducing, reusing, and recycling to minimize all environmental impacts. Businesses typically analyze the environmental impacts associated with only the production of the product. However, the life-cycle approach requires a full environmental audit of all environmental impacts. It includes a full life cycle from the acquisition of raw materials, through production, to final disposal of the product, packaging, and other by-products. This includes upstream factors, such as pollution or environmental depletion caused by mining of raw materials, refining or manufacturing by the supplier, and transport. The impacts of the actual manufacture and production of the product at the company are analyzed. Finally, the downstream factors such as distribution and disposal of the product and packaging after customer use are also included. Low-cost factors such as water use, energy use, and clean air may be also considered.

Adopting a life-cycle approach can result in financial gains for the company by shedding costs and gaining from recycling. Because the relationships among business processes are complex, life-cycle analysis requires a sophisticated understanding of material flows, resource reuse, and product substitution. Shifting to an approach that considers all resources, products, and waste as an interdependent system will take time, but government can facilitate the shift by encouraging the transition to a systems approach.

For many products and systems, design and development costs are relatively well known; however, the costs associated with system operation and maintenance support are somewhat hidden. Furthermore, it has been determined that a major portion of the projected life-cycle cost for a given system stems from the consequences of decisions made during the early phases of advance planning and system conceptual design. We may define the life-cycle cost (LFC) as

$$LFC = CoA + CoU + CoD + PDC,$$

where CoA = cost of acquisition, CoU = cost of use, CoD = cost of disposal, and PDC = post-disposal costs.

8.6 Risk, Safety, and Health Factors

The management of technological systems has become increasingly complex. Engineers and scientists are being called on to engage in technical as well as managerial activities. These technology managers are required to direct technical project groups, cross-disciplinary work teams, and task forces, evaluate and compensate group effectiveness, and solve the problems associated with complex work environments. A major issue facing today's products and processes is how to reduce the risks associated with the **hazards** of products, work processes, industrial and occupational settings, and work environments.

A **risk** involves loss of health or life. Risk may be defined as a measure of the probability and severity of adverse effects (loss). A hazard is an endangerment, an obstacle possibly leading to an accident.

The overall costs of occupational accidents in the United States are more than $150 billion. These costs include direct losses such as medical expenses, equipment and material, insurance, lost wages, property damages, legal fees, regulatory fines, and indirect losses such as lost time of production, equipment obsolescence, loss of public confidence, loss of prestige, degradation of employee morale, loss of market share, and loss of company reputation for quality and safety.

Generally speaking, products or processes are often designed without an extensive evaluation for their potential hazards. For example, in 1996, more than 10,400 work-related deaths were reported in the United States. The multidisciplinary field of safety and health contains a number of tools and techniques suitable for hazard identification and analysis of products, tools/equipment, vehicles, and technological subsystems. Unfortunately, product and process design teams do not make full use of "safety" resources commonly recommended by the safety community. Little standardization exists for using these techniques across the science and engineering fields. Moreover, as the system under study becomes more technologically complex the hazards and risk are difficult to evaluate. An estimate of the risk for several specific exposures is summarized in Table 8.3.

Table 8.3 **Examples of Risk Exposures**

	Risk estimate (annual deaths per exposed population)	Lives saved (annually)
Underground construction	1.6 in 10^3	8
Crane-suspended platform	1.8 in 10^3	5
Asbestos	6.7 in 10^6	75

Source: Adapted from Y. Haimes, *Risk Modeling, Assessment and Management,* New York: Wiley, 1998.

Consider the following example of an accident on the London Railway illustrating the limits of safety systems:

> October 5th, a horrific collision just outside Paddington station, on one of the most congested sections of Britain's rail network, killed at least 70 people, making it the worst crash for 42 years. Another 150 were injured.
>
> The Paddington crash has come in the midst of a public inquiry into an accident two years ago at Southall, on the same stretch of line, which claimed seven lives and injured 150. Quite

rightly, this week's calamity has revived questions about the safety of Britain's railways.[2]

Presumably, a risk–benefit analysis was done when the London Railway was designed. Many projects, especially public works and regulations, are justified on the basis of a risk–benefit analysis. The questions answered by such a study are the following: Is the product worth the risks connected with its use? What are the benefits? Do they outweigh the risks? We are willing to take on certain levels of risk as long as the project (the product, the system, or the activity that is risky) promises sufficient benefit or gain. If risk and benefit can both be readily expressed in a common set of units (say, lives or dollars as in cost–benefit analyses), it is relatively straightforward to carry out a risk–benefit analysis and to determine whether we can expect to come out on the benefit side.

Cost–benefit analysis, while seen as a tool to assist in running a business, is less attractive to environmentalists as a tool of environmental policy. They believe that the costs of environmental protection are easier to quantify and express in dollars than are the benefits of environmental protection. After all, what are the economic values of feeling healthier, living longer, being able to see farther and more clearly on a summer day, knowing that clean groundwater will be available for use by future generations, and protecting endangered species from extinction? There is also a suspicion that the costs of preventing or controlling pollution are not as large as industry and regulators suggest.

In the next section, Professor William Samuelson provides a description of cost–benefit analysis.

A COST-BENEFIT RESULT

The economic value of the public health and environmental benefits that Americans obtain from the Clean Air Act Amendments of 1990 exceeds their costs by a margin of four to one, according to a new study. The report by the Environmental Protection Agency projects that the law and its associated programs prevent thousands of premature deaths and millions of asthma attacks related to air pollution every year.

The study shows that by the year 2010, the amendments to the Clean Air Act will prevent 23,000 Americans from dying prematurely each year, and avert more than 1,700,000 incidences of asthma attacks and aggravation of chronic asthma.

SOURCE: Environmental Protection Agency, Nov. 17, 1999, *www.epa.gov/oar/sect812.*

---

[2] *The Economist,* Oct. 9, 1999, p. 65.

8.7 Cost–Benefit Analysis[3]

By William Samuelson
Professor
School of Management
Boston University

> *My way is to divide a half sheet of paper by a line into two columns; writing over the one Pro, over the other Con. . . . When I have got them altogether in one view, I endeavor to estimate their respective weights; and where I find two, one on each side, that seem equal, I strike them both out. If I find a reason Pro equal to two reasons Con, I strike out the three, . . . and thus proceeding, I find where the balance lies. . . . I have found great advantage from this kind of equation, in what might be called moral or prudential algebra.*
> — Benjamin Franklin, September 19, 1792

Cost–benefit analysis is a method of evaluating public projects and programs. Accordingly, cost–benefit analysis is used in planning budgets, in building dams and airports, devising safety and environmental programs, and in spending for education and research (Musgrave, 1969). It also finds a place in evaluating the costs and benefits of regulation: when and how government should intervene in private markets. In short, almost any government program is fair game for the application of cost–benefit analysis. The logic of cost–benefit analysis is simple. The fundamental rule is undertake a given action if and only if its total benefits exceed its total cost. (The rule applies equally well when taking one action forecloses pursuing another. Here, the action's cost is an opportunity cost: the foregone benefits of this next best alternative.) Applying the cost–benefit rule involves three steps: (1) identifying all impacts (pro or con) on all affected members of society, (2) valuing these various benefits and costs in dollar terms, and (3) undertaking the program in question if and only if it produces a positive **total net benefit** to society.

The aim of cost–benefit analysis is to promote economic efficiency (Gramlich, 1992). While there is little controversy concerning the need to use resources wisely, there is criticism of the way cost–benefit analysis carries out this goal. Some critics point out the difficulty (perhaps impossibility) of estimating dollar values for many impacts. How does one value clean air, greater national security, unspoiled wilderness, or additional lives saved? The most difficult valuation problems arise when benefits and costs are highly uncertain, nonmarketed, intangible, or occur in the far future. Proponents of cost–benefit analysis do not deny these difficulties; rather they point out that any decision depends, explicitly or implicitly, on some kind of valuations. For instance, suppose a government agency refuses to autho-

---

[3]This section has been reprinted with permission from *The Technology Management Handbook* (Richard Dorf, ed.), pp. 4-35–4-39, Boca Raton, FL: CRC Press, 1998. Copyright CRC Press, Boca Raton, Florida.

rize an $80 million increase in annual spending on highway safety programs, which is projected to result in 100 fewer highway deaths per year. The implication is that these lives saved are not worth the dollar cost, that is, the agency reckons the value of such a life saved to be less than $800,000. All economic decisions involve trade-offs between benefits and costs. The virtue of the cost–benefit approach is in highlighting these trade-offs.

A second point of criticism surrounds step three, whereby only total benefits and costs matter, not their distribution. A program should be undertaken if it is beneficial in aggregate, i.e., if its total dollar benefits exceed total costs. However, what if these benefits and costs are unequally distributed across the affected population? After all, for almost any public program, there are gainers and losers. (Indeed, citizens who obtain no benefit from the program are implicitly harmed. They pay part of the program's cost either directly via higher taxes or indirectly via reduced spending on programs they would value.) Shouldn't decisions concerning public programs reflect distributional or equity considerations? Cost–benefit analysis justifies its focus on efficiency rather than equity on several grounds. The first and strongest ground is that the goals of efficiency and equity need not be in conflict, provided appropriate compensation is paid between the affected parties. As an example, consider a public program that generates different benefits and costs to two distinct groups, A and B. Group A's total benefit is $5 million. Group B suffers a loss of $3 million. The immediate impact of the project is clearly inequitable. Nonetheless, if compensation is paid by the gainers to the losers, then all parties can profit from the program. A second argument for ignoring equity is based on a kind of division of labor. Economic inequality is best addressed via the progressive tax system and by transfer programs that direct resources to low income and other targeted groups. According to this argument, it is much more efficient to use the tax and transfer system directly than to pursue distributional goals via specific public investments. Blocking the aforementioned project on equity grounds has a net cost: foregoing a $5 million dollar gain while saving only $3 million in cost. Finally, though it is not common practice, cost–benefit analysis can be used to highlight equity concerns. As step one indicates, the method identifies, untangles, and disaggregates the various benefits and costs of all affected groups. This in itself is an essential part of making distributional judgments. The method's key judgment about equity comes when these benefits and costs are reaggregated: all groups' benefits or costs carry equal dollar weight.

8.7.1 APPLYING COST–BENEFIT ANALYSIS: A SIMPLE EXAMPLE

A city planning board is considering the construction of a harbor bridge to connect downtown and a northern peninsula. Currently, residents of the peninsula commute to the city via ferry (and a smaller number commute by car taking a slow, "great circle" route). Preliminary studies have shown that there is considerable demand for the bridge. The question is whether the benefit to these commuters is worth the cost. The planning board has the following information. Currently, the ferry provides an estimated 5 million commuting trips annually at a price of $2.00 per trip; the ferry's average cost per trip is $ 1.00, leaving it $1.00 in profit. The immediate construction cost of the bridge is $85 million dollars. With proper maintenance, the bridge will last indefinitely. Annual operating and maintenance costs are estimated at $5 million. Plans are for the bridge to be toll-free. Since the bridge will be a perfect substitute for the ferry, the ferry will be priced out of business. The planners estimate that the bridge will furnish 10 million commuting trips per year. The discount rate (in real terms) appropriate for this project is 4%.

Figure 8.1 shows the demand curve for commuter trips from the peninsula. The demand curve shows that at the ferry's current $2.00 price, 5 million trips are taken. Should a toll-free bridge be built, 10 million trips will be taken. Furthermore, the planning board believes that demand is linear.

Currently, the ferry delivers benefits to two groups: the ferry itself (its shareholders) and commuters. The ferry's annual profit is (2.00 – 1.00)(5) = $5 million. In turn, the commuters' collective benefit takes the form of **consumer surplus** — the difference between what consumers are willing to pay and the actual price charged. The triangular area between the demand curve and the $2.00 price line (up to their point of intersection at 5 million trips) measures the total consumer surplus enjoyed by ferry commuters. The area of this triangle is given by (0.5) (4.00 – 2.00) (5) = $5 million. Thus, the sum of profit plus consumer surplus is $10 million per year. Supposing that this benefit flow is expected to continue indefinitely at this level, the resulting net present value is 10/0.04 = $250 million. Here, the

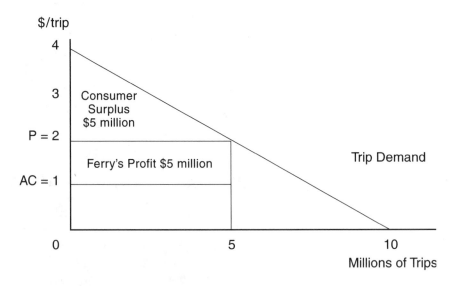

Figure 8.1 **Benefit–cost analysis of building a bridge.**

annual net benefit (in real terms in perpetuity) is capitalized by dividing by the appropriate (real) interest rate.

Now consider the cost–benefit calculation for the bridge. First note the adverse effect on the ferry: it is put out of business, so its profit is zero. Second, there is a burden on taxpayers. They must foot the bill for the construction and maintenance costs of the bridge. Since the bridge charges no toll, it generates no revenue. Third, the entire benefit of the bridge takes the form of consumer surplus (the triangle inscribed under the demand curve and above the zero price line). The dollar value is $(0.5)(4.00)(10) = \$20$ million per year. In present value terms, this benefit comes to $500 million against a total cost (also in present value terms) of $85 + 125 = \$210$ million. Thus, the bridge's net benefit is $290 million. Since this is greater than that of the ferry, the bridge should be built.

The decision to build the bridge depends crucially on charging the "right" toll. In the present example, no toll is charged. Here, the right price is zero because there is a negligible cost (no wear and tear or congestion) associated with additional cars crossing the bridge. The general principle behind **optimal pricing** is simple: the optimal price should just equal the marginal cost associated with extra usage (Layard, 1977). For instance, since large tractor trailer trucks cause significant road damage to highways, they should be charged a commensurate toll. Here, a zero price insures maximum usage of the bridge and maximum benefit (with no associated cost). Setting any positive price would exclude some commuters and reduce the net benefit.

Regulating the Ferry

Before concluding that public provision is warranted, government decision makers should consider another option: regulating the private ferry market. Here, regulation means limiting the price the ferry operator can charge. From a cost–benefit point of view, what is the optimal regulated price? Basic economic principles provide a direct answer: the price that would prevail in a perfectly competitive market. If free entry of competitors were feasible, the ferry's price would be driven down to the zero-profit point: $P = AC = \$1.00$. Thus, this is the price that the government should set for the (natural monopolist) ferry operator.

At a $1.00 price, the ferry makes 7.5 million in trips and makes a zero profit. Commuters realize total consumer surplus that comes to $(0.5)(4.00 - 1.00)(7.5) = \$11.25$ million per year. (As always, consumer surplus is given by the area under the demand curve and above the price line.) The present value of the net benefit from ferry regulation is $11.25/0.04 = \$281.25$ million. Building the bridge, with a discounted net benefit of $290 million, has a slight edge over the regulatory alternative and continues to be the best course of action.

8.7.2 VALUING BENEFITS AND COSTS

The main issues with respect to valuing benefits and costs concern (1) the role of market prices, (2) ways of valuing nonmarketed items, and (3) the choice of appropriate discount rate. In most cases, market prices provide the correct values for benefits and costs. In perfectly competitive markets, the price of the good or service is an exact measure of its marginal benefit to consumers and its marginal cost to producers, $P = MB = MC$. In some circumstances, however, correct valuation of benefits and costs based on market prices may require modifications. One instance occurs when "inframarginal" effects are important. We know that the current market price reflects the valuation of marginal units (i.e., the last ones consumed). However, if the output impact is large, a significant amount of consumer surplus will be created above and beyond revenue generated. In the previous example of the bridge, the benefit consisted entirely of consumer surplus. With a zero toll, the bridge generated no revenue.

A second problem concerns price distortions. Taxes are one source of distortions. For instance, suppose the government sets a tariff on the import of an agricultural good with the intent of protecting domestic farmers from foreign competition. The result is that the domestic price for the good is $1.00 above the world price. How should one value the new crops grown using water from a federal dam? The difficulty is that, instead of a single price reflecting marginal benefit and marginal cost, there are two prices. If all of the crop is sold on the world market, then that is the appropriate price, and similarly for the home market. If the crop is sold on both markets, then the appropriate value is a weighted average of the separate prices. A similar difficulty stems from the presence of monopoly, where the monopoly price is greater than marginal cost. Depending on circumstances, output could be valued at P or MC. A final example involves the employment of labor. Consider the cost of labor used to build the bridge. If undertaking the project has no effect on overall employment (construction workers are bid away from other jobs), the labor cost is measured by the going market wage. Alternatively, suppose that there is widespread unemployment so that new jobs are created. Then the analysis should include the wage paid (net of the cost of the worker's lost leisure time) as an additional benefit for the newly hired worker.

Nonmarketed Benefits and Costs

Several difficulties occur when valuing nonmarketed items. For instance, how might one measure the dollar benefits provided by public elementary and secondary schools? This is a tough question. The difficulty is that public education is provided collectively (i.e., financed out of local tax revenues); there is no "market" value for this essential service. Parents do not pay market prices for their children's education, nor are they free to choose among public schools. By contrast, valuing education provided by private schools is far easier to pin down. It is at least as much as these parents are willing to pay in tuition. If a private school fails to deliver a quality education, parents will stop paying the high market price.

This same point about valuation applies to all nonmarketed goods: national security, pollution, health risks, traffic congestion, even the value of a life. In the absence of market prices, other valuation methods are necessary. Roughly speaking, there are three approaches to valuing nonmarketed goods and so-called "intangibles." One method is to elicit values directly via survey, that is, ask people what they really want. Surveys have been used to help ascertain the benefits of air quality improvement (including improved visibility in Los Angeles from smog reduction), the benefits of public transport, the cost of increased travel time due to traffic, and the value of local public goods. A second approach seeks to infer values from individual behavior in related markets. For instance, the benefit of a public secondary school education might be estimated as the expected difference in labor earnings (in present-value terms) between a high school graduate and a ninth-grade dropout. Appealing to labor markets provides a ready measure of the economic value of these years of schooling (though not necessarily the full personal value). One way to measure the harm associated with air pollution is to compare property values in high-pollution areas vs. (otherwise comparable) low-pollution areas. Finally, society via its norms and laws places monetary values on many nonmarketed items. Workman's compensation laws determine monetary payments in the event of industrial injuries. Judges and juries determine the extent of damages and appropriate compensation in contract and tort proceedings.

The Discount Rate

Evaluating any decision in which the benefits and costs occur over time necessarily involves discounting. The discount rate denotes the trade-off between present and future dollars, that is, one dollar payable a year from now is worth $1/(1+r)$ of today's dollar. In general, the net present value (NPV) of any future pattern of benefits and costs is computed as

$$NPV = [B_0 - C_0] + [B_1 - C_1]/(1 + r) + [B_2 - C_2]/(1 + r)^2 + \ldots [B_T - C_T]/(1 + r)^T$$

As always, future benefits and costs are discounted relative to current ones. Moreover, the public-sector manager employs the present value criterion in exactly the same way as the private manager. The decision rule is the same in each case: undertake the investment if and only if its NPV is positive. For most public investments, costs are incurred in the present or near-term, while benefits are generated over extended periods in the future. Consequently, lower discount rates lead to higher project NPVs, implying that a greater number of public investments will be undertaken.

What is the appropriate discount rate to use in evaluating public investments? The leading point of view is that they should be held to the same discount rate as private investments of comparable risk. For instance, consider a municipal government that is considering building a downtown parking garage. Presumably, this project's risks are similar to those of a private, for-profit garage, so its NPV should be evaluated using a comparable discount rate. A fundamental principle of modern finance is that individuals and firms demand higher rates of return to hold riskier investments. Thus, setting the appropriate discount rate means assessing the riskiness of the investment.

A second point of view holds that the choice of the discount rate should be a matter of public policy. Proponents of this view argue that the overall rate of investment as determined by private financial markets need not be optimal. They contend that left to their own devices private markets will lead to underinvestment. For instance, there are many investments that generate benefits in the distant future — benefits that would be enjoyed by future generations. However, unborn generations have no "voice" in current investment decisions, so these investments (beneficial though they may be) would not be undertaken by private markets. The upshot is that private markets are too nearsighted and private interest rates are too high, thereby discouraging potentially beneficial investment. According to this view, public investments should be promoted by being held to lower discount rates than comparable private investments.

8.7.3 CONCLUDING REMARKS

When used appropriately, cost–benefit analysis is a useful guide to decisions concerning public programs and government regulation. Inevitably, some aspects of the analysis — specific benefits, costs, or the discount rate — may be uncertain or subject to error. The appropriate response is to utilize sensitivity analysis. For instance, if the discount rate is uncertain, the project's present value should be computed for a range of rates. If the NPV does not change sign over the range of plausible rates, the optimal investment decision will be unaffected.

Defining Terms

Consumer surplus: The benefit enjoyed by consumers, equal to the difference between what they are willing to pay and the price they actually pay for a good or service.

Optimal pricing: The price set for a public project should equal the marginal cost associated with additional usage.

Total net benefits: The objective of cost–benefit analysis is economic efficiency, that is, maximizing the sum of net benefits measured in dollar terms.

Further Information

The two following articles discuss the potential uses and abuses of cost–benefit analysis. The textbook reference offers an extensive treatment from which parts of this condensed survey are adapted.

Arrow, K. *et al.* Is there a role for benefit-cost Analysis in environmental, health, and safety regulation?, *Science,* 272(12): 221–222, 1996.

Dorfman, R. Why benefit–cost analysis is widely disregarded and what to do about it, *Interfaces,* 26: 1–6, 1996.

Samuelson, W., and S. Marks. *Managerial Economics,* chap. 14, Fort Worth, TX: Dryden Press, 1995.

Plans for the Metcalf Energy Center, a natural gas-fired power plant that will provide 600 megawatts to San Jose, California, have received strong support from environmental groups such as the Loma Prieta Chapter of the Sierra Club and the Center for Energy Efficiency and Renewable Technologies (CEERT). This "anti-sprawl" project will not require any new transmission towers and includes the restoration of nearby wetlands, planting 800 new trees, and using recycled wastewater for cooling towers—a good example of an ecologically sound, clean-burning plant that benefits, rather than diminishes, its natural surroundings. ARCHITECTURAL RENDERING COURTESY OF THE DEVELOPERS, CALPINE CORPORATION AND BECHTEL ENTERPRISES HOLDINGS, INC.

CHAPTER 9

Energy, Environment, Economy, and Society

9.1 Energy

If the machine has freed man from slavery to toil, then the fuel of this change has been the abundant energy stored in the natural resources of the planet. In the 19th century man developed a capability not only to manufacture materials and goods, but also to manufacture and control power. More correctly, mankind learned to release the vast amounts of stored energy contained within the fossil fuels of the world and to control this energy and put it to use in the technological processes of industry, commerce, and the home. A century after the invention of the first successful oil well, only 1% of America's physical work is done by people. The rest is done by machines fueled with the energy released from natural resources.

An adequate, affordable energy supply and efficient energy use are indispensable ingredients of the economic well-being of individuals and nations. In the United States and worldwide, energy accounts for about 7% of GDP and a similar share of international trade; global investments in energy-supply technology (oil refineries and pipelines, electric power plants and transmission lines) total hundreds of billions of dollars per year; and annual global expenditures on items whose energy-using characteristics are potentially important to their marketability (automobiles, aircraft, buildings, appliances, and industrial machinery) run into the trillions. When and where energy becomes scarce or expensive, recession, inflation, unemployment, and the frustration of aspirations for economic betterment are the usual results.

Energy is no less crucial to the environmental dimensions of human well-being than to the economic ones. It accounts for a striking share of the most troublesome environmental problems at every geographic scale — from wood smoke in Third World village huts, to regional smog and acid precipitation, to the risk of widespread radioactive contamination from accidents at nuclear energy facilities, to the buildup of carbon dioxide and other greenhouse gases in the global atmosphere. The growth of energy use, driven by the combination of population increase and economic development, has increased some of these problems to levels variously disruptive of human health, property, economic output, food production, peace of mind, and enjoyment of nature in many regions. And all of these aspects of human well-being could eventually be impacted over substantial areas of the planet by the kinds of global climatic changes widely predicted to result from continued buildup in the atmosphere of carbon dioxide from fossil fuel combustion.

Improvements in energy technology and the widespread penetration of these improvements in the marketplace in the 21st century are needed to enhance the positive connections between energy and economic well-being and to ameliorate the negative connections between energy and environment. Such improvements in technology can lower the monetary and environmental costs of supplying energy, lower its effective costs by increasing the efficiency of its end uses, reduce overdependence on oil imports, slow the buildup of heat-trapping gases in the atmosphere, and enhance the prospects for environmentally sustainable and politically stabilizing economic development in many of the world's potential trouble spots.

The problem with this vastly increasing use of energy resources may be the finite nature of the energy resources of the world — or at least our ability to tap only a limited amount of them. Energy is the capacity for doing work. The fuel resources of our world contain the capacity to accomplish work, such as moving an automobile. The unit of energy is the joule (J) and several other units of energy are used as shown in Table 9.1. When energy is put to use, it is consumed at some rate over a period of time. The rate of use of energy is defined as power with units of joules per second.

Table 9.1 **Units of Energy**

1,055 joules = 1 British thermal unit (Btu) = 0.252 kilocalories
3,412 Btu = 1 kilowatt-hour = 3,600,000 joules (J)
1 quad = 1×10^{15} Btu = 1 Q Btu = 1 quadrillion Btu

Table 9.2 **Units of Power**

1 watt (W) = 1 joule per second
1 kilowatt (kW) = 1.341 horsepower
1 horsepower (hp) = 746 W

Table 9.3 **Prefixes of Scientific Units**

Prefix	Abbreviation	Power of 10
kilo	k	10^3
mega	M	10^6
giga	G	10^9
tera	T	10^{12}
peta	P	10^{15}
exa	E	10^{18}

Table 9.2 lists the commonly used units of power. Table 9.3 shows the prefixes for common scientific units. For example, using Tables 9.2 and 9.3, we can describe an automobile with 140 hp as equivalent to 10.4 kW.

About 90% of the energy consumed in the United States is derived from fossil fuels — coal, oil, and natural gas. All fossil fuels required millions of years to be formed. At the rate we consume them, they will not replace themselves. Thus fossil fuels are a finite or limited source. We may define Q_R as the proven reserves of a fossil fuel and Q_P as the total accumulated quantity of the fuel removed from the ground up to any time. The cumulative discoveries, Q_P, represent the quantities removed from the ground plus the recoverable quantities remaining in the ground. Therefore,

$$Q_D = Q_P + Q_{R.} \tag{9.1}$$

If Q_D is eventually limited or finite, we encounter the effect portrayed in Figure 9.1. As time progresses, the cumulative discoveries, Q_D, reach the finite limit. As production, Q_P, increases, eventually the proved reserves, Q_R, reach a maximum and begin to reduce. Of course, the reserves are never fully exhausted, because as the reserves become very low and scarce, the price of this fuel will rise and consumption will decrease. As Figure 9.1 shows, the reserves are reduced slowly as time progresses toward the far right of the curve for Q_R.

In 1998, the 5.9 billion people on earth consumed inanimate energy forms at a rate of about 427 Q Btu with 75% of it derived from fossil fuels. Table 9.4 records the world energy use. The United States consumed about 94 Q Btu in 1998 or about 22% of the world's consumption, with fossil fuels accounting for 85% of the total used.

Much of the total energy used by a nation is in the form of electricity. Electrical power is readily distributed to homes and factories and is a valuable source of energy to an industrialized society. Electrical energy is primarily generated by using fossil fuels to generate steam, which then drives electric generators. In 1998 the world consumed about 13.0 peta-watt-hours (PW hr) while the United States consumed about 3.5 PW hr — about 27% of the total world electric energy consumption. Worldwide, about 63% of electric power is generated using fossil fuels.

The dramatic increase in the scale of energy consumption is illustrated in Table 9.5. For the period from 1850 to 1990, total global energy use

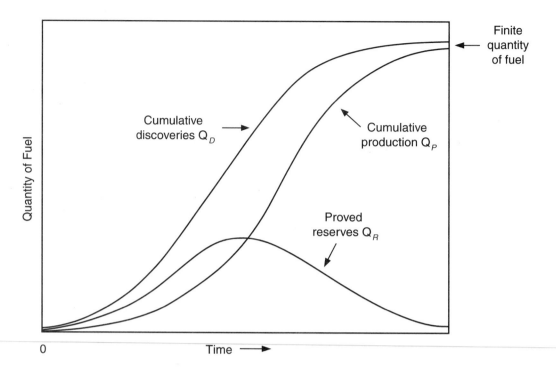

Figure 9.1 **The consumption of a limited fuel over the lifetime of the fuel.**

Table 9.4 **World and U.S. Energy Use in 1998**

Fuel source	Percent of total world use	Percent of total United States use
Petroleum	33.0	38.1
Coal	21.2	22.3
Natural gas	20.0	24.1
Biomass[a]	13.0	3.1
Hydropower	6.1	4.4
Nuclear	6.2	7.6
Solar, wind, geothermal	0.5	0.4
Total:	100.0	100.0

[a]Biomass = wood, charcoal, crop wastes, and manures.

increased by more than a factor of 16. Total energy can be decomposed into the product of population and energy use per capita. Both of these factors have contributed to total energy increase. As shown in Table 9.5, population increased by nearly a factor of 5 and energy use per capita by a factor of 3.5, from 19 to 66 GJ per capita. The total energy increased by a factor of 16.6 with about 17% of the total energy generated by nuclear power and another 19% by hydropower.

Table 9.5 **Global Energy Use in 1850 and 1990**

	1850	1990	Ratio of 1990 to 1850
Population (billion)	1.1	5.3	4.8
Energy per capita (GJ)	19.0	66.0	3.5
Total energy (EJ)	21.0	349.0	16.6

Table 9.6 **Energy Uses in the United States in 1997**

Sector	Subsectors	Percent of primary use	Percent of sector use
Residential Buildings		12	
	Space heat		50
	Water heat		20
	Air conditioning		5
	Appliances		25
Commercial Buildings		24	
	Space heat		35
	Water heat		16
	Air conditioning		8
	Lighting		21
Transportation		26	
	Passenger cars		55
	Truck freight		25
	Aircraft		7
Industry and Agriculture		38	
	Fuels		18
	Chemicals		15
	Metals		8
	Pulp and paper		8

Energy uses, by sector, are shown in Table 9.6 for the United States in 1997. Energy is most intensely used by industry and agriculture.

Figure 9.2 shows the forms of energy consumed in the United States since 1850. Prior to 1850, wood accounted for most of the energy consumption. Wood is biomass and can be continually grown and replaced. After 1860, the United States switched to using coal for its furnaces and engines. Starting early in the 1900s, oil became a significant part of the fuel supply, and later natural gas became heavily distributed and utilized. Nuclear energy became important after 1970, but reached a peak and leveled off as safety issues became apparent.

What about the energy future for the United States and the world? A panel on energy research and development created a report for the Office of the President of the United States (1997). This report prepared a scenario for the energy future which they called "business as usual." This scenario

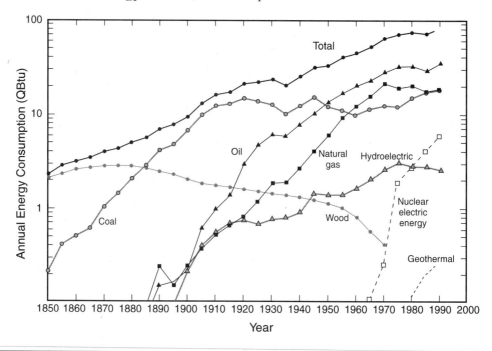

SOURCE: From U.S. Energy Information Administration.

Figure 9.2 **Forms of energy consumed in the United States since 1850.**

indicates that world energy demand could be four times as large in 2100 as the 1995 demand, and fossil fuels would be providing about two-thirds of the total. Energy demand in the United States would be about two times as large in 2100 as in 1995 and fossil fuels would account for 88% of U.S. energy use in 2100.

U.S. oil imports in 1995 were $60 billion on the deficit side of the U.S. trade balance. The U.S. import bill could reach $108 billion by 2015.

Energy consumption is highest in industrialized nations. Table 9.7 shows the energy consumption in six large nations. The United States of America is the largest consumer of energy and also has the highest energy consumption per capita.

Table 9.7 **Largest Energy Consuming Nations in 1995**

Rank	Nation	Consumption (EJ)	Population (millions)	Energy per capita (GJ per person)
1	United States	96.1	272	353
2	China	36.4	1,250	29
3	Russian Federation	39.7	145	205
4	Japan	18.8	125	150
5	Germany	13.6	82	166
6	India	13.5	970	14

EJ = 10^{18} joules; GJ = 10^9 joules.

9.2 Energy and the Environment

Energy use and environmental problems are strongly interrelated. Energy systems are critical to the economy of nations and are slow to change. At the same time environmental issues are critical to energy planning since environmental constraints and the costs of coping with them are as important as resource scarcity or the monetary costs of energy technology. Environmental considerations may turn out to be the most important considerations in society's choices about how much energy should be supplied from what sources.

In the emerging nations, at the local level, air pollution results from fossil fuel and biomass fuel consumption. Many countries use biomass fuels (wood, charcoal, crop wastes, and manures) for heating and cooling in dwellings.

Among the world's many local water pollution problems, those produced by coal mine drainage, oil refinery emissions, oil spills from pipelines and tankers, and leakage into groundwater from underground fuel storage tanks are significant to the energy sector.

Fossil fuels used in power plants and vehicles cause air-basin smog. Acid rain from emissions of oxides of nitrogen and sulfur can impact forests, fish, and soils. At the global level, the emission of heat-trapping carbon-dioxide gas from fossil fuel combustion is the largest contributor to the possibility that the atmosphere's greenhouse effect will significantly change the global climate.

Of course, ameliorating the environmental problems caused by energy supply will be partly a matter, in many circumstances, of putting in place appropriate combinations of incentives and regulations that effectively incorporate environmental costs into the decision making of both energy producers and consumers. Improvements in energy technology itself are an essential part of any strategy for addressing environmental problems.

The wider environmental challenge to energy technologies is to provide energy options that can substantially mitigate the local, regional, and global environmental risks and impacts of today's energy supply system. They should do so at affordable costs and without incurring new environmental risks as serious as those that have been ameliorated. Further, they should be applicable to the needs and contexts of developing countries as well as industrialized ones.

Expansion of the use of nuclear energy plants would reduce the dependence on fossil fuels. Nevertheless, safety and health concerns would need to be addressed before this approach could be adopted.

Another means of reducing the impact of fossil fuel use is a significant increase in the efficiency of energy technologies. Significant innovations could result in important savings. For example, consider the efficacy of electric lamps as shown in Table 9.8. If most electric light was shifted from incandescent light to fluorescent, a reduction of energy used to achieve a specified light level could result in an energy use reduction of 50%.

Table 9.8 **Efficacy of Electric Lamps**

Type	Efficacy[a] (lumens/watt)
Low-pressure sodium	180
High-pressure sodium	130
Metal–halide	100
Fluorescent	80
Mercury-vapor	56
Incandescent	23

[a]Efficacy = light output in lumens per watt input.

During the 1980s and 1990s, great progress has been made in reducing the environmental impacts of fossil fuel use — particularly of coal use in electric power production — in cost-effective ways. In the United States natural gas has become the fuel of choice for new electric generation because of its low cost, low environmental impact, relatively small scale, and advancing turbine technology, and because of the competitive pressures of electric industry restructuring. This trend to natural gas is likely to continue for several decades and contributes positively to environmental improvement, particularly by reducing CO_2 emissions to the extent that gas replaces coal.

9.3 Energy Technology, Efficiency, and Efficacy

The global consumption of fossil fuels is likely to increase throughout the 21st century. Because fossil resources are limited, the world's use of fossil fuels will most likely peak at some time during the next 200 years as shown in Figure 9.3 by the dashed line labeled "Current technologies" (as of 2000). On the other hand, a significant increase in the efficiency and efficacy of energy systems could result in a future as shown by the solid line in Figure 9.3. By efficiency we mean the ratio of output to input. Thus, we seek high energy efficiencies.

Similarly, we seek efficacious results, that is, results that produce the desired goal. Overall, we seek high efficiency (good results for the given inputs) while obtaining the right (desired) results. Efficient and efficacious results are those that are aligned with our economic and environmental goals and provide an output that is close to the input. In formula terms we consider efficiency to be

$$\eta = \frac{\text{Energy output}}{\text{Energy input}}$$

and efficacy to be

$$\epsilon = \frac{\text{Actual output consequences}}{\text{Desired output consequences}}.$$

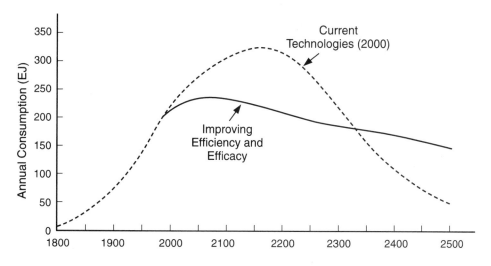

Figure 9.3 **Two scenarios for the consumption of fossil fuels throughout the world.**

The best system has a high return r where

$$r = \eta \times \epsilon.$$

Table 9.9 shows the energy efficiencies of several devices. Of course, it is the energy return, r, of a system that is important. Consider the simplified diagram of an energy system for a passenger vehicle as shown in Figure 9.4. First, consider the internal combustion engine vehicle that uses gasoline. The fossil fuel, petroleum, is refined to yield the energy carrier gasoline with an efficiency, η_1, of 90%. The vehicle's engine has an efficiency η_2 of 25%. Therefore, the efficiency of the gasoline car is

$$\eta = 0.25 \times 0.90 = 0.225$$

or the system efficiency is 22.5%.

The electric vehicle is potentially more efficient using an electric generator and electric motor for the components of Figure 9.4. Therefore, today we use electric energy for street trolleys, underground railways, and light transit vehicles. The electric energy is delivered to the vehicle by means of a set of overhead wires or a third rail in a subway. However, it is difficult to deliver electric energy to an automobile. The current system uses a plug converter that charges a storage battery in the electric car. This system is limited by the storage capacity of the battery. The energy density of common fuels is given in Table 9.10. Note that gasoline has about four times the energy density of a storage battery per kilogram of weight added to the vehicle.

Table 9.9 **Energy Efficiencies of Several Processes**

Device	Efficiency, η (percent)
Electric generator	98
Electric motor	90
Steam boiler	88
Home gas furnace	85
Home oil furnace	65
Steam–electric power plant	40
Diesel engine	38
Automobile engine	25
Fluorescent lamp	20
Photovoltaic cell	10
Incandescent lamp	5

Table 9.10 **Energy Density of Common Fuels**

Fuel	Density (MJ/kg)
Electric storage battery	12
Wood	18
Coal	24
Ethanol	27
Natural gas	35
Petroleum	42
Gasoline	46
Hydrogen	110

An example of a very efficient energy system is shown in Figure 9.5. This combined heat and power system is often called a **cogeneration system**. Since we are using most of the energy contained in the fossil fuel, we can achieve a system efficiency of 90%. These plants are built near a large user such as a university or factory. They provide steam for heating, chilled water for cooling, and electric power.

Investments in energy efficiency are the most cost-effective way to simultaneously reduce the risks of climate change, oil import interruption, and local air pollution, and to improve the productivity of the economy. Improvements in the use of energy have been a major factor in increasing the productivity of U.S. industry throughout the 1980s and early 1990s. Between 1973 and 1986, the U.S. consumption of primary energy stayed at around 75 quads, whereas the GNP grew by more than 35%.

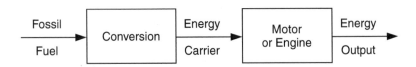

Figure 9.4 **Simplified energy system for a vehicle.**

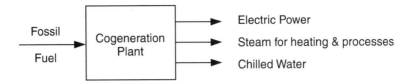

Figure 9.5 **Model of a cogeneration system.**

The energy intensity of the U.S. economy, measured in energy use per dollar of GDP, has declined from its high in 1920 to its current low as shown in Table 9.11. To reduce the nation's energy intensity, the efficiency of major energy users must increase. Improvements in energy efficiency reduced the energy intensity of economic activity in the United States by nearly one-third between 1975 and 1995, an improvement that is now saving U.S. consumers about $170 billion per year in energy expenditures and is keeping U.S. emissions of air pollutants and carbon dioxide about one-third lower than they would otherwise be. Further major increases in efficiency can be achieved in every energy end-use sector: in transportation, for example, through use of much more fuel-efficient cars and trucks; in industry through improved electric motors, materials-processing technologies, and manufacturing processes; in residential and commercial buildings through high-technology windows, superinsulation, more efficient lighting, and advanced heating and cooling systems.

Table 9.11 **Energy Intensity of the U.S. Economy, 1900–2000**

Year	1900	1920	1940	1960	1980	1990	2000[a]
Energy intensity[b]	26	35	25	23	20	14	13

[a]Estimated.
[b]Energy intensity = thousand Btu/dollars of GDP for constant 2,000 dollars.
SOURCE: U.S. Department of Energy.

9.4 Energy and Government

The energy system of an industrialized nation is part of the integrated social and political system as shown in Figure 9.6. The social processes of an industrialized nation include the important interaction of energy, economics, and the environment. Energy and economics interact directly through supply and demand, prices, and capital investments. The energy marketplace is a result of the population, investment capital, government policies, technology, energy sources, and energy demand. The marketplace balances all of these factors and yields an equilibrium of supply and demand by means of prices and controls. As the world's energy marketplace changes, a redistribution of economic wealth occurs.

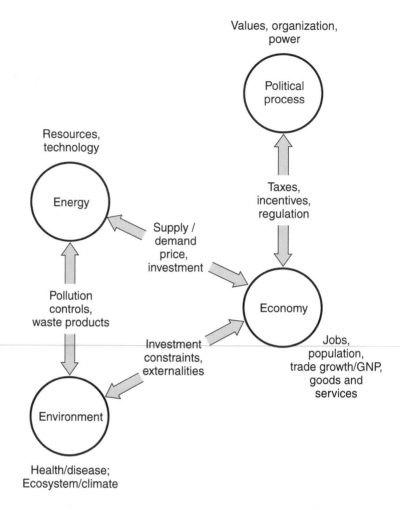

Figure 9.6 **The interactive social processes of an industrialized nation.**

Research and development (R&D) is a systematic means for creating the technical improvements needed for the energy system. Technology that is useful today is the result of R&D that was done in the past, and what will be deployable in the future depends on the R&D that is being done now and that will be done tomorrow. It is important to understand that while some kinds of energy R&D can bring quite rapid returns, the timescales on which most kinds of energy R&D exert a significant influence on deployed energy systems are longer. This is related not only to the time required to complete the R&D but also to the long turnover times of most energy supply and energy end-use equipment: for example, three to five decades for electric power plants and oil refineries; five decades or more for residential and commercial buildings; and a decade or even more for automobiles and household appliances.

These long timescales are one of the reasons why energy R&D is not left entirely to the private sector. It is in society's interest to investigate — as part of its strategy for preparing for an uncertain future — some high-potential-payoff energy alternatives for which the combination of a long time horizon for potential economic returns, uncertainty of success, and cost of the R&D makes this pursuit unattractive to private firms. Another rationale for a government role in R&D is that some of the most badly needed improvements in energy technologies relate to externalities such as environmental impacts that are not valued in the marketplace and hence do not generate the market signals to which firms respond. Still another is that the fruits of some kinds of R&D are difficult for any one firm or small group of firms to appropriate, even though these innovations may be highly beneficial to society as a whole.

9.5 The Energy Systems Model of Society[1]

BY JOHN PEET
Senior Lecturer
Department of Chemical and Process Engineering
University of Canterbury

The conventional macroeconomic model is of production and consumption linked in a circle of flows of exchange value. This picture can be modified by acknowledging the essential part played by the material flows of raw resources of materials and energy from the environment, through the economy and out to waste. Figure 9.7 represents the economy as a pathway by which energy resources are used for conversion of material resources into the goods and services needed for the economic system's normal functioning.

In practice, primary resources of energy (such as fuels in the earth's crust) or materials (such as minerals) have to be transformed to make them usable by an economy. The energy sector — which Gilliland (1977) refers to as an **energy transformation system** — normally originates at a coal mine or oil/gas well, where the resource is physically extracted from its geological deposit. This stage can then include, for example, a power station, which takes the fuel and transforms it into electricity; or an oil refinery, which takes crude oil and produces refined petrol, diesel, and other products. Another example is a hydroelectricity station, which takes water in an elevated lake or river and produces electricity from the potential energy of that water. Figure 9.7 shows the energy sector as existing outside the mainstream economy; in economic convention it is usually incorporated within the production sector.

---

[1]An earlier version of the material presented here was first published in Chapter 6 of the author's book, *Energy and the Ecological Economics of Sustainability*, Washington, D.C.: Island Press, 1992.

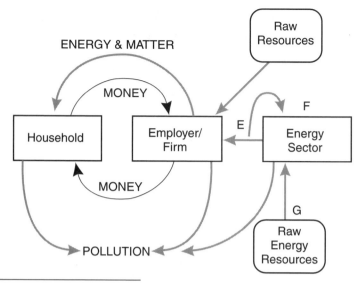

SOURCE: After Gilliland, 1977.

Figure 9.7 **Energy supply to an economy.**

One important characteristic of an energy (transformation) sector is that it uses capital, operating, maintenance, and other material input flows from the economy, represented in Figure 9.7 by the feedback flows F of energy embodied in those goods and services, to convert "raw" primary flows of energy G into "useful" consumer energy E. It is only when the flow of E is greater than F that there is a supply of **net energy** to the economy. The net energy criterion (which can take a number of forms; see Hall *et al.,* 1986) is a means of indicating the **physical accessibility** of a resource. It is the amount of effort (measured in energy units) required to extract the resource and deliver it in the appropriate form, to the consumer.

During the last century or so in most industrializing countries, economies have gone from an almost complete dependence on biomass-based fuels (mainly wood) to coal, then to oil and natural gas. Each transition was characterized by change to a more accessible energy source, but that change was not possible until technology had developed to the stage at which the greater level of accessibility was achievable. In each case, the accessibility of the new resource (measured by both net energy and economic methods) was better than that from the previous one. This meant costs were lower, so substitution could proceed rapidly, once started.

There is increasing evidence, however, that energy is becoming more difficult to obtain and process into readily usable forms. Worldwide, oil discoveries per meter of well drilled are declining, and exploration is being done in ever more inhospitable places. Many countries have substantial reserves of fossil fuel (mainly coal) but the majority is often of low quality. In other words, resources are not homogenous, and humankind has (quite reasonably) exploited the easily accessible, high-quality resources first. Lower quality, less accessible resources are left for future generations.

9.5.1 NET ENERGY OF RESOURCES OF DECLINING ACCESSIBILITY

As an example of the type of thing that could happen, consider a simple coal mine. Let us assume that the coal seam is vertical, starting at the surface. (While exaggerated to make a point, this is broadly what will happen in the long term, because resources that are closest to the surface will normally be extracted first, leaving deeper deposits until later.)

At the start of the mining operation, only simple materials and resources are needed to mine the coal and deliver it to trucks or railway wagons. As the coal face goes lower, the total work that has to be done to get the coal to the surface increases, because of the increased vertical height up which it has to be lifted. The greater the height, the heavier and more complex the lifting mechanism has to be; and the faster it has to run, just to maintain a constant rate of delivery of coal to the surface. Soon, a substantial proportion of the energy in the coal delivered to the surface will be needed, just to build, run, and maintain the machinery needed to get it out. This is quite separate from the effort expended in other activities, such as pumping water to keep the mine dry, and the precautions necessary to ensure safety at the coal face, where the working conditions are becoming ever more inhospitable.

As the depth increases, the proportion of net energy delivered to the surface declines, because more and more of the mine's output is needed, just to drive the mining machinery itself. To ensure a constant supply of net energy at the surface (as is normal for economies), the rate of mining will have to accelerate. The faster the miners try to get the coal out, and the greater the demands placed on the mine by the economy, the faster the process approaches the stage at which the whole operation may become energetically pointless. Clearly, such an end result would be stalemate, in that the mining operation does not actually do anything for the economy, other than supply employment. Beyond the point where the net energy yield is zero, the operation is in fact taking more than it supplies. *This will happen irrespective of the amount of coal left in the mine.* The resource will then be economically and energetically useless.

The lower the quality of the energy resource, the sooner this will happen. Peat or lignite will be accessible to shallower depths than bituminous coal or anthracite. The direct and indirect energy requirements of the other parts of the mining activity will ensure that, in practice, the depth that is mineable will be very much shallower than that which is calculated from the energy requirements of lifting alone.

Many people coming from the economic viewpoint rely on improvements in technology at this stage, to ensure that the economy will always get what it wants. Such a view is not supported by the physics. It is quite possible that lifting machinery and water pumps will be improved, but there is an absolute minimum energy requirement for lifting a ton of coal to the surface and pumping a liter of water that is absolutely unalterable by technology, being a consequence of the second law of thermodynamics. The most that can be hoped for, from technology, is that it will enable mining to go a little deeper; the constraint remains.

9.5.2 NEW ENERGY TECHNOLOGIES

The further question may then be asked of whether we can envisage making more radical changes in technology. For example, would underground gasification enable the coal resource to be made available with a lower expenditure of feedback energy? The answer is not clear, because of so many unknowns. However, we know that such methods are unsuitable for many coal seams (e.g., those with geological faulting), and also incur large increases in depletion rate, waste, and pollution, due to the low efficiency with which the coal is converted into gas. The capital structures needed to carry out such operations will also be substantial.

The "ultimate" unlimited form of energy is believed to be nuclear fusion. In this, deuterium, a naturally occurring isotope of hydrogen, would serve as fuel for processes that duplicate the reactions that occur in the interior of the sun — at temperatures of millions of degrees. Although claims of breakthroughs in this area have been with us since I left school (more than 40 years ago), fission reactor programs have so far produced nothing more than the ability to absorb vast amounts of money and energy, in construction and operation of ever more elaborate experimental machines. Whether they will ever show an economic or an energetic benefit nobody knows, but even if they do, the results will only be available to a few rich nations; fusion power will almost certainly be too expensive (and too dangerous) for anybody else.

9.5.3 ENERGY AND SYSTEMS

It is a basic tenet of neoclassical economics that substitution is always possible, which effectively makes scarcity only relative. But from biophysical systems reasoning, substitution of a resource by another of lower accessibility is not one in which the two resources can be considered separately — they exist in a systems relationship, in which the one affects the other. The increased energy requirements (from current resources) of investment in access to new energy resources give strong indications that the cost of delivered energy will rise at an accelerating rate in the future, and that substitution cannot be taken for granted.

The universality of the process of consumption of high-quality available energy and its rejection as low-quality unavailable energy by physical systems makes energy *one* logical means of studying the structure and dynamic behavior of economic systems. It also provides *one* means of combining concepts from a range of disciplines, and thus of assisting in the development of a more unified understanding of the behavior of complex physical systems.

Thermodynamics is concerned with the properties of systems, not of things on their own. In energy analysis, the expenditure of energy and time to do work is a linking mechanism between the natural environment system and the human socioeconomic system. This link may enable us to gain a more unified, more complete understanding of the combined environment–economy system.

9.5.4 AIMS OF THE ENERGY SYSTEMS WORLD VIEW

In the above discussion, it is clear that a perspective gained from looking at the economy as a system, in which primary resources are physically transformed into goods and services, has little to contribute to a discussion of how people are likely to react in social situations. That is because it is based on the physical world, not the social. Most physical scientists would agree that this perspective is basically applicable to the physical constraints and long-term *limits* to social activities.

The picture that comes from the energy systems view is that (1) the long-term economic costs of energy resource development are likely to be much higher than currently believed; and (2) that these costs are likely to increase at an accelerating rate over time. If these insights alone were to be incorporated into resource economics and the treatment of subjects such as depletion, then I believe the economics of sustainable alternatives such as energy demand management ("energy conservation") and renewable energy sources (the family of solar options) would be seen to be much more favorable. The approach makes it clear that there are strong indications of limits to growth of economies in any sense that involves the production and consumption of material goods and services.

While the immediate aims of the energy systems perception are modest, the outcomes of acceptance of its findings would be far reaching, in policy discussions on the use of energy resources in society.

Further Information

A valuable collection of papers in energy analysis and related fields is available in Sergio Ulgiati (Ed.), "Proceedings of the International Workshop, Advances in Energy Studies: Energy Flows in Ecology and Economy," Porto Venere, Italy, 26/30 May 1998, published by MUSIS, Museum of Science and Scientific Information, Rome. (S. Ulgiati, Department of Chemistry, University of Siena, Pian dei Mantellini, 44-53100 Siena, Italy. Email ulgiati@unisi.it)

Coal mining is one of the most hazardous methods of exploiting a natural resource. Efforts to recover land disrupted by mining are increasingly, and successfully, pursued by mining corporations, but the image of men working deep underground for low pay remains in the forefront of the public's consciousness of coal. PHOTOGRAPH COURTESY OF CORBIS.

CHAPTER 10

Coal

10.1 History of Coal Use

After decades of declining production and increasing disfavor, coal, the most abundant fossil energy resource in the world, has been making a strong comeback. It is one of the ironies and dilemmas of our environmentally aware age that in the future, we may use more, not less, of this relatively heavily polluting fuel.

Coal, the fossil fuel extensively used by man, was initially used by the Chinese at the time of Marco Polo or earlier. The use of coal as a major source of energy began in England in the 12th century when pieces of black rock called "sea coles" were discovered to be combustible. Eventually, it was deduced that these rocks could be dug from strata of rock along the cliffs in England and then from holes sunk to the strata. In 1234 King Henry III granted Newcastle-upon-Tyne the right to mine coal. By the late 13th century, coal smoke was already a source of pollution in London. Coal was used as a domestic fuel, as a fuel for lime burning, and by blacksmiths and for other metallurgical processes. By 1658 the production of coal at Newcastle had reached 529,000 tons per year. By 1750 the annual production in England reached 7 million tons.

Coal in many ways was the fuel of the Industrial Revolution. Coal was used during the Industrial Revolution for metallurgical processes, glassmaking, fuel for railroads, and, in general, for the steam engine. By 1860, world production of coal reached 150 million metric tons. From 1860 to 1910, world annual production of coal grew from 150 to 1,100 million metric tons, at an annual growth rate of 4.4%.

During the period from 1910 to 1940, world coal production grew at the relatively low rate of 0.75% per year. However, after 1940, the growth rate of world coal production again rose to 3.6%. World coal production for several time periods is given in Table 10.1.

Coal is actually a family name for a variety of solid organic fuels. The origin of coal was plants, which were accumulated in a bog and became a soggy mass of plant debris we call peat. When peat was compressed and burned more than 300 million years ago, it became lignite. Successive invasions of the sea, piling up layer upon layer of material, resulted in the deep burial of the lignite. Deep burial often results in a rise in temperature and an expelling of the moisture, and thus lignite became bituminous coal. In some areas the layers of coal were subjected to large compressive forces, resulting in "hard coal" or anthracite. The main constituent of coal is carbon and hydrogen, with small added quantities of sulfur, oxygen, and nitrogen.

The production of coal in the United States started about 1820, when 14 tons were reported to have been mined. From 1820 to 1900, the mining of coal increased rapidly to 280 million tons annual production, at an annual growth rate of 6.6% per year. By 1910, the annual production rate was 500 million tons per year. During World War II, the annual production of coal rose to more than 600 million tons. However, after 1947, the development of gas pipelines resulted in a shift of fuel use from coal to gas and petroleum. Therefore, by 1960, the annual production of coal had dropped to 440 million tons. By 1975, the use of coal had revived to 640 million tons. As of 2000, coal consumption in the United States has leveled off at about 800 million tons.

Coal provided the world 62% of its energy in 1910 and that percentage dropped to 23% by 2000. Nevertheless, world use of coal has increased as shown in Table 10.1. As shown in Table 10.2, many nations use coal for a high percentage of their industrial and home energy uses. The five nations that provide the largest coal production are listed in Table 10.3. Nevertheless, coal is increasingly recognized as a leading threat to human health, and one of the most environmentally disruptive human activities.

Table 10.1 **World Coal Production**

Year	Production (millions of metric tons)
1860	150
1880	320
1900	780
1920	1,200
1940	1,700
1960	2,100
1980	3,000
2000	3,500[a]

[a]Estimated.

Table 10.2 **Large Countries Using Coal**

	Percentage (%) of nations' energy use	Percentage (%) of electric energy use
China	73	75
United States	25	55
India	57	73
Russia	40	60
Poland	68	97

Table 10.3 **Nations with the Largest Coal Production**

Rank	Nation	Production (EJ)[a]
1	China	20.5
2	United States	26.0
3	Russian Federation	7.1
4	India	6.7
5	Australia	5.6

[a]Exajoules = EJ = 10^{18} J.

Coal has been linked to air pollution, smog, respiratory problems, and heart disease. Coal mining and extraction also causes safety and health problems.

More than one-half of the coal consumed in the United States is used by electric utilities to fuel the boilers to generate electricity. Coal is also important for making industrial steam and in the manufacture of steel. While coal was used as fuel for railroad engines from 1850 to 1950, locomotives now use diesel fuel or electric power.

The reserves of coal available are variously estimated at 8 to 16 trillion tons worldwide. One estimate of the world reserves is a proven reserve of 8 trillion tons (189×10^{18} Btu) and an estimated total resource of 16 trillion tons.

Coal has provided the majority of the world's electricity for many years, and will continue to do so until the green forms of energy — photovoltaic arrays, wind turbines, tidal hydroelectric plants, and fuel cells — are ready to meet all the world's energy needs. Meanwhile, the preferred course of action is to optimize every aspect of coal consumption, from mining to burning to cleanup of emissions and disposal of particulates.

There are no near-term limits on the global availability of coals. Measured readily recoverable reserves surpass 700 billion metric tons, which is about equivalent to 250 years at the rate of the 1995 extraction.

10.2 Coal and the Environment

In the future, the major markets for advanced coal power and fuel technologies will not be in the United States but in coal-intensive developing countries such as China and India, where gas is not widely available for these purposes. Providing attractive coal technologies that are much more efficient with greatly reduced CO_2 and other emissions is an important environmental objective. To take advantage of this environmental opportunity, the industrialized nations should maintain technological leadership in coal power technologies and develop a strong international program including collaborative R&D and commercialization activities. This will require a shift away from the current focus on the U.S. and European markets and toward a focus on coal-intensive developing countries.

China and India rely on coal for their industrial sectors. China is the world's largest producer and consumer of coal and has larger coal reserves than any other country. In fact, it depends on coal for 75% of its energy and uses it for everything from power generation to home cooking. Of the world's ten most polluted cities, nine are in China.

How can China meet its growing need for energy without further degrading its air and water? Over time China and India might shift toward a balanced portfolio of energy sources that include hydro, clean coal, natural gas, and wind power. Coal peaked in its use in the United States in 1920 and may peak in use in Asia in 2020. Coal accounts for about 22% of energy use in the United States in 2000. Perhaps China and India can limit the use of coal after 2020 to 50% of their energy use.

As an example of wide-ranging environmental impacts of coal, in 1999 New York State launched a lawsuit against 17 coal-burning plants in upwind states such as Ohio and Indiana. It cited the plants for pollution and failure to update their emission control systems. This lawsuit may lead to changes in coal use in the Midwest.

10.3 Coal Technology

The current system of coal use is shown in Figure 10.1. The coal is extracted from a mine and transported to a power plant where its energy is converted to steam and eventually to electricity. The electricity is then transmitted to homes, offices, and factories.

Research and development focuses on technologies that increase efficiency and reduce emissions. For new electric generation capacity, coal cannot compete with natural gas environmentally or economically at this time in the United States. Gas power technologies are less expensive and they emit far less CO_2 per unit of electricity produced than the best coal technologies. However, the cost of coal is likely to remain low, and the cost of gas may rise as demand for it increases. So, at some time in the future advanced coal technologies may be less expensive to use if CO_2 emissions can also be controlled economically, assuming control will be required. The same CO_2 requirement would pertain to gas, of course, although emissions are less intense.

An alternative, lower impact approach is to use coal gasification. The Southern California Edison (SCE) Mojave desert plant is the world's first commercial-scale integrated coal gasification combined-cycle (IGCC) plant. The IGCC design includes a new 120-MW electrical generating unit. The project's industrial sponsors viewed coal gasification as a conduit to use the world's vast coal resources in a way that would meet or surpass environmental performance requirements without add-on pollution controls. By taking advantage of rapidly improving gas turbine technology, a significant increase in conversion efficiencies could be achieved, thus reducing CO_2 emissions as well. This project provides a commercial-scale process to better understand operational dynamics, coal suitability, and environmental performance.

A clean coal-fired plant is planned for construction in Kentucky. The 400-MW plant will use IGCC. The plant will combine municipal solid waste with coal to form fuel briquettes for the gasification process. The synthesis gas will be burned in a combustion turbine to generate electricity and exhaust heat will be used to boil water to drive a steam turbine. The combination of the two types of power-generating turbines accounts for the name "combined cycle." Some of the synthesis gas will be directed to a 1.25-MW molten carbonate fuel cell in the plant's power-generating section. When operations begin in 2002, electricity from the plant will be sold to East Kentucky Power Cooperative under a 20-year contract. This alternative, coal gasification process, is illustrated in Figure 10.1.

Figure 10.1 The current coal system and an alternative coal gasification system.

	Steps	Current System	Coal Gasification System	
1	Extraction	Coal Mine	Coal Mine	
2	Resource	Coal	Coal	
3	Conversion Process	Heat	Gas	Heat
		Steam	Heat	Steam
		Turbine	Turbine	Turbine
4	Conversion Output	Electricity	Electricity	Electricity
5	Distribution	Electric Grid	Electric Grid	
6	Final Energy	Electric Power	Electric Power	
7	User	Home, Industry	Home, Industry	

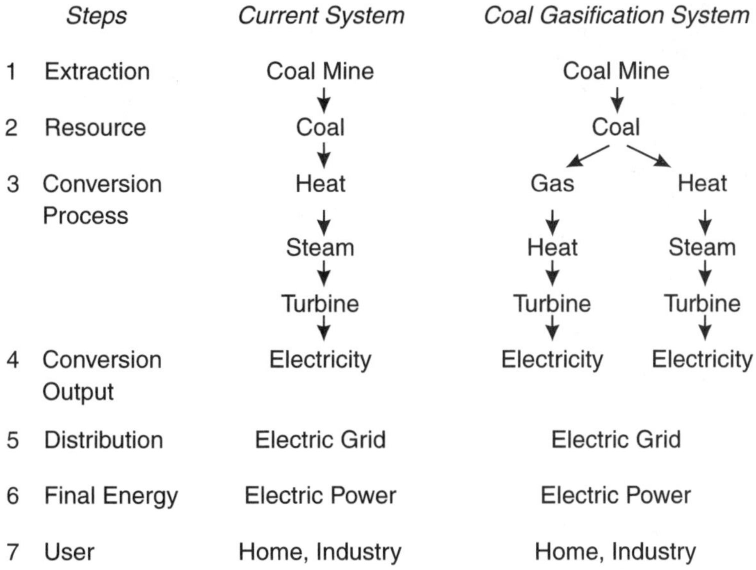

An aerial view of Texaco's UK North Sea Captain Field. Gas turbines, using the field's associated gas as fuel, power this platform. Heat-recovery units take hot exhaust from the turbines and redistribute it to the production process areas. PHOTOGRAPH BY CHRIS SANDERS. COURTESY OF TEXACO, INC.

CHAPTER 11

Petroleum and Natural Gas

11.1 Petroleum

By the 10th century, oil lamps were being used in Egypt. Crude oil was oozing out of rocks or was available by digging a shallow well in the Middle East. By the 1800s, the demand for good lighting fuel was growing and was dependent on whale oil. In August 1859, in the heart of Pennsylvania, Edwin Drake struck an oil source while drilling a well at Titusville. This oil find led to commercial oil drilling activity. By 1870, John D. Rockefeller had founded the Standard Oil Company and started to consolidate control of the oil market.

Petroleum has assumed an important and growing role as an energy source in the world. Its superior qualities and the ease with which it can be transported made petroleum the preferred fuel after 1920. The result is that the U.S., European, and Japanese economies are heavily dependent on the availability of oil. The world economy is rapidly approaching the same dependence that industrialized nations now experience.

Nevertheless, we understand that oil, like natural gas and coal, is finite in supply. The increased use of petroleum around the world has caused an accelerating problem of supply and demand and resulted in price fluctuations. The United States has become increasingly reliant on oil as a primary fuel, as well as a supply for the production of chemicals and other materials.

The annual production and consumption rates of petroleum in the United States since 1920 are shown in Table 11.1. Consumption of oil has seen an average annual growth of 5% per year. U.S. consumption grew by a factor of 16 over the 80-year period from 1920 to 2000.

Table 11.1 **Annual Consumption and Production of Petroleum in the United States**

Year	Consumption (millions of barrels)	Domestic Production (millions of barrels)	Imports as percent of consumption (%)
1920	434	443	2
1930	862	898	4
1940	1,285	1,353	5
1950	2,375	1,974	17
1960	3,611	2,575	29
1970	3,365	3,517	34
1974	5,900	3,500	40
1980	6,250	3,500	45
1990	6,500	3,400	48
2000	6,900	3,400	51

As domestic production of oil in the United States fell short of meeting the demand, imports have grown to 50% of our current use of oil, thus causing us to be even more heavily dependent on other nations.

The annual consumption of petroleum in the world is growing at a rate of 4% per year. Oil and oil products are heavily used in all industrialized nations, and other less developed nations are also becoming dependent on oil. Europe and Japan are heavily dependent on imported oil for their source of a primary energy supply. Currently, these regions and the United States consume two-thirds of the world's production of petroleum.

The annual world production of oil from 1910 to 2020 is shown in Table 11.2. The world production of petroleum, by region, is shown in Table 11.3. The nations that produce the most oil are listed in Table 11.4. The largest source of oil is the Middle East, 37.4% of world total, which includes Saudi Arabia, the nation that produces the most oil. The United States produces only about one-half of its needs.

Petroleum is the primary fuel for the transportation systems of the industrial nations, since the automobile is totally dependent on the availability of gasoline, a refinery product of petroleum. Transportation uses more than 55% of the petroleum consumed in the United States. Industry uses petroleum for heating, for diesel engines, and in chemical processes. Petroleum is a most important raw material in the chemical industry.

An oil refinery turns crude oil into gasoline, diesel oil, lubricating oil, kerosene, and jet fuel. Gasoline is the primary product of refineries (45% average) and fuel oil of various grades ranks second in importance. Fuel oil is used to heat homes and buildings and to supply power for factories and railroads. Jet fuel contains a mixture of gasoline, kerosene, and oils with low freezing points. Asphalt for roads is another product of crude oil.

Petrochemicals are chemicals made from petroleum; examples are plastics, nylon, fertilizers, and medical drugs.

Table 11.2 **Annual World Production of Oil**

Year	Production (million metric tons)	(billions of barrels)
1910	50	0.4
1950	500	4
1970	1,500	11
2000	3,500[a]	26[a]
2020	5,000[a]	36[a]

[a]Estimated.

Table 11.3 **Petroleum Production by Region in 1995**

Region	Production (EJ)[a]
Africa	14.4
Europe	25.9
North America	21.0
Central and South America	18.7
Middle East[b]	37.4
Asia and Oceania	21.2
World Total	138.5

[a]Exajoules = EJ = 10^{18} J.
[b]Middle East = Iran, Iraq, Jordan, Kuwait, Saudi Arabia, Syria, United Arab Emirates.

Table 11.4 **Nations with the Largest Oil Production in 1995**

Rank	Nation	Oil production (EJ)[a]
1	Saudi Arabia	17.8
2	United States	16.4
3	Russian Federation	12.8
4	Iran	7.7
5	China	6.3
6	Norway	5.9

[a]Exajoules = EJ = 10^{18} J.

11.2 Petroleum Resources and Prices

The availability of petroleum as a world resource and its annual production depend on several parameters, some of which relate to the geology of the earth and to the techniques for oil production, but many of which are dependent on economics, governmental regulations, material and equipment supply, and similar factors. A prediction of future oil supplies can be found in a composite assessment of several questions: (1) How much oil is there to be found? (2) How effectively and rapidly can new oil deposits be located? (3) How much of the oil that has been found or will be found can be recovered? (4) How fast can known oil be processed? The answer to every one of these questions is at best an estimate, and the uncertainties in the estimates arise from both nonphysical and physical factors.

Obtaining accurate estimates of world petroleum and natural gas resources and reserves is difficult and uncertain. Terminology used by the oil industry to classify resources and reserves has no broadly accepted standard classification. Such classifications have been a source of controversy in the international oil and gas community and confusion persists. **Recoverable resources** include discovered and undiscovered resources. **Discovered resources** are those resources that can be economically recovered.

Discovered resources include all production already out of the ground and reserves. Reserves are further broken down into proved reserves and other reserves. Therefore,

Reserves = Discovered resources − Cumulative production.

World petroleum reserves were estimated to be about one billion barrels with 60% of the reserves in the Middle East. Table 11.5 shows the oil reserve estimates for various regions of the world. At the rate of current oil production, 25 billion barrels per year, the world's oil reserves should last about 40 years.

Table 11.5 **Estimated Oil Reserves by Region in 1995**

Nation	Oil reserves (billions of barrels)
Africa	62
North America	88
Europe	89
Middle East	61.0
Asia–Pacific	58
South America	70
World Total	977

Recoverable resources, which include undiscovered resources, are estimated to exceed 1 trillion barrels. Of course, much of these recoverable resources may never be discovered or be economically or environmentally recoverable.

The price of crude oil has a strong influence on the exploration for new crude oil and thus the discovered resources. As oil prices rise people reduce their use, thus lowering demand. As oil companies increase their output, prices drop. The cost of finding a new barrel of oil is about $8. Thus, as the price of crude oil drops toward $10, the exploration effort declines sharply.

The average price of domestic crude oil in the United States in 1995 was $14.65 per barrel, as compared to $13.30 per barrel in 1960 (1995 dollars). Costs of imported oil in 1995 were between $15 and $17 per barrel. In 1981, the cost of domestic oil in the United States averaged $52 per barrel and imported oil cost between $57 and $62 per barrel (1995 dollars), about four times costlier than in 1995. In late 1999, oil prices rose to $25 from a low of $11 a year earlier.

The price of crude oil is quite volatile as demand wanes and surges and new sources come on line. In the next section we discuss the potential limited supply of oil as worldwide demand grows.

11.3 Limits to Oil Resources

The two major views of the limit of the oil supply are that (1) it is limited and very soon we will hit that limit or (2) it is essentially unlimited and we will not approach a limit in this century.

Let's start with the concept that there is a limited supply of economically recoverable crude oil. Assuming this worldwide limit is 2,500 billion barrels, a peak production rate of 36 billion barrels is reached in 2020 (see Table 11.2) after which we start to experience a decline in the annual production of oil as shown in Figure 11.1. Under this scenario the price of oil will continually climb as is also shown in Figure 11.1.

On the other hand, many assert that oil will never run out because technology is making it possible to find, produce, and refine oil so efficiently that its supply, at least for practical purposes, is basically unlimited.

First, let's start by reviewing the potential for a limit of the oil resource. Campbell and Laherrere (1998) state the case cogently by critiquing the viewpoint that oil will remain plentiful for 43 or more years:

> Unfortunately, this appraisal makes three critical errors. First, it relies on distorted estimates of reserves. A second mistake is to pretend that production will remain constant. Third and most important, conventional wisdom erroneously assumes that the last bucket of oil can be pumped from the ground just as quickly as the barrels of oil gushing from wells today.

Basically, this view is that if demand keeps rising as production tapers off, then this imbalance will result in a severe economic shock. The disagreement centers on the real reserves of oil and the economic accessibility of this oil in a timely manner. Campbell and Laherrere's view is that the peak shown in Figure 11.1 will be reached in 2010 or at least by 2020. As Campbell and Laherrere note, price rises may curb demand, thus delaying the occurrence of the peak, but the peak will simply be delayed by a few years. They call for a shift to natural gas and work on safer nuclear energy so that the oil peak will be indefinitely delayed.

One of the reasons reserves grew during the 1990s is improving exploration technology. For example, 3-D seismic technology has enabled geologists to see promising reservoirs as a cavern in the ground rather than as a line on a piece of paper. Geologists have improved seismic imaging of

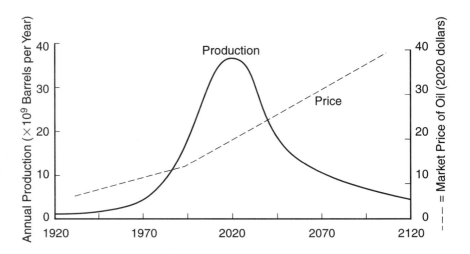

Figure 11.1 A petroleum production scenario based on an estimated value for the total worldwide oil resource of 2,500 billion barrels over time.

seafloor geology with the hope of finding vast new sources of oil. Furthermore, drilling technology has improved significantly. Wells are being drilled in water and reaching more than 8,000 feet below the surface. Since oil is expensive to locate and develop, exploration companies do not aggressively search for new fields until they start to exhaust their resources. Those who see an optimistic scenario for oil discoveries state that (Taylor, 1999):

> When undiscovered oil, efficiency improvements, and the exploitation of unconventional crude oil resources are taken into account, it is difficult not to be optimistic about the long-term prospects for oil as a viable energy source well into the future.

The substitution of one fuel for another over time has been encouraging: coal for wood, oil for coal, and recently natural gas for oil. Technology and ingenuity tend to produce new solutions. Figure 11.1 shows a peak in 2020 for oil production assuming a total oil resource of 2,500 billion barrels. If the actual total resource is 3,000 billion barrels, the peak might occur later in the century — say, 2050. The world had a cumulative production (CP) of about 900 billion barrels by 2000. If we use the simple assumption that the peak occurs at one-half the total resource (TR) and the annual use (AU) is about 30 billion barrels (bb) over the next several decades, we can calculate the years from the current year (2000) until we reach the expected peak (T_p). Then, T_p is

$$T_p = \frac{\frac{1}{2}\text{TR} - \text{CP}}{\text{AU}}.$$

If TR = 2,500 bb, CP = 900 bb, and AU = 30 bb, then

$$T_p = \frac{1,250 - 900}{30} = \frac{350}{30} = 11.3 \text{ years}$$

or the peak year is 2011. On the other hand, if TR = 3,000 bb, then

$$T_p = \frac{1500 - 900}{30} = \frac{600}{30} = 20 \text{ years}$$

and the peak year is 2020. This calculation is sensitive to the estimated total resource and the annual use. As the price of oil climbs it is probable that the annual use would decline.

In summary, it is difficult to estimate a scenario for the next 50 years since the variables are price, total resource, world annual use, and substitution of alternative fuels. One shift in fuel use is to transfer many of our uses to natural gas. In the next section, we consider natural gas.

11.4 Natural Gas

For years, natural gas has been used primarily for generating electricity and fueling kitchen stoves and some home furnaces. In the Alaskan oil fields the gas is pumped back into the ground to maintain pressure in the oil wells. In Nigeria and the Middle East, it is simply burned off (often called **flaring**). But such waste is soon to become a thing of the past. The use of natural gas will increase due to its lower environmental impacts. Figure 11.2 shows one scenario for future uses of coal, oil, and natural gas. By 2050, natural gas would supply one-third of the world's energy and oil would supply another one-third. Coal would decline to 20% of the world's energy, and hydropower and renewable energy sources and nuclear power would supply the remaining 14% of the world's total energy use.

Natural gas and gas manufactured from coal or oil have been used as fuels for lighting, cooling, and heating for more than a century. In 1880 approximately 200 billion cubic feet of fuel gas was used in the world.

The use of natural gas and gaseous fuel generated from coal developed in the United States after 1890. In 1900 the United States used approximately 200 billion cubic feet of gas as energy fuel. Today, about 22 trillion cubic feet of gas is used as fuel in the United States.

Natural gas was discovered in the United States along with petroleum and became available as petroleum wells were developed. Natural gas, which is CH_4 (methane), was not generally used at first, since it was not as easily stored or transported as liquid petroleum. However, as natural gas was recognized as a valuable fossil fuel, methods of storage and transportation began to develop.

Natural gas can be found together with oil in associated gas fields or independently from oil in unassociated gas fields. Some gas has been generated in

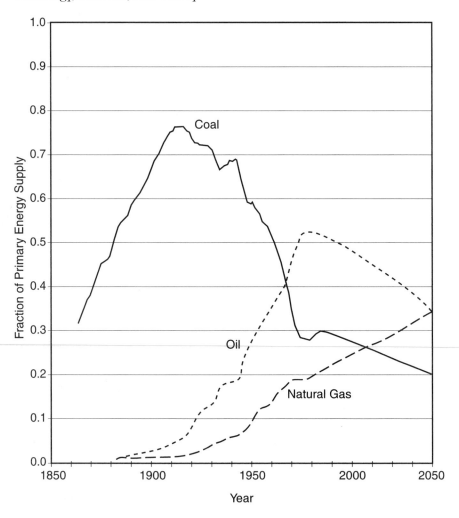

Figure 11.2 **One possible scenario with natural gas supplying one-third of the world's energy.**

the past from oil and some from coal. Gas is the cleanest (in terms of pollutants) and most flexible natural fuel, but it is more difficult to store and transport than liquid or solid fuels, especially when the distribution system must extend from one continent to another. Natural gas emits 40% less carbon dioxide than coal and 25% less than oil. It is also far more abundant than oil, at least in the United States. In associated gas fields, the production of gas is directly linked to the production of oil and its flow cannot be modified separately from that of oil. Gas that is not used can be reinjected, destroyed, or liquefied. Reinjection helps to maintain pressure in the oil field and preserves the gas for future use. However, much natural gas that is produced together with oil is still destroyed through flaring.

The six largest producers of natural gas are listed in Table 11.6. The United States and Russia have large resources of natural gas and are fully exploiting it with a well-developed pipeline distribution system.

Table 11.6 **Six Largest Producers of Natural Gas in 1995**

Rank	Nation	Production (EJ)[a]
1	United States	24.0
2	Russian Federation	22.7
3	Canada	6.1
4	United Kingdom	2.9
5	Netherlands	2.8
6	Algeria	2.4
	World Total	88.7

[a]Exajoules = EJ = 10^{18} J.

One estimate of the world's natural gas reserves is about 60 EJ or the equivalent of 900 billion barrels of oil. This natural gas, however, is not readily accessible to users and the lack of easy distribution reduces ready sources to about 30 EJ.

Because of the forces of competition loosed by deregulation and advances in the technology of finding and producing gas from ever more difficult formations, the price of natural gas at the well head is at less than \$4 per million Btu and is expected to remain at such levels for the next 20 years even with a one-third increase in consumption during that period. This shift to gas is in effect a continuation of a long-term trend to ever more efficient, less carbon-intensive fuels — part of the gradual "decarbonization" of the world energy system. Figures for the United States show the broader trends, though recent patterns are distorted somewhat by the disruption in gas markets that occurred during the 1970s. As natural gas supplies level off or are voluntarily kept in the ground in order to reduce carbon emissions, a substitute will be needed in the latter half of this century.

Natural gas is the fuel of choice for electric utilities in the United States and Canada. In the wake of the summer 1998 power shortages, utilities stepped up orders for new gas turbine power plants.

Output from such plants should triple by 2002 from 1995 levels. The increased use of gas in the residential heating and cooking market should continue as well. By the year 2020, the United States could consume nearly 30 trillion cubic feet of gas a year. The consumption of natural gas in the United States from 1920 to 2020 is given in Table 11.7. By 2020, the energy share of the U.S. economy should be about equal for oil and natural gas.

Table 11.7 **Natural Gas Consumption in the United States, 1920–2020**

Year	Consumption (trillions of cubic feet)	Consumption (EJ)[a]
1920	0.8	0.9
1940	2.7	3.0
1960	12.8	14.0
1980	20.0	21.8
2000	22.0	24.0
2020	30.0[b]	32.7[b]

[a]Exajoules = EJ = 10^{18} J; 1 cubic foot of natural gas = 1.09 MJ.
[b]Projected.

China uses natural gas for home heating and to make fertilizer. China has about 20 trillion cubic feet of gas resources. The reserves of natural gas in the United States are about 1,500 trillion cubic feet. At a consumption rate of 25 trillion cubic feet annually, the United States could exhaust its known reserves after 60 years. Of course, new technologies and better exploration and storage mechanisms could extend the reserves for at least another 40 years to 2100. Others have asserted that the reserves in the United States and Canada exceed 2,500 trillion cubic feet and that natural gas could supply a significant fraction of the U.S. requirements well into the 22nd century.

Table 11.8 lists estimates of the world's natural gas resources. Of course, much of this resource may not be discovered nor recovered.

Table 11.8 **Estimates of World Natural Gas Resources (Trillions of Cubic Feet)**

North America	2,500
Middle East	3,500
Russia	3,000
World Total	15,000

The reserves of natural gas in the world may exceed 15,000 trillion cubic feet or about 16,000 EJ. The estimated oil reserves are about 1,000 billion barrels or about 6,000 EJ. Thus, there may be about 2.5 times as much energy reserves in natural gas as in oil. Oil, however, is readily transportable as a liquid while it is difficult to gather and transport natural gas from remote, scattered locations.

11.5 Alternative Sources of Natural Gas

Because natural gas is available in many regions of the world, it is attractive to consider the transporting of this gas to consumer locations to meet the demand for this fuel. Because natural gas converted to a liquefied form occupies only about 1/600th the volume of normal gas, moving it by special insulated tank-ship is possible.

Natural gas can be liquefied by lowering its temperature until a liquid state is achieved. It can be transported in refrigerated ships. The process of using ships and providing special-handling facilities adds significantly to the final LNG cost. If oil prices stay low, prospects for LNG development will remain low in the future. However, LNG projects planned by OPEC member countries may become significant during the next 20 years with shipments of LNG exports ultimately accounting for up to 25% of all gas exports.

Many companies hope to produce clean diesel fuel from natural gas at remote locations and then ship the liquid fuel to the users. Diesel fuel derived from natural gas has no polluting sulfur or aromatics. It has a high cetane number, meaning it burns more thoroughly. It takes about 10,000 cubic feet of gas to make one barrel, or 42 gallons, of synthetic fuel. One factor spurring the recent flurry of activity is the rising price of crude oil.

It is estimated that the world has about 5,000 trillion cubic feet of stranded (remote) gas reserves, and drillers flare off about 3,700 billion cubic feet per year. For example, the United States has about 4 trillion cubic feet of gas reserves on the Alaskan North Slope and no way to get it to market.

Another possible source of liquids would be the tar sands of Canada. The presence of the Athabasca tar sands in Canada has been known of for many years. These tar sand deposits cover an area of 9,000 square miles in the province of Alberta. The tar sands are beds, or layers, of a mixture of sand, water, and bitumen. The water and bitumen form a film around each tiny grain of sand. When a handful of sand is compressed, it leaves a discernible oily stain and smell. The deposits lie under the earth's surface and are estimated to contain approximately 600 billion barrels of oil. Perhaps half of this could be eventually recovered. Only 10% of the deposits, however, have a thin enough overburden — up to 300 feet — to permit recovery by open-pit mining operations. These near-surface deposits might yield up to 20 billion barrels of oil. Obtaining oil from the deeper deposits will require *in situ* processes. Among methods proposed to liquefy the oil are controlled underground fire, steam injection, and emulsion injection. In all of these plans, the idea is to heat the reservoir and apply pressure sufficient to cause the heavy oil to migrate to drilled recovery wells.

Methane hydrates are deposits under the ocean floor made up of nodules of methane bearing ice. Underground hydrate layers are found throughout the world and may constitute 250×10^{18} cubic feet of natural gas. The goal is to work out methods for tapping these methane hydrates. This is in the experimental stage.

Power lines from nearby Hoover Dam bring electricity to Southern California. PHOTOGRAPH COURTESY OF CORBIS.

CHAPTER 12

Electric Energy

12.1 Electricity

Electricity serves as a **carrier** of energy to the user. Energy present in a fossil fuel or a nuclear fuel is converted to energy in the form of electricity in order to transport and readily distribute it to customers. By means of transmission lines, electric power is transmitted and distributed to essentially all residences, industries, and commercial buildings in the United States.

Electricity is a relatively useful energy carrier because it is readily transported with low attendant losses, and improved methods for safe handling of electricity have been developed during the past 80 years. Furthermore, methods of converting fossil fuels to electric power are well developed, economical, and reliably safe. Means of converting solar and nuclear energy to electric energy are currently in various stages of development or of proven safety. Geothermal energy, tidal energy, and wind energy may also be converted to electric energy. The kinetic energy of falling water may also be readily used to generate hydroelectric power.

The United States consumed about 3.7 billion kilowatt-hours (kWh) of electric energy in 1900, for a per capita consumption of 49 kWh per person per year. By 1974 total consumption had grown to 1,862 billion kWh, with a per-capita consumption of 8,612 kWh/year. By 1995 total consumption had increased to 3,345 billion kilowatt-hours with a per-capita consumption of 12,250 kWh (see Table 12.1). As shown in Table 12.1, China consumes about 1,000 billion kWh, but the per-capita figure is only 800 kWh.

Table 12.1 **The Eight Nations That Consumed the Most Electric Energy in 1995**

Rank	Nation	Consumption (billion kWh)[a]	Consumption per capita (kWh per person)
1	United States	3,345	12,250
2	China	1,008	800
3	Japan	990	6,730
4	Russia	860	5,900
5	Canada	537	17,900
6	Germany	535	6,500
7	France	493	8,360
8	India	415	420

[a]1 kWh = 1 kilowatt-hour = 3.6×10^6 joules.

The growth in the consumption of electric energy in the United States is shown in Table 12.2. The average annual growth rate averaged 7.0% per year during the period from 1900 to 1995. This growth rate is expected to decrease during the next 25 years as population levels off, conservation practices are developed, and more efficient use of electricity is achieved.

Table 12.2 **Electric Energy in the United States**

Year	Consumption (billion kWh)	Consumption per capita (kWh)
1900	3.7	49
1930	115.0	935
1950	329.1	2,161
1970	1,530.0	7,469
1990	3,140	12,200
2000	3,400[a]	12,200

[a]Estimated.

Ninety percent of the U.S. industrial economy gets its power as electricity in the office and in the factory. About 70% of the growth in U.S. energy demand has been met via electricity.

The largest portion (60%) of electric energy is consumed by the commercial and industrial sector. Electricity is heavily used to heat, cool, and light commercial buildings. It is also consumed in large amounts for industrial processes such as for the production of aluminum and other metallurgical processes.

The greatest use of electricity is for industrial drives (such as motors in steel mills and on assembly lines). Refrigeration is the next largest use, and includes home refrigeration.

12.2 Generation and Transmission of Electric Power

Power plants for producing electricity may use the energy of falling water, geothermal steam, steam produced from nuclear reactors, or fossil fuel boilers. Fossil fuels provide the largest part of the energy used to generate electricity, about 70% in 2000. The types of power plants and the percent of electricity generated by each type is given in Table 12.3. Because fossil fuel boilers convert only about one-third of the input energy to output electricity, the energy attributable to fossil fuel plants is relatively high.

The overall efficiency of converting a fossil fuel to energy rose slowly from 1920 to 1980 and started to level off at about 35%. By 1990 the end of thermal efficiency and scale improvements contributed to the reversal of a trend toward productivity improvements — improvements that previously made electric utilities the marvel of American industry. As can be seen in Table 12.3, coal still accounts for a large share of electric generation. Oil was used for 14% of all electric generation in 1972 but declined to 2.4% in 1997.

Table 12.3 Sources of Generation of Electric Energy in the United States in 1997

	Percent of total (%)
Coal	56.7
Natural gas	9.4
Petroleum	2.4
Hydroelectric	10.9
Geothermal	0.5
Nuclear	20.0
Wind and solar	0.1
Total	100.0%

Electric power plants are not often located at the point of end use; they are generally situated away from the metropolitan areas at dam sites or near sources of cooling water. The electric power generated is then transmitted to the user by means of transmission lines. Energy losses occur at each stage of energy movement (which are shown in Figure 10.1 for a coal-fired system). The average efficiency of each step is given in the Figure 12.1. The overall efficiency is the product of the efficiency at each step, so that overall $\eta = 0.25$.

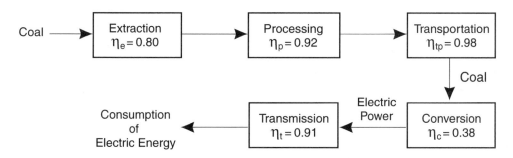

Figure 12.1 **The movement of energy through the various stages of a coal-fired electric power plant.**

12.3 Fuels Generation and Transmission of Electricity

Electricity has become an essential ingredient of business, industry, and everyday living in the developed countries, and the future trend is toward greater dependence on it. During this decade, most of the world's electricity will be generated by the combustion of fossil fuels. In the United States (the biggest single consumer of electricity worldwide), the cheapest fuel is coal, from which 57% of power was generated in 1997. Other contributors to the supply were natural gas (9%) and oil (2%). Nonfossil sources included nuclear power (20%) and hydroelectric power (11%). Other developed countries are also heavily dependent on fossil fuels for power generation. The people of less developed countries have become increasingly aware of the desirability of enhanced availability of reliable electricity. China and India, for example, have substantial reserves of coal, and have been burning them at an increasing rate to generate more electricity. One strategy for the future would be to embark on greatly expanded generation of electricity from renewable resources and from nuclear reactors. Even if methods of reducing CO_2 emissions from power plants are economically practical, many years would be required to achieve a significantly reduced level of emissions.

Generating electricity from renewable sources is receiving much interest. One of the handicaps of wind power and solar power, though, is intermittent generation. This factor can be minimized through the use of energy storage devices, but the net effect is to increase the overall cost of wind and solar power. These sources of electricity are also facing strong economic competition from very efficient equipment powered by natural gas. Those power generators can be located close to major customers, avoiding power transmission losses. Air pollution is minimal, and emissions of CO_2 per kilowatt-hour are less than those from coal-fired equipment.

The potential availability of cheaper power has led to the deregulation of electric utilities. Entrepreneurs are able to use existing transmission lines to deliver electricity to the highest bidder. About half of all U.S. electricity generation is now sold on the wholesale market before it reaches customers. Much of this power travels long distances between source and user. Growth in these activities comes at a time when many parts of the North American transmission system are already operating close to their limits. If the United States and the rest of the world are to have a sustainable and prosperous future, numerous steps must be taken. Many of them will involve changes in the means of generating and distributing electricity. New research is needed to decrease the cost of renewable power; to develop hydrogen fuel cells that generate competitive electricity; to provide safe and cheaper nuclear power; and to control the supply, transmission, and delivery of reliable electricity.

In 1950, a 200-MW plant was cheapest to build but by the early 1980s a 1,000-MW plant was the best economic choice. By 1990, the size of the optimal plant started declining as gas turbine plants became competitive. By 2000, the optimal gas turbine plant is in the range of 100 to 200 MW.

A primary objective is to increase the efficiency of electric generation. One goal is to increase the efficiency of gas-turbine generators operating in the range of 100 to 500 MW to greater than 50%. Natural gas turbines are expected to make up more than 80% of the power generation capacity to be added to the United States in the next decade.

The passage of the Energy Policy Act of 1992 requires open access to utility transmission lines and deregulation of the power industry by states. The result has been a greatly increased volume of bulk power sales — a development that stresses the reliability of today's transmission systems, which were not designed to support a wholesale electricity market of this magnitude. About one-half of all domestic generation is now sold on the wholesale market before it is delivered to customers.

The future of power delivery will include the rapidly developing area of small, affordable electricity generation and storage units often called distributed resources. With capacities in the range of 1 kW to 10 MW, these distributed units will move generation closer to the point of use, enabling improved power quality and reliability and providing the flexibility to meet a wide variety of customer and distribution system needs.

By 2005, we can expect fuel cells and small gas turbines (20 to 200 kW) to be available for the generation of distributed power.

12.4 Power Prices and Conservation

The retail price of electricity, in real terms, rose 53% from a low 5.7 cents/kWh in 1972 to 8.7 cents/kWh in 1982. Also during that period the nominal price rose even more dramatically, from 1.9 to 6.1 cents/kWh, a staggering 220%, almost tripling the price. This price increase was primarily due to the increase in oil prices between 1972 and 1982. The nominal price continued to rise after 1982, but at a much more gradual rate. Meanwhile, the real price showed a steady decline of about 30% by 1996. By the summer of 2000, electric prices started rising again as a shortage of electric power emerged.

One method of reducing the use of fossil fuels is to reduce the use of electric energy through conservation. Conservation can be achieved through the use of efficient products such as refrigerators, washers, and all electric appliances. Energy-efficient refrigeration presents a solid cost saving potential for supermarkets since refrigeration systems account for a majority of their electric costs.

Electric vehicles (EVs) represent one of the great opportunities for energy efficiency and environmental improvement in the 21st century. EVs could be over 40% more efficient for congested urban driving than conventional automobiles, and significant air quality benefits are anticipated from widespread EV use, even if today's fossil-fuel-heavy electricity generation mix is maintained.

12.5 The Future of the Electric Power Industry

Though its importance has drawn less attention, electrification is increasingly being acknowledged as a key element of sustainable development. Electrification is increasingly seen as an enabling technology that can address many of the basic needs of the world's poor. It provides a way to apply efficient, advanced (low-carbon and no-carbon) generating technologies and efficient end-use technologies that can change people's lives in many ways. Electricity-powered motors, pumps, refrigerators, and control systems for water purification and wastewater treatment can play an essential role in meeting these needs.

There appears to be an emerging consensus that the widespread deployment of distributed, off-grid, renewables-based generating technologies could be very important in accelerating the advance of electrification into most of the developing world now lacking power grids and delivery infrastructures. The technologies include photovoltaics for home and commercial facilities.

To bring electric power to all peoples of the world, power plants of varying sizes that are environmentally sound are required. For the developing countries, this could mean tripling their electricity use, on a per-capita basis, by increasing from 1,000 kWh in 2000 to 3,000 kWh per person in 2050 (Moore, 1999).

New technologies such as the Internet and information technologies may accelerate efficiency gains and reduce reliance on fossil fuels by substituting directly for fossil fuel technologies, by performing old tasks in new ways, or by providing new functional capabilities. The goal is to move from dependence on coal power plants to new, lower impact methods of using coal such as a coal power system combining electricity generation, hydrogen separation, chemical production, and CO_2 sequestration.

The use of solar photovoltaics (PV) could be beneficial, but for PV to become a major source will require approximately another 5-fold improvement in cost and performance beyond the 100-fold improvement attained during the 20 years ending at 2000.

The renewable technologies of wind turbines and biomass fuels are already deployed around the world at about 10 times the currently installed PV capacity, and both have significant potential for further contributions to global electrification. Breakthroughs in low-cost, practical electricity storage technologies could greatly increase the value of renewable resources for dispatchable electricity generation. Perhaps the greatest opportunity for distributed power will be in bringing electricity to rural regions of the developing world.

During the 2000s, research in technologies such as fuel cells and biomass gasification could help broaden the array of available energy sources for distributed power applications, both stationary and mobile. Fuel cells are an important distributed technology, with half a dozen types under development for powering vehicles or for use as stationary generators.

Table 12.4 **Carbon Intensity of Fossil Fuels and Wood**

Fuel	Intensity (tons of carbon per ton of oil equivalent)
Wood	1.25
Coal	1.08
Oil	0.84
Natural gas	0.64

To lower the impact of CO_2 emissions, coal and oil will decline in use as a shift toward natural gas occurs. Table 12.4 provides the carbon intensity of wood and fossil fuels. A goal during the next 20 years would be to move toward an electricity-hydrogen economy with globally abundant, clean, low-cost energy for use in fuel-cell-powered vehicles and in distributed generating units in homes and businesses.

To achieve an overall low-impact economy, we will need to increase the rate of productivity and electric energy efficiency while reducing emissions and water consumption for the electric energy system.

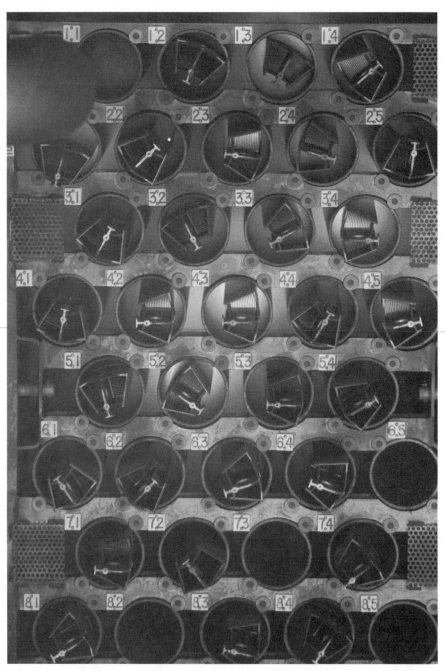

Nuclear reactor fuel rods, containing uranium oxide pellets. When the rods are assembled, the critical mass necessary for a chain reaction is reached. Safe disposal of radioactive spent fuel rods remains one of the controversial issues surrounding nuclear power plants. PHOTOGRAPH COURTESY OF CORBIS.

CHAPTER 13

Nuclear Power

13.1 Nuclear Energy Technology

Two distinct technological processes involving the nuclei of atoms can be harnessed, in principle, for energy production. **Fission** is the splitting of a nucleus and **fusion** is the joining together of two nuclei. For any given mass or volume of fuel, nuclear processes generate more energy than can be produced through any other fuel-based approach. Another attractive feature of these energy-producing reactions is that they do not produce greenhouse gases or other forms of air pollution directly. Nuclear fission is a mature though controversial energy technology where electricity is generated from the energy released when heavy nuclei break apart. In the case of nuclear fusion, much work remains in the effort to sustain the fusion reactions and then to design and build practical fusion power plants. Fusion's fuel is abundant, namely, light atoms such as the isotopes of hydrogen. The most optimistic timetable for fusion development is initial commercial plants by about 2050, because of the extraordinary scientific and engineering challenges involved. However, fusion's benefits are so globally attractive that fusion R&D is an important component of today's effort.

13.2 Nuclear Fission Power

The development of nuclear fission power dates back to the 1940s and the achievement of the first nuclear explosion at Alamogordo in New Mexico in July 1945. After World War II, the United States formed the Atomic Energy Commission (AEC) in 1946 to oversee the development of nuclear power reactors as well as nuclear weapons. The successor to the AEC is the Department of Energy.

Nuclear energy can be generated by fission (the splitting of a nucleus). Fission reactions do not generate greenhouse gases or the air pollutants that produce urban smogs and regional acid precipitation. Fission power, as of 1999, provides about 17% of the world's electric power with 429 nuclear power reactors operating in 30 countries and 26 more plants under construction.

A nuclear reactor power plant essentially consists of a uranium fuel core, a converter loop, a turbine-generator, and a condenser. The critical size and arrangement of the reactor depends on several factors, including (1) the geometry of the fuel elements, (2) the purity of the fissionable material, (3) the converter for extracting the energy from the reaction, and (4) the control mechanism.

Most fission reactors include the following components: the core, a coolant, control rods, and a moderator. The core is made up of bundles of fuel rods, which contain uranium oxide pellets. When a number of bundles of rods are assembled, a critical mass is reached and the chain reaction starts. Individual fuel rods do not contain sufficient fuel for a critical mass.

The coolant, either gas or liquid, flows over the fuel core, removing heat from the fission process. The coolant does not come in contact with the actual fuel, since the radioactive material itself is sealed within the fuel rods.

The control rods are made of material that readily absorbs neutrons. These rods are usually cylinders or sheets of metal (boron steel or cadmium), positioned inside the fuel assembly. If the rods are pulled out of the bundle, more neutrons are available to cause fissioning of the fuel, so the rate of reaction increases. If the rods are inserted into the fuel bundle, they act as a neutron sponge, so that fewer neutrons are available to the fuel. Thus the chain reaction slows or may be stopped completely. This makes it possible to produce heat at a desired rate and to shut down the reactor completely.

The first nuclear power plant went into operation in 1957 at Shippingport, Pennsylvania. The Shippingport plant is a 90-MW unit. By the end of 1975, the United States had 58 operating nuclear power plants, with a combined capacity of 39,900 MW, which was about 9% of the total U.S. electric generating capacity in 1975. By 1998, nuclear electric plants had about a 100-GW capacity.

Nuclear fission electric power plants cause issues to arise regarding operating hazards, particularly the chances of a serious reactor accident, the difficulties of safeguarding the fissionable materials used as reactor fuels, and the still unresolved problem of long-term storage for radioactive wastes. The possibility of technological failures, earthquakes, and other unforeseen natural disasters, and of human actions ranging from carelessness to deliberate sabotage, appears to be particularly significant with nuclear power systems. Because of the consequences to human health and to the environment of any large release of radioactive substances, nuclear fission is viewed by many people as a hazardous process for providing electric power. Extremely safe and reliable systems for control of the reactor are required in order to ensure public acceptance of nuclear fission reactors.

Nevertheless, nuclear fission has substantial advantages over traditional sources of energy. Nuclear plants do not emit particulates or sulfur oxides, as do fossil fuel plants. The fuel requires less mining and thus results in less disruption of the environment. If fossil plants were used to produce the amount of electricity generated by 450 nuclear plants, more than an additional 300 million metric tons of carbon would be emitted each year. Nuclear power can play a significant role in mitigating climate change. A nuclear expansion must be performed under very high safety standards. Additionally, capital cost reductions from advanced designs and production methods will be required (Sailor, 2000).

Some countries depend on nuclear power. In 1996, nuclear power provided 77% of the electricity in France, 33% in Japan, 26% in the United Kingdom, and 20% in the United States. The United States has the largest number of operating nuclear reactors (109) and the largest nuclear capacity (about 100,000 MW) of any nation. Nuclear fission power is a widely used technology with the potential for further growth, particularly in Asia.

Several problems with the expansion of nuclear power include disposal of spent nuclear fuel, concerns about nuclear weapons proliferation, concerns about the safe operation of plants, and uncompetitive economics. But given the projected growth in global energy demand as developing nations industrialize, and given the desirability of stabilizing and reducing greenhouse gas emissions, it is important to reconsider fission energy as a widely viable and expandable option if this is at all possible. A properly focused R&D effort to address the problems of nuclear fission power — economics, safety, waste, proliferation — may be useful.

A significant event occurred in April 1986 when operators of a nuclear reactor in Chernobyl, in the Ukraine, lost control of the chain reaction. The reactor itself was consumed in a partial nuclear meltdown, and clouds of radioactive emissions spewed forth and were carried by winds across vast stretches of the European continent.

Although the United States has the largest number of operating reactors of any country in the world, the outlook is that no new nuclear plant will be built in this country in the next 10 to 20 years. The decline of nuclear power in the United States has resulted from many factors: a drop in annual electricity consumption growth rates, low natural gas prices, improved efficiency of gas-fired combined-cycle plants, escalation of nuclear plant construction costs, the unresolved problems of waste disposal and storage, and concerns about proliferation and safety. These factors, combined with the current trend toward deregulation of the electric utilities, may lead to early shutdown of some operating nuclear plants in the United States.

The current market for new nuclear reactors is primarily in Asia, where developing economies are buying and installing diversified electric generation capacity.

13.3 Fusion Energy

Whereas nuclear fission gets its energy from large nuclei as they break up, thermonuclear fusion reactions yield energy when two light nuclei fuse together to form a heavier one. It is fusion that powers the sun and the stars, and the challenge is to establish a workable fusion reactor in practice.

Two light nuclei are brought together to yield energy upon fusion within a high-temperature gas. The very hot gas at a relatively low density is known as a **plasma**. Within the plasma are free electrons and ions that flow like a current. Fusion power requires much additional work in the quest to make the fusion reaction self-sustaining and to design and build practical fusion power plants; as mentioned earlier, the most optimistic timetable for fusion to reach commercialization is another half century. But the potential benefits of fusion are so large that fusion R&D is an important component of current energy R&D portfolios in the United States and internationally.

In total, the United States, through the Department of Energy and its predecessors, had invested $14.7 billion (1997 dollars) in fusion science and technology through fiscal year 1997. Results and techniques from fusion plasma science have had fundamental and pervasive impact for many other scientific fields, and they have made substantial contributions to industry and manufacturing. Since 1970, fusion power achieved in experiments has increased from less than 0.1 W to 12 MW. Recent experiments may be approaching the break-even threshold, where the amount of fusion power produced exceeds the power used to heat and confine the plasma.

The fusion reactor, if achieved, has many advantages. The first of these is the ready and inexpensive availability of fuel. The fusion reactor is expected to present fewer environmental hazards than do the fission reactors. No long-lived heavy-element isotopes are produced, and there is no production or handling of the biologically dangerous and radioactive plutonium. There is no need to transport radioactive fuel elements, and all processing would be done on site. Also, there is probably less danger of accident. Estimates for the efficiency of the reactor range from 40 to 50%.

Practical fusion reactors are not yet available because the physics of containing and heating plasmas to thermonuclear conditions in a controlled manner has proved extraordinarily difficult. Large-scale fusion experiments, however, have been conducted in various countries, and the necessary conditions of plasma temperature and heat insulation have been largely achieved, suggesting that fusion energy for electric power production is now a reasonable possibility.

From a practical viewpoint, the initiation of nuclear fusion in a hot plasma is but the first step in a whole sequence of steps required to convert fusion energy to electricity. The intervening steps in this process would transform the fusion energy into heat at conditions appropriate, for example, to generate steam that can be employed in a Carnot cycle to drive a turbine and produce electricity.

In order for fusion to become commercially viable, certain breakthroughs are necessary, particularly in plasma chemistry and physics, development of adequate containment fields, particle beam accelerators, inertial confinement techniques, triggering devices (such as lasers), and refinement of raw materials. But these obstacles are not insurmountable. Time must also be allowed for regulatory approval processes, construction, and other start-up measures. The launch of nuclear fusion facilities may begin as early as 2050, with a significant number being online by 2100.

13.4 Nuclear Power and the Large Environment[1]

By David Bodansky
Professor of Physics, Emeritus
Department of Physics
University of Washington

13.4.1 INTRODUCTION

The development of nuclear energy has come to a near halt in the United States and in much of the rest of the world. The construction of new U.S. reactors has ended and although there has been a rise in nuclear electricity generation in the past decade, due to better performance of existing reactors, a future decline appears inevitable as individual reactors reach the end of their economically useful lives.

An obstacle to nuclear power is the publicly perceived environmental risk. During this development hiatus, it is useful to step back and take a look at nuclear-related risks in a broad perspective. For this purpose, we categorize these risks as follows:

- *Confined risks.* These are risks that can be quantitatively analyzed, and for which the likelihood and scale of possible damage can be made relatively small.

---

[1]This section is reprinted with permission from *Physics and Society,* Vol. 29, No. 1, January 2000, pp. 4–6 (with some updates and editorial changes).

- *Open-ended risks.* These are risks that cannot be well quantified by present analyses, but which involve major dangers on a global scale.

As discussed below, public concern has focused on risks in the confined category, particularly reactor safety and waste disposal. This has diverted attention from more threatening, open-ended risks of nuclear weapons proliferation, global climate change, and potential scarcity of energy in a world with a growing population. The rationale for this categorization and the connection between nuclear power and these open-ended risks are discussed below.

13.4.2 CONFINED RISKS

Nuclear Reactor Accidents

The belief that reactor accident risks are small is based on detailed analyses of reactor design and performance, and is supported by the past safety record of nuclear reactors, excluding the accident at Chernobyl in 1986. Defects in the design and operation of the Chernobyl reactor were so egregious that the Chernobyl experience has virtually no implications for present reactors outside the former Soviet Union. Chernobyl is a reminder, however, of the need for careful, error-resistant design if there is to be a large expansion of nuclear power in many countries.

At the end of 1998 there had been over 8,000 reactor-years of operation outside the former Soviet Union, including about 2,350 in the United States. Only one accident, that at Three Mile Island (TMI), has marred an otherwise excellent safety record. Even at TMI, although the reactor core was severely damaged, there was very little release of radioactivity to the environment outside the reactor containment. Subsequently, U.S. reactors have been retrofitted to achieve improved safety and, with improved equipment and greater attention to careful procedures, their operation has become steadily more reliable.

A next generation of reactors can be even safer, either through a series of relatively small evolutionary steps that build directly on past experience or through more radical changes that place greater reliance on passive safety features — such as cooling water systems that are directly triggered by pressure changes (not electrical signals) and that rely on gravity (not pumps). It would in fact be remarkable if the accumulated past experience, both good and bad, did not improve the next generation.

Nuclear Waste Disposal

The second dominant public concern is over nuclear wastes. Current plans are to dispose of spent fuel directly, without reprocessing, keeping it in solid form. Confinement of the spent fuel is predicated on its small volume, the ruggedness of the planned containers, the slowness of water movement to and from a site such as Yucca Mountain, Nevada, and the continual decrease in the inventory of radionuclides through radioactive decay.

Innumerable studies have been made to determine the degree to which the radionuclides will remain confined. One way to judge the risks is to examine these studies as well as independent reviews. An alternate perspective on the scale of the problem can be gained by considering the protective standards that have been proposed for Yucca Mountain.

Proposed standards were put forth in preliminary form by the Environmental Protection Agency (EPA) in 1985. These set limits on the release of individual radionuclides from the repository, such that the attributable total cancer fatalities over 10,000 years would total less than 1000. This target was thought to be achievable when the only pathways considered for the movement of radionuclides from the repository were by water. However, the development of the site was put in jeopardy when it was later recognized that escaping ^{14}C could reach the "accessible environment" relatively quickly in the form of gaseous carbon dioxide. A release over several centuries of the entire ^{14}C inventory at Yucca Mountain would increase the worldwide atmospheric concentration of ^{14}C by about 0.1%, corresponding to an annual average dose of about 0.001 mrem/year for hundreds of years. The resulting collective dose to 10 billion people could be sufficient to lead to more than 1,000 calculated deaths.

It is startling that ^{14}C might have been the showstopper for Yucca Mountain. It appeared that this could occur, until Congress took the authority to set Yucca Mountain standards away from the EPA pending future recommendations from a panel to be established by the National Academy of Sciences (NAS). The panel issued its report in 1995. It recommended that the period of concern extend to up to one million years and that the key criterion be the average risk to members of a "critical group" (probably numbering less than 100), representing the individuals at highest risk from potentially contaminated drinking water. It was recommended that the calculated average risk of fatal cancer be limited to 10^{-6} or 10^{-5} per person per year. According to the estimates now used by federal agencies to relate dose to risk, this range corresponds to between 2 mrem/year and 20 mrem/year.

Taking the NAS panel recommendations into consideration, but not fully accepting them, the EPA in August 1999 proposed a standard whose essential stipulation is that for the next 10,000 years the dose to the *maximally* exposed future individual is not to exceed 15 mrem/year. This may be compared to the dose of roughly 300 mrem/year now received by the *average* person in the United States from natural radiation, including indoor radon.

Attention to future dangers at the levels represented by any of these three standards can be contrasted to our neglect of much more serious future problems, to say nothing of the manner in which we accept larger tolls today from accidents, pollution, and violent natural events. While we have responsibilities to future generations, the focus should be on avoiding potential disasters, not on guarding people thousands of years hence from insults that are small compared to those that are routine today.

Fuel Cycle Risks

Risks from accidents in the remainder of the fuel cycle, which includes mining, fuel production, and waste transportation, have not attracted as much attention as those for reactor accidents and waste disposal, in part because they manifestly fall into the confined-risk category. Thus, the September 1999 accident at the Tokaimura fuel preparation facility resulted in the exposure of many of the workers, including two eventually fatal exposures. It involved an inexcusable level of ignorance and carelessness and may prove a serious setback to nuclear power in Japan and elsewhere. However, the effects were at a level of harm that is otherwise barely noticed in a world that is accustomed to coal mine accidents, oil rig accidents, and gas explosions. The degree of attention given the accident is a measure of the uniquely strict demands placed on the nuclear industry.

13.4.3 OPEN-ENDED RISKS

Nuclear Weapons Proliferation

The first of the open-ended risks to be considered is that of nuclear weapons proliferation. A commercial nuclear power program might increase this threat in two ways:

- A country that opts for nuclear weapons will have a head start if it has the people, facilities, and equipment gained from using nuclear power to generate electricity. This concern can explain the U.S. opposition to Russian efforts to help Iran build two nuclear power reactors.

- A terrorist group might attempt the theft of plutonium from the civilian fuel cycle. Without reprocessing, however, the spent fuel is so highly radioactive that it would be very difficult for any subnational group to extract the plutonium even if the theft could be accomplished.

To date, the potential case of Iran aside, commercial nuclear power has played little if any role in nuclear weapons proliferation. The long-recognized nuclear weapons states — the United States, the Soviet Union, the United Kingdom, France, and China — each had nuclear weapons before they had electricity from nuclear power. India's weapons program was initially based on plutonium from research reactors and Pakistan's on enriched uranium. The three other countries that currently have nuclear weapons, or are most suspected of recently attempting to gain them, have no civilian nuclear power whatsoever: Israel, Iraq, and North Korea.

On the other side of the coin, the threat of future wars may be diminished if the world is less critically dependent on oil. Competition over oil resources was an important factor in Japan's entry into World War II and in the U.S. military response to Iraq's invasion of Kuwait. Nuclear energy can contribute to reducing the urgency of such competition, albeit without eliminating it. A more direct hope lies in stringent control and monitoring of nuclear programs, such as those attempted by the International Atomic Energy Agency. The United States' voice in the planning of future reactors and fuel cycles and in the shaping of the international nuclear regulatory regime is likely to be stronger if the United States remains a leading player in the development of civilian nuclear power.

In any event, the relinquishment of nuclear power by the United States would not inhibit potential proliferation unless we succeeded in stimulating a broad international taboo against all things nuclear. A comprehensive nuclear taboo is highly unlikely, given the heavy dependence of France, Japan, and others on nuclear power, the importance of radionuclides in medical procedures, and the wide diffusion of nuclear knowledge — to say nothing of the unwillingness of the nuclear weapons states to abandon their own nuclear weapons.

Global Climate Change

The prospect of global climate change arises largely from the increase in the atmospheric concentration of carbon dioxide that is caused by the combustion of fossil fuels. While the extent of the eventual damage is in dispute, there are authoritative predictions of adverse effects impacting many millions of people due to changes in temperature, rainfall, and sea level. Most governments profess to take these dangers seriously, as do most atmospheric scientists. Under the Kyoto agreements, the United States committed itself to bringing carbon dioxide and other greenhouse gas emissions in the year 2010 to an overall level that is 7% lower than the 1990 level. Given the 10% increase from 1990 to 1998, this will be a very difficult target to achieve.

Nuclear power is not the only means for reducing CO_2 emissions. Conservation can reduce energy use, and renewable energy or fusion could in principle replace fossil fuels. However, the practicality of the necessary enormous expansion of the most promising forms of renewable energy, namely, wind and photovoltaic power, has not been firmly established. Additionally, we cannot anticipate the full range of resulting impacts. Fusion is even more speculative, as is the possibility of large-scale carbon sequestration. If restraining the growth of CO_2 in the atmosphere warrants a high priority, it is important to take advantage of the contribution that nuclear power can make — a contribution clearly illustrated by French reliance on nuclear power.

Global Population Growth and Energy Limits

The third of the open-ended risks to be considered is the problem of providing sufficient energy for a world population that is growing in numbers and in economic aspirations. The world population was 2.5 billion in 1950, has risen to about 6 billion in 1999, and seems headed to some 10 billion in the next century. This growth will progress in the face of eventual shortages of oil, later of gas, and still later of coal.

The broad problem of resource limitations and rising population is sometimes couched in terms of the "carrying capacity" of the Earth or, alternatively, as the question posed by the title of the 1995 book by Joel Cohen: *How Many People Can the Earth Support?* As summarized in a broad review by Cohen, recent estimates of this number range from under 2 billion to well over 20 billion, centering around a value of 10 billion.

The limits on world population include material constraints as well as constraints based on ecological, aesthetic, or philosophical considerations. Perhaps because they are the easiest to put in "objective terms," most of the stated rationales for a given carrying capacity are based on material constraints, especially on food supply, which in turn depends on arable land area, energy, and water.

Carrying capacity estimates made directly in terms of energy, in papers by Pimentel *et al.* (1994) and by Daily *et al.* (1994), are particularly interesting in the present context as illustrations of the possible implications of a restricted energy supply. Each group concludes that an acceptable sustainable long-term limit to global population is under 2 billion, a much lower limit than given in most other estimates. They both envisage a world in which solar energy is the only sustainable energy source. For example, in the Pimentel paper the authors conclude that a maximum of 35 quad of primary solar energy could be captured each year in the United States, which, at one-half the present average per-capita U.S. energy consumption rate, would suffice for a population of 200 million. For the world as a whole, the total available energy would be about 200 quads, which Pimentel *et al.* conclude means that "1 to 2 billion people could be supported living in relative prosperity."

One can quarrel with the details of this argument, including the maximum assumed for solar power, but it dramatically illustrates the magnitude of the stakes, and the centrality of energy considerations.

13.4.4 CONCLUSIONS

If a serious discussion of the role of nuclear power in the nation's and world's energy future is to resume, it should focus on the crucial issues. Of course, it is important to maintain the excellent safety record of nuclear reactors, to avoid further Tokaimuras, and to develop secure nuclear waste repositories. But here — considering probabilities and magnitudes together — the dangers are of a considerably smaller magnitude than those from nuclear weapons, from climate change, and from a mismatch between world population and energy supply.

The most dramatic of the dangers are those from nuclear weapons. However, as discussed above, the implications for nuclear power are ambiguous. For the other major areas, the picture is much clearer. Nuclear power can help reduce the severity of predicted climate changes and can help ease the energy pressures that will arise as fossil fuel supplies shrink and world population grows. Given the seriousness of the possible consequences of a failure to address these matters effectively, it is an imprudent gamble to let nuclear power atrophy in the hope that conservation and renewable energy, supplemented perhaps by fusion, will suffice.

It is therefore important to strengthen the foundations on which a nuclear expansion can be based, so that the expansion can proceed in an orderly manner — if and when it is recognized as necessary. Toward this end, the federal government should increase support for academic and industrial research on nuclear reactors and on the nuclear fuel cycle, adopt reasonable standards for waste disposal at Yucca Mountain, and encourage the construction of prototypes of the next generation of reactors for use here and abroad. Little of this can be done without a change in public attitudes toward nuclear power. Such a change might be forcibly stimulated by a crisis in energy supply. It could also occur if a maverick environmental movement were to take hold, driven by the conclusion that the risks of using nuclear power are less than those of trying to get by without it.

Upstream arch barrels at the Mountain Dell Dam, Salt Lake City. John S. Eastwood developed these inclined, reinforced concrete barrel vaults, which transfer the weight of the impounded water to the ground. PHOTOGRAPH BY JACK E. BOUCHER. COURTESY OF THE LIBRARY OF CONGRESS, PRINTS AND PHOTOGRAPHS DIVISION.

CHAPTER 14

Renewable Energy

14.1 Renewable Energy Technology

Renewable energy is that which is regenerated at the same rate as it is used. **Renewable energy technologies** (RETs) can provide electricity, fuels for transport, heat and light for buildings, and power and process heat for industry. These technologies generally have little or no emissions of greenhouse gases, air pollutants, or other environmental impacts. RETs can also offset imports of foreign oil and offer important economic benefits; for example, growing biomass energy crops on excess agricultural lands would increase farm income while potentially allowing a reduction in federal farm income support programs.

Properly managed, these technologies generally have very little environmental impact, with little or no emissions of greenhouse gases (GHG) or air pollutants, water contaminants, or solid wastes. Through the use of these technologies, the risk of global warming, the most difficult environmental challenge is reduced, many of the regulatory controls on air emissions that are in place today become irrelevant, and health is improved. The inherent cleanliness of these technologies minimizes decommissioning costs and virtually eliminates long-term liability for possible environmental or health damages. RETs can also offset imports of foreign oil and offer important direct economic benefits.

Renewable energy resources include biomass, geothermal energy, hydropower, ocean energy, solar energy, and wind energy. Each of these has unique characteristics that require different approaches to R&D and system integration. The specific characteristics of RETs are listed in Table 14.1.

Table 14.1 Characteristics of Renewable Technologies

- Site specificity
- Variable availability
- Diffuse energy flow
- Low fuel costs

Most RETs vary by region and site. For example, they vary by how strong the sun shines or from which direction the wind blows. At most locations, however, there are one or more high-quality resources available. Ascertaining the optimal mix requires careful regional and site-specific evaluations of the resources over long periods as well as some degree of matching the system to the site. In some cases, relatively long-distance transport or transmission is required to get the energy generated at the best resource sites to where the people actually are.

Renewable energy resources often vary in their availability. For example, geothermal and biomass energy are available on demand, while solar energy varies with the time of day and degree of cloud cover. Thus, careful integration of intermittent resources like the sun and wind with other energy supplies or energy storage is needed to provide power when people need it.

Most of these resources are diffuse, requiring large areas for energy collection, and concentration or upgrading to provide useful energy services. This increases up-front capital costs and encourages strategies to control costs, for example, by integrating systems into building roofs, walls, or windows. The diffuseness of the resource often leads to energy conversion at capacities much smaller than for conventional energy and to modular system designs. While such systems are not well suited for exploiting economies of scale in capacity, they are well suited for factory mass production, which may permit rapid reduction in costs with cumulative production experience. For such modular technologies a rapid rate of incremental improvement is more easily achieved as experience grows than with large-scale technologies.

Many RETs involve collecting natural flows of energy. Once the capital investment in the collection system is made, there are no recurring fuel costs. In effect, these systems pay upfront for energy collected over the lifetime of the system. This eliminates the risk of fuel cost increases but raises the upfront capital cost and risk if the system does not perform as expected.

The basic RETs — biomass, geothermal, hydropower, ocean solar, and wind energy — are described in Table 14.2.

Table 14.2 **Renewable Energy Sources**

- **Biomass** includes the full range of organic plant materials, such as trees, grasses, and even aquatic plants. It can be burned to produce heat or converted into liquid or gaseous fuels. The heat may be used to produce electricity.
- **Geothermal energy** is the accessible thermal energy or heat content of the earth's crust. It can be used to produce electricity, process heat, or to heat/cool buildings. Geothermal resources can be utilized locally.
- **Hydropower** is the energy drawn from water falling or flowing downhill.
- **Ocean energy** resources include energy conversion processes utilizing the temperature difference between surface and deep waters, recovery of potential energy from the rise and fall of the tides, and the recovery of kinetic energy from wave motion.
- **Solar energy** or sunlight is used to generate electricity directly using photovoltaic cells or to produce heat that can then be used directly or converted into electricity in a thermal power plant.
- **Wind energy** is used to turn a wind turbine to generate electricity. It is also used directly to power equipment such as water pumps.

Renewable energy technologies have made remarkable progress during the 1980s and 1990s. Prices for energy from RETs such as wind turbines and photovoltaics (PVs) have come down by as much as 10 times. Prospects for bringing RETs to broad market competitiveness are good. With continuing R&D coupled to carefully targeted demonstration and commercialization, RETs are now poised to become major contributors to U.S. and global energy needs during the next several decades. RETs may, by 2025, provide one-half as much as fossil fuels do at present.

The technologies that tap renewable energy resources are quite diverse. Biomass power technologies collect organic plant material, agricultural or forest product residues, or dedicated crops, and burn it in systems similar to coal-fired power plants but smaller in scale. Geothermal power systems use naturally trapped underground hot water to power electric generators. Their biggest challenge is identifying and tapping hydrothermal resources, similar in many respects to oil and gas exploration and production. Photovoltaic devices consist of thin layers of semiconductors that generate electricity when sunlight hits them. They use many of the technologies of the electronics industry, but also can employ different materials. Solar thermal-electric systems concentrate sunlight to produce electricity in a thermal power plant. These systems face serious materials constraints due to the high temperatures required and thermal cycling. Wind energy technologies convert the kinetic energy in wind flows using three-dimensional aerodynamic principles to optimize energy capture by the turbine blades. Biomass fuels can be produced by genetically engineering enzymes to convert plant fiber (cellulose) into sugars and then ferment the sugars into ethanol.

Intermittent RETs, such as wind and solar, can be only used by electric utilities for power needs when they are available. Thus, they are typically used as a base load source. Thus, the emphasis is given to gas turbines to supply peak power needs. If intermittent RETs are to make very large contributions to electricity supplies in the longer term (50% or more), technologies are needed that would make it possible to store energy for many hours at attractive costs.

Many RETs can be sited in distributed configurations much closer to customers than is the case with conventional power-generating technologies. Distributed electric generation refers to small (10-W to 100-kW) power plants at or near the loads operating in a stand-alone mode or connected to a local power grid. When such systems are integrated into electric grids, new control technologies and strategies and new power management entities (e.g., the "distributed utility") will be needed to exploit optimally the potential economic benefits offered by such systems and integrate them into the grid in ways that ensure high-quality electric service. Similarly, new control technologies and management techniques will be needed for integrating RETs with fossil fuel technologies for use in applications remote from utility grids.

The prospects are perhaps best for solar, wind, and hydro, somewhat less bright for geothermal and biomass, and least attractive for ocean energy.

Solar and wind energy technologies need research on the properties of semiconductors for photovoltaics, high-temperature materials and long-life reflectors for solar thermal systems, fatigue-resistant materials for wind turbine blades, geochemical characterization tools for geothermal reservoirs, and computational aerodynamic models for wind turbines.

At present prices for natural gas, gas-fired systems provide electricity at lower costs than RETs. Given the competition from natural gas, all RETs face great difficulty capturing sufficient market and production scale to drive costs down. Further, it is often difficult for a company to attract financing for continued R&D and commercialization when it might not have a positive net return for 10 years or more. RET energy system costs will decline during the next decade and many may compete satisfactorily with oil.

Many RETs are well suited for developing countries. The small scale and modularity of most RETs make them good fits to energy systems in developing countries. PV technology is already competitive for household lighting and other domestic uses in rural areas of developing countries where some two billion people do not have access to electricity. Wind turbines are used for pumping applications in many areas. Modern small-scale biomass power systems may offer farmers new income-generating opportunities while providing an electricity base for rural industrialization. The environmental attractions of RETs will have a special appeal in developing countries. Rapid industrial growth and high population densities are creating increasingly severe environmental problems in many developing regions, especially in urban centers.

New RET technologies face the chicken-and-egg problem of generally having high costs, and thus being limited to low market volumes, but needing large market volumes to drive costs down. Making this transition is difficult given the low costs of energy today. One possible time line for the adaptation of renewable energy technologies is given in Table 14.3.

The potential sources of renewable energy are illustrated in Figure 14.1, which shows only part of the energy flow diagram for the earth. The rate of energy flow from the sun can be calculated to be equal to 1.73×10^{17} W. The tidal energy input is 3×10^{12} W. Therefore, the sun supplies *essentially all* of the energy flowing to the earth. When the solar power enters the earth, a large part of it, 30%, is reradiated or reflected back into space as short-wave-length radiation (flow ①). Another part, in flow ②, sets up differences of temperature in the atmosphere and oceans in such a manner that the convective currents produce the winds, ocean currents, and waves. This mechanical energy is eventually dissipated into heat and radiated into space. Flow ③ follows the evaporation, precipitation, and surface runoff channel of the hydrologic cycle. A fraction of the incident solar energy is captured by the leaves of plants in the process of photosynthesis (flow ④). In this process, solar energy is stored as chemical energy.

Table 14.3 **Possible Time Line for Renewable Energy Technologies**

Renewable energy technology	Possible date for achieving 10% of the potential
Hydro	2010
Wind	2020
Geothermal	2030
Solar	2040
Biomass	2050

14.2 Hydroelectric Power Generation

Energy flows in the hydrologic cycle of evaporation, precipitation, and surface runoff of water can be harnessed to generate electric power. Worldwide, as we noted in Figure 14.1, this flow is 4×10^{16} W, and only a small fraction is used by man to turn water turbines and electric generators.

The earliest use of water power was in water mills several thousand years ago. By the 16th century, water mills had to be adapted to many processes. The use of water falling through a distance, in order to turn a turbine and later an electric generator, began after the general introduction of electric power in the United States and Europe at the end of the 19th century. In 1882 a hydroelectric station was placed in operation on the Fox River in Appleton, Wisconsin. A waterwheel drove two Edison-type generators with a total output of 25 kW, which was sold to two paper mills and one residential customer for lighting incandescent lamps.

In 1895 the first large hydroelectric power plant was built at Niagara Falls, New York. A large central power station was used at the falls and the power was distributed over a wide area. Large water turbines were used to

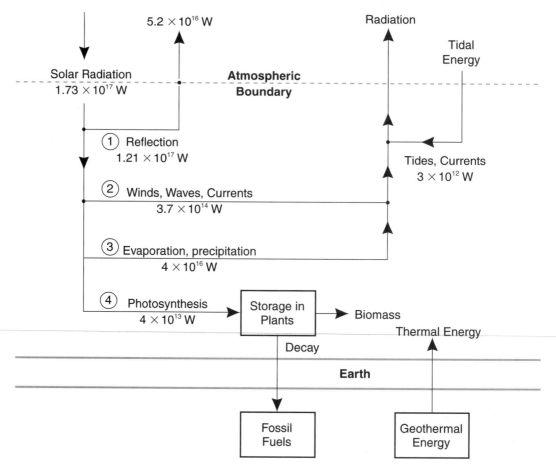

Figure 14.1   **Energy flow diagram for the earth.**

harness the falling water. The height of the falls is about 200 feet and an estimated 210,000 cubic feet (5.95×10^6 liters) of water per second flows over it. Three million dollars of capital was raised to build the power plant, and it was decided to generate alternating-current electric power to be transmitted over a transmission line to Buffalo. Approximately 4 MW was generated at the original Niagara power plant.

The water, often stored behind a dam, falls through a height (or head) of distance h. The water's potential energy is converted to kinetic energy, and the flowing water turns a water turbine. The rotating shaft of the turbine turns the electric generator, which yields the electricity.

Water turbines are highly efficient, converting most of the kinetic energy to rotational energy of the shaft. The outlet speed of the water leaving the turbine is sufficiently small so that the efficiency of the turbine is very high (95%). If the efficiency of the turbine is 95% and the efficiency of the electric generator is 95%, the overall efficiency of these two devices is $(0.95)^2 = 0.90$.

Hydroelectric projects often are combined with flood control and irrigation projects and therefore must involve the evaluation of the total environmental effects of the project. Furthermore, the dams often provide recreational areas with a lake and beaches. Undesirable effects include upstream flooding of river valleys, downstream water-flow reduction, impact on the area required for the lake, and the effects of long electric transmission lines from the project site to the area where the electric power is used. The environmental impacts of a hydroelectric project must be thoroughly analyzed since, after it is completed, they are essentially irreversible.

Hydropower is a relatively mature and competitive renewable energy technology with 92,000 MW installed capacity in the United States. As of 2000, some 1,200 hydropower plants generate approximately 10% of U.S. electricity and provide annual revenue in excess of $20 billion. There is the potential to install from 35,000 to 50,000 MW of new hydropower capacity in the United States, in large part using existing dam structures and reservoir systems. This approach avoids the potential environmental impacts associated with new dam structures, including land inundation, silting, and water quality.

Although hydropower emits little pollution, a variety of environmental issues have been raised, including its impacts on water quality, river flows, and aquatic ecology.

The worldwide potential for hydropower is very large, but low-cost fossil fuels, environmental concerns, and the capital-intensive nature of major hydro installations will slow development. As of 2000, 71 large-scale projects are under way worldwide with plant capacities of 1,000 MW or more. There is an opportunity for the United States to provide more efficient, more environmentally sustainable turbomachinery to a growing global market. Most of the world's potential hydropower sites are either low-head (2 to 20 m) or medium-head (20 to 100 m). These are good sites for distributed, local power plants.

China and India are planning large hydropower projects. The Chinese Three Gorges Dam is designed both to generate some 20 GW of electricity and to help control the floodwaters of the temperamental Yangzi River.

As a renewable and sustainable form of energy already providing one-fourth of the world's electricity, there is little doubt of hydropower's continued growth; only a small percentage of its potential has been tapped in Asia, Africa, and Latin America. However, there is increasing recognition that hydro development under many circumstances may indeed be a poor choice for providing energy. The amount of land flooded per unit of power generated, the effect on displaced indigenous peoples, changes in ecosystems (both upstream and downstream), health effects on local populations, and silting are only a few of the factors that must be considered. In Latin America, hydropower provides 60% of its electricity.

To sustain and increase the U.S. hydropower capacity of about 95,000 MW, new turbine technologies are needed that are less environmentally damaging to fish and ecosystems.

The Tunoe Knob offshore wind farm in Denmark provides electric power for approximately 4,000 households. PHOTOGRAPH COURTESY OF VESTAS WIND SYSTEMS.

CHAPTER 15

Wind and Solar Energy

15.1 Wind Energy

A large amount of power is contained in the movement of air in the form of the wind. The mechanical devices that are used to convert kinetic energy in the wind into useful shaft power are known as **windmills** (the earliest machines were used to mill grain), wind machines, or wind turbines.

The Persians built the first known windmills as early as 250 B.C., several of which are still in use today. Use of the horizontal windmill, whose sails rotate around a horizontal axis, spread throughout the Islamic world after the Arab conquest of Iran. The windmill in Europe in the 11th century was also a horizontal-axis type; and by the 17th century the Netherlands had become the world's most industrialized nation through extensive use of wind power in ships and the familiar Dutch windmill.

During the 18th century, many important improvements were made in horizontal windmills. Windmills were built with 8 to 12 sails constructed from sailcloth stretched over a wooden frame. Later 18th-century windmills were characterized by four to six arms that supported long rectangular sails. In 1750, the Netherlands had 8,000 windmills in operation. In Germany there were over 10,000 windmills by the mid-19th century. The designs of windmills continued to be improved and refined. Sails gave way to hinged shutters, like those of Venetian blinds, and the shutters in turn gave way to the propeller or airscrew.

The use of wind power for sailing ships has prevailed for several thousand years. However, the use of sailing ships for commercial trade began to decrease in the early 1800s. The demise of the clipper ship era was begun when the steamship the S.S. *Savannah* was introduced. In 1819 the *Savannah* was the first steamship to cross the Atlantic Ocean. An American, Daniel Halladay, constructed a self-governing windmill in Connecticut in 1854. This improvement and those developed by others resulted in a windmill with many short sails and self-governing speed. Factories throughout the country produced and sold millions of such machines in the next half-century. Wealthy people bought them to provide running water in their homes. Union Pacific used them to fill trackside tanks with water for the first locomotives that chugged across the continent, and cattlemen and homesteaders bought them to help turn the great American plains into the source of much of our food today.

By 1850, the use of windmills in the United States provided about 1×10^9 kWh each year. By 1870, the amount of power provided by windmills began to decrease as the steam engine became predominant. Up until the 1930s windmills provided power for water pumping and small (2-kW) electric generators for rural locations. By 1950, windmills were largely replaced by rural electric distribution systems.

The advantages of using wind energy are (1) it does not deplete natural resources; (2) it is nonpolluting, making no demands upon the environment beyond the comparatively modest use of land area; and (3) it uses cost-free fuel. These advantages must be weighed against the disadvantages: (1) that wind is an intermittent source of energy and (2) that total system costs are high when an energy storage system is included.

The total potential production of electric energy from the wind in the United States is estimated to be 1.5×10^{12} kWh/year. Of course, to realize this potential by the year 2020, a vast capital investment would be required.

Wind power grew in use worldwide during the 1990s. The cost of wind-generated electricity is now comparable to the average cost of generation from conventional sources. One reason is that advances in wind turbine technology have made generating electricity at lower wind speeds more profitable; another is that the number of sites with good wind resources is greater than originally projected. Wind power provides less than 1% of the world's energy. Wind power capacity is about 10,000 MW worldwide and the U.S. capacity is about 1,700 MW.

The cost of wind energy is about 4.2 cents/kWh, making it cost competitive. Wind energy is intermittent, highly variable, and site specific; it exists in three dimensions and is the least dependent on latitude among all of the renewable resources. The power density (in watts per unit area) in moving air (wind) is a cubic function of wind speed and therefore even small increases in average wind speeds can lead to significant increases in the capturable energy. Wind sites are typically classified as good, excellent, or outstanding, with associated mean wind speeds of 13, 16, and 19 mph, respectively.

A wind-electric system (WES) is a wind machine and an electric generator. Large-scale harnessing of wind energy will require hundreds or even thousands of WESs arranged in a wind farm with spacings of about 2 to 3 diameters crosswind and about 10 diameters apart downwind. The power output of an individual WES will fluctuate over a wide range, and its statistics strongly depend on the wind statistics.

At around 1,700 MW, nearly 90% of all the WESs installed in the world are in California. They are expected to generate nearly 3 billion kWh of electricity per year for the state's utilities to which they are interconnected. Because of their lack of control and the intermittent nature of wind-derived energy, they are not embraced enthusiastically by electric utilities. This gap is expected to be bridged very soon with appropriate computer controls and operating strategies. Wind energy is already an economical option for remote areas endowed with good wind regimes. The modularity of WES, coupled with the associated environmental benefits, potential for providing jobs, and economic viability, points to a major role for wind energy in the generation mix of the world in the decades to come. For example, India has about 900 MW of wind energy capacity.

The price of wind-generated electricity has dropped to about 5 cents/kWh in 2000. Although wind turbines generate electricity without causing any air pollution or creating any radioactive wastes, they have an impact on the environment. Wind turbines require a lot of land, but most of that land will remain available for other uses such as farming or livestock grazing. Roads must be built to facilitate installation and maintenance of the wind turbines. Turbines generate noise, but newer turbines are much quieter than earlier models. Current industry standards call for characterization of turbine noise production and rate of decay with distance from the turbine as part of the turbine-testing process. In addition, turbines are large structures that will significantly change the landscape where they are installed, creating visual impacts. Another environmental issue facing the wind industry is that of bird deaths as a result of collisions with wind turbines.

Wind turbines will generate about 20 billion kWh of electrical energy in 2000. This is about 1% of the world's electric energy use. However, the wind energy industry is growing. With government incentive programs, WES usage could grow to supply 2% of the world's electric power by 2020. Note in Table 15.1 that wind energy use grew at an annual rate of 22% between 1990 and 1998 (Brown, 2000).

Table 15.1 **Energy Use by Source, 1990–1998**

Source	Wind	Solar PV	Geothermal	Hydro
Annual rate of growth	22%	16%	4%	2%

The European Wind Energy Association projects that up to 40 GW of wind power capacity could be installed in Europe by 2010. Danish wind turbine manufacturers supply about half of the generating equipment in the world. The European wind energy companies accounted for about 90% of worldwide sales in 1999. India, China, and other developing countries have considerable wind energy potential and could generate substantial employment by building a strong indigenous base. India already has 14 domestic turbine manufacturers.

The value of wind power can be increased if wind power is integrated with energy storage. Energy storage is discussed in the next chapter.

15.2 Wind Power: Where Will It Grow?

By Martin J. Pasqualetti
Professor
Department of Geography
Arizona State University

Renewable energy has a certain cache and so, by inclusion, should wind power. Indeed, wind power is attracting attention, some of it sharply worded, both from developers and from the public. In this section, I summarize the current status of wind power, and suggest one element of a wind energy future.

The reason that wind power is newsworthy, even though we have been using it for thousands of years, is that it has been reborn in a new form, this time not just moving ships or lifting water, but generating electricity — and doing so on an impressive scale. At present, it is the quickest growing renewable energy resource in the world, faster than any direct type of solar, geothermal, tidal, or, at present, hydropower. Two billion dollars were invested in wind power in 1998 alone, and its generating capacity is more than 14,000 MW worldwide (Figure 15.1). For every hour this capacity is available and operating, enough electricity is produced to illuminate 140 million 100-W light bulbs, a string of lights at 1-foot intervals all the way around the equator of our planet.

Wind power is growing most quickly in Europe, particularly Germany, Denmark, and Spain (Figure 15.2). In Schleswig Holstein, in northern Ger-

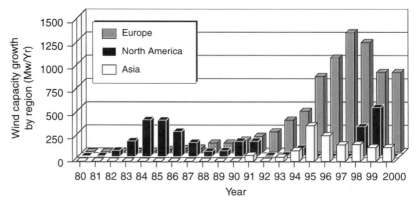

Figure 15.1 **Wind generating capacity, 1980–2000.**

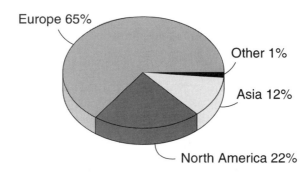

Figure 15.2 **Major users of wind power.**

many, for example, wind power now provides 15% of the electricity, enough to replace two of Germany's large coal-fired power plants, according to Christopher Flavin of the World Watch Institute in Washington, D.C. Danish researchers have been talking about providing up to 50% of their electricity from wind power, and they estimate that perhaps as much as 10% of the world's electricity can come from the wind within the next few decades. In the United States, the wind potential in North Dakota, presumably with comparable environmental advantages regarding CO_2, is adequate to provide about one-third of the electrical needs of the country (Figure 15.3).

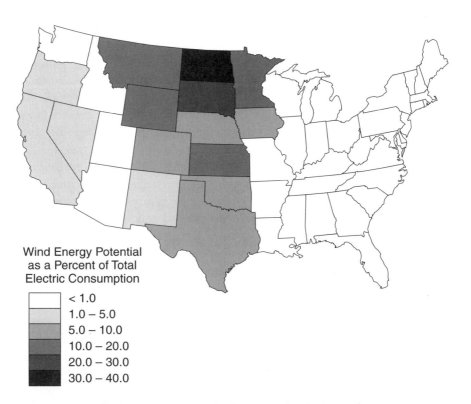

Figure 15.3 **Wind energy potential in the United States.**

The potential in many other parts of the world, also large, is found in places such as Patagonia and Central Asia. In the long run, wind power — both onshore and off — could eventually exceed hydropower (currently supplying 20% of the world's electricity) as an energy source.

Going along with the fast growth and great potential of wind power, it is now cost effective (about $800 per kilowatt installed), and a low producer of the environmental impacts that so cloud the future of coal and uranium as fuels. Wind power produces electricity free of the sullying impacts of atmospheric pollution; it requires no water for cooling. In addition, there is no short- or long-lived radioactive waste, nor permanent dedication of land, air, or water resources associated with wind energy installations.

The enthusiasm for wind power in Denmark is leading the Danes to count on its expansion to help reduce CO_2 emissions by 20% in 2005 compared to 1988. This would make the environmental benefits of wind power so significant that from a socioeconomic point of view wind power and natural gas power are equal. Also of significance, and owing to its modularity, wind turbines can be constructed as the need arises, in small increments, rather than as huge centralized stations that dominate today's pattern of electricity generation.

Ironically, given all that is good about wind power, it is inevitably, abruptly, and unavoidably visual, even intrusive. The question is whether wind's conspicuous and quick landscape changes will curtail the contribution it will ultimately make. In an era that seems to be dominated by the desire for energy that is free of all environmental costs, wind energy cannot be hidden. It is a vertical element in a horizontal world. By its very nature it cannot be camouflaged by vegetation or other structures. Indeed wind tends to be strongest and most consistent in the same areas that are by their nature the most obvious, such as grassy plains, ridges, seashores, and mountain passes that are prized by transportation engineers. And, the turbines can be quite large, with blades as long as 40 m and hub heights as high as 65 m.

There are only two solutions to the dilemma that has developed from the convergence of wind power sites with public intolerance of visual infractions. First, wind power can be relegated to relatively unpopulated areas, where it might be easier for people to balance wind's visibility against its various supply and environmental benefits. This option may be feasible in a few places in densely settled Europe, but the relatively large lightly inhabited central plains of the United States are particularly appealing sites for wind energy development.

The second option is not one that can be accommodated quickly, but is nevertheless promising. It entails an educational and acclimatization process, something that can take most of a generation. Such a program also usually requires the establishment of regulatory steps, mandates, and procedures in order to bolster public confidence. This option requires identifying and repeating the advantages of wind power, and putting in place the system for broadcasting that message.

The message that wind advocates want to advertise is one of clean abundance in comparison with few drawbacks. Wind turbines produce some sounds, but properly engineered turbines now produce noise that is less than a nearby highway or breaking surf. Turbines may also pose a threat to birds, although this problem is now thought to be smaller than was earlier assumed. Wind power produces no long-term wastes, requires no cooling water, emits no air pollutants, has no effect on surface or subsurface water supplies, and it can be removed from the landscape with no lasting effects.

Wind power's most obvious drawback, its visibility, can be a serious deterrent to public acceptance. Visibility has, for example, drawn strong public rebukes of wind power in several places, including California (most notably, near Palm Springs), Denmark, Germany, and other countries. In England, at present, aesthetic concerns have wind power development at a standstill.

The question is how best to deliver a balanced message about wind power. The options include media advertising, information tours, public presentations, publications, wind "fairs," modules in school curricula, visitor centers, and the like. All of these approaches are being used in one form or another, but such public education is usually a slow process. And it can be tenuous; one dead eagle can erase years of public instruction.

The inherent barrier to vibrant public encouragement of wind power, at least in countries of the Organization of Economic Cooperation and Development (OECD), is that these countries are relatively prosperous. They tend to have other, more familiar, energy options that can fill energy demands more completely than can wind power, even if the aggregate impacts of these conventional approaches would be greater.

Wind power's economic viability and its intrusive but otherwise light environmental touch on the land support some positive suppositions. First, the development of wind power will meet least resistance in locations where it satisfies, in addition to the obvious requirement of viable wind concentration, various combinations of the following siting criteria:

- Rural lands
- Offshore
- Noncherished lands (e.g., national parks)
- Lands currently unconnected to a power grid
- Areas of low economic prosperity.

The last-named criterion is perhaps the most important and this is where wind energy can grow fastest and do the most good. Such areas will have the fewest options, most desperate need, and the least amount of capital. In contrast, where countries have available to themselves fossil and nuclear fuel, or even ample opportunities for energy efficiency and conservation, the opposition that wind power has already encountered is likely to continue, despite the availability of funds to increase wind power capacity. Such opposition is likely to continue at least until public acclimatization can advance toward acceptance. In developing countries, especially in areas on the fringes of settlement, the first and most dominating motivation will be simply obtaining supplies of electricity.

Wind power, where available, can be developed in small increments with very little existing infrastructure. As such, it is, at present, the most suitable source of energy for these areas. This argument for wind power could produce a slowing of construction of wind power in Europe, and a quickening pace in such places as Mongolia, western China, central India, and the windy treeless portions of Africa and South America. Wind power could be a salvation in such places, and these places can return the favor for the entire wind energy industry.

15.3 Solar Energy

All our food and fuel (exclusive of nuclear energy) has been made possible by the sun through the photosynthetic combination of water and atmospheric carbon dioxide in growing plants. Solar energy is the basic energy support for life and underlies the wind, the climate, and fossil fuels. The energy flow diagram of Figure 14.1 portrays the flow of the sun's energy into the atmosphere and then ultimately into heat, wind, evaporation of water, and photosynthesis.

In this chapter we discuss the direct use of the sun's energy as it impinges on the earth. Thus, we are interested here in the tapping of solar energy in the form of electromagnetic energy prior to its conversion to wind energy, fossil fuels, and plants. Immediately outside the earth's atmosphere the sun provides energy at the rate of $1,353 \text{ W/m}^2$ normal to the sun. Since the area of the diametral plane of the earth is $1.27 \times 10^{14} \text{ m}^2$, the solar input to the earth is $1.73 \times 10^{17} \text{ W}$.

The average solar radiation received at the earth's surface is approximately 630 W/m^2. This is an *average* figure; the actual radiation received at a specific location can be significantly higher or lower. Of course, this radiation is received only at a time when the point on the earth is normal to the sun's rays. Thus, no radiation will be received at night, and a reduced amount is received at times other than noon. The solar energy received also varies significantly with latitude as well as with time of year. The solar energy received in a given location in the United States during a clear day in December is about one-half of that received in July.

Solar radiation at the earth's surface, averaged over the day and over the total year, has been recorded for the United States and is called the **average annual insolation**. The region consisting of the southern two-thirds of New Mexico and Arizona and the bordering desert regions of Nevada and California receive an average insolation of about 260 W/m^2. This is about 40% more than the insolation of New York or the New England states.

The use of direct power from the sun's radiation has many advantages. Solar power is abundantly available, even in regions remote from the source of fossil fuels. It is essentially a nondepletable source of energy in comparison with fossil fuels or nuclear fission power, and it is cost free in its original radiation form. Of course, there is a significant cost for the capital plant required for converting solar energy to other forms of energy. If solar energy is utilized locally then, the need for transporting the energy is avoided. Also, solar power can be used in small units, as for an individual building or home. Solar devices hold promise for the developing world as well as for the economically developed world. Because solar power burns no fuel, it causes no air or water pollution.

A disadvantage of the utilization of solar radiation is its dispersal over the surface area of the earth. The average insolation in the United States is 180 W/m^2 and the land area required to collect significant amounts of energy is large. For example, the land required to collect the equal of the output of a 1000-MW power plant would be 5.6×10^6 m^2 or 2.15 square miles assuming a collector efficiency of 100% (actual efficiencies are 10%).

Solar radiation has been used by people since the beginning of time for heating their domiciles, for agriculture, and for personal comfort. Solar heating has been utilized in various forms since ancient times, when the sun's rays were focused for heating and cooling. In 1774 Joseph Priestly used lenses to concentrate solar rays to decompose mercuric oxide into mercury and oxygen. In 1872, in the desert of North Chile, a solar distillation unit covering 4750 m^2 of land was built to provide freshwater from saltwater. This plant was operated for 40 years, producing 6,000 gallons of water per day.

At an exhibition in Paris in 1878, sunlight was focused onto a steam boiler that operated an engine which drove a printing press. During the period 1901 to 1915, several solar collectors used with steam engines of several horsepower were constructed in California and Pennsylvania. While solar energy arriving at the land surface of the United States is about 500 times the present rate of consumption of energy, it is also both diffuse and intermittent. Gathering sunlight and providing it in useful form when needed at a competitive cost is the principal challenge to solar energy technology.

The first stage in the process of utilizing solar energy is collecting it by some means. Solar energy can be collected by three means: (1) thermal collectors, (2) photovoltaic collectors, and (3) biological collectors. In addition, the collectors can be either focusing or nonfocusing. Focusing (or concentrating) collectors can use lenses, mirrors, or selective films to concentrate the solar radiation. An example of a photovoltaic system is a collector consisting of solar electric cells. Biological collection is facilitated by means of photosynthesis. We do not include ocean thermal gradients, hydroelectric power, or wind energy here (they are discussed elsewhere), although they are inherently a natural means of collecting the sun's radiant energy.

The thermal collector operates on the basis of absorbing the solar radiation at a surface and thus raising the temperature of the collector surface. The photovoltaic collector directly uses the photons in the solar beam.

High-temperature solar thermal technologies use mirrors to concentrate the sun's rays onto receivers, in which the solar heat is recovered. Solar thermal electric (STE) systems make electricity from the recovered solar heat in conventional thermal power cycles. The quality of thermal energy employing STE conversion necessitates concentrated sunlight. Parabolic troughs, parabolic dishes, and central receivers are used to generate temperatures in the range of 400 to 500°C.

The only STE technology with commercial experience uses parabolic trough collectors coupled to steam turbine power units and natural gas backup. Some 354 MW was installed in California from 1984 to 1991, and as a result of this experience capital costs for the solar portion fell from $4,500 to $2,900/kW.

Solar cells that directly generate electricity by means of photovoltaic processes are the predominant source of power for space satellites. These cells convert sunlight directly into electricity without an intermediate thermodynamic cycle. A photovoltaic device requires a material in which mobile charge carriers can be generated by absorption of light, and a built-in potential barrier by which these charge carriers can be separated from the region generating them. Semiconductor materials with *p-n* junctions fulfill these requirements. Semiconductors manufactured from silicon (Si), gallium arsenide (GaAs), cadmium sulfide (CdS), and copper sulfide (Cu_2S) can be used for photovoltaic conversion. Existing silicon cells develop about 0.5 V, so that large number of cells must be arranged in series in order to achieve high voltages. The output of the cells is a direct current, which is converted to alternating current.

The photovoltaic effect consists of the generation of an **electromotive force** as a consequence of the absorption of radiation; that is to say, a current will flow across the junction of two dissimilar materials when light falls on it. A photovoltaic cell (PV cell), also known as a solar cell, is simply a large-area semiconductor *p-n* junction diode with the junction positioned very close to the top surface. PV technologies convert sunlight directly into electricity using solid-state devices. Most technologies use flat-plate collectors that convert diffuse as well as direct sunlight into electricity. At a typical 10% efficiency, 600 ft^2 of collectors are needed to generate electricity at the average U.S. household use rate at a site with average insolation. Global PV production was 89 MW in 1996 and is estimated to be 150 MW in 2000.

PV is typically the least costly electric technology for small-scale remote applications ranging from callboxes along U.S. freeways to lighting for rural households in developing countries. Market strategy consists of identifying niches where the technology is cost effective today against conventional energy; aggregating these demands to scale up production; and driving down prices by technology improvements and increases in the scale of production. These cost reductions expand the range of market niches where the technology is cost competitive. This market strategy may allow PVs to penetrate ever larger markets, in the following progression: rural remote, village minigrid, distributed grid-connected, peaking, and finally intermediate and base load power applications.

Installed PV costs have fallen from $9,000 to $6,000 per kW in 1996. It is estimated that the installed cost will be $3,000 per kW in 2004. It may be possible to reduce this cost to $1,000 per kW by 2020, which would be competitive with other services.

At $1,000 per kW, it may be possible to mount panels on roof tops up to 5,000 km² and they could generate 25% of the electric power used in the United States. If a typical building roof is 10×25 m or 250 m², then this system would require 20 million buildings.

Table 15.2 shows the typical size of an electric power plant using different types of sources. Note that 10 solar panels on a house roof could supply twice the power required for a typical home when the sun is shining. Using energy storage, say, in batteries, the PV system meets the home's needs.

Table 15.2 **Typical Size of Electric Power Plant**

Type	*Size(kW)*
Coal utility plant	1,000
Gas turbine plant	200
Ten wind turbines	50
Ten solar panels	20

15.4 The Potential for Photovoltaics

BY ROGER MESSENGER
Professor
Department of Electrical Engineering
Florida Atlantic University

Since the atmosphere of the earth is receiving roughly 1.3×10^{17} W on a 24-hr basis from the sun, and since the present electrical generating capacity of the planet is about 2.7×10^{12} W, and since the sun will probably shine for another 4 billion years or so, the sun presents a reasonable source of energy for the earth, at least in the short term. Furthermore, since Hubbert's model suggests the peaking of world petroleum production within the next decade (Hubbert, 1971) and since coal and nuclear appear to be less-than-attractive replacements for petroleum as fuels if environmental effects are considered, the wind and the sun deserve serious consideration for large-scale power production. In particular, photovoltaics (PVs) present a nearly classical example of the need to develop sustainable business, engineering, and polit-ical models to facilitate the transition to a sustainable energy supply.

PV cells are made of semiconductor materials such as silicon, gallium arsenide, cadmium telluride, and copper indium diselenide. The cells are capable of converting up to 30% of incident sunlight directly to electricity, depending on the technology. Production-scale multicrystalline silicon cells, for example, convert about 15% of incident sunlight to electricity, while efficiencies of 18.6% have been reported for laboratory cells (Rohatgi *et al.*, 1996).

The test for sustainability for PVs is the net energy production of the PV system. No energy system can be sustainable, unless over its lifetime it will produce more useful energy than was used in the production, operation, and decommissioning of the source. Although this may appear to contradict the second law of thermodynamics, the implication is that the primary energy source must be nearly infinite.

The degree to which a source can be considered sustainable can be con-veniently termed the energy return on investment, where ROI = E_{out}/E_{in}, which represents the ratio of the energy produced by a source over its life-time (E_{out}) to the energy required to obtain or produce the source (E_{in}). In other words, it takes energy to get energy. To convert petroleum to a usable form, for example, it must be found, drilled, pumped, transported, refined, and transported again. All of these steps take energy.

For PVs the ROI is presently estimated to be in the range of 4:1. Hence, for each kilowatt-hour used to produce a PV system, the system will gener-ate 4 kWh. What this means is that if a PV cell produces energy for 5 hr/day for 30 years, it will operate for 54,750 hr. To generate 4 kWh in this amount of time, the cell must have an output power of 7.306×10^{-5} kW = 0.073 W. For a silicon PV cell, this amount of power will require a cell area of approx-imately 5 cm². As efficiency improvements are made in the manufacture and operation of silicon and other emerging PV technologies, such as thin films, the energy ROI for PVs should continue to increase. Furthermore, each kilowatt-hour of fossil electricity that is replaced by a kilowatt-hour of PV electricity involves a reduction in CO_2 emissions by nearly 90% (Markvart, 1994). If the energy used to produce a PV cell is produced by other PV cells, then the CO_2 emissions associated with PV-generated elec-tricity become almost negligible.

Having established that PV sources are sustainable, the challenge, then, is to overcome the business, technical, and political barriers to large-scale deployment. While the wholesale cost of electricity from the wind has now dropped to approximately $0.05/kWh in areas with sufficient wind (Moore, 1999), the cost of PV electricity is still close to $0.25/kWh in areas with sufficient sunlight (Messenger and Ventre, 1999). But as the cost of PVs has continued its historic decline, the cost of fossil- and nuclear-generated electricity has remained relatively stable, but is likely to begin to increase during the 2000s. One might logically ask, then, when will PV electricity become cost effective? The answer is now, for many applications, some of which are stand-alone and others of which are grid-connected (utility interactive).

15.4.1 STAND-ALONE APPLICATIONS

When a few hundred kilowatt-hours per month are needed at a location not conveniently served by the utility grid, it is likely that the extension of the grid, if at all possible, may cost tens of thousands of dollars. Since the annual payment on an 8%, $10,000 loan over a 20-year period is $1,018, the cost per kilowatt-hour for each $10,000 of grid extension cost is in the neighborhood of $0.50. As a result, many stand-alone PV systems are currently used in remote locations for pumping water, operating communication repeaters, powering highway sign lighting, and powering cathodic protection systems, to name a few. The life-cycle costs of these systems are less than the fossil alternatives, including small fossil generators, even when environmental degradation factors are ignored. If a price is assigned to environmental degradation, then the PV alternative becomes even more attractive.

15.4.2 UTILITY INTERACTIVE SYSTEMS

The wholesale cost of electricity to the grid may range from a few cents per kilowatt-hour to more than $0.20/kWh. Baseload electricity from fossil or nuclear plants that are built for $2/W or less, with fuel costs of pennies per kilowatt-hour and high capacity factors, is relatively inexpensive. But a peaking generator having fuel costs of close to $0.10/kWh and a capacity factor of less than 10% must recover the capitalization costs with a relatively low amount of generation, and, hence, a high cost per kilowatt-hour. The capacity factor of a generation facility is the percent of rated capacity obtained from the source, generally measured on an annual basis.

For example, suppose a 1-MW gas turbine generator can be installed for $500,000. Suppose also that the generator is only used to meet summer peak demands on the worst of days, so that the capacity factor of the unit is 5%. With this capacity factor, the unit will only operate for $0.05 \times 8760 = 438$ hr/year. During these 438 hr, the unit will generate 438,000 kWh of electricity.

If the plant is built with money borrowed at 8% for 20 years, the annual cost of repayment of the loan will be $50,939. Thus, the amortization cost is $0.116/kWh. Adding a jet fuel cost of approximately $0.10/kWh, plus maintenance and profit, the cost approaches $0.25/kWh. As the capacity factor of the system is increased, the fuel cost remains high, so the cost per kilowatt-hour asymptotically approaches about $0.15/kWh as the capacity factor approaches unity. Since utilities purchase power from the lowest priced source, the peaking generator capacity factor is essentially automatically limited by the price of electricity from base load sources. Hence, use of peaking generators is only cost effective when base load generation is at full capacity.

Now consider a PV system built at $6000/kW. If the system faces west, in some locations it will be able to generate as much as 6 kWh/day/kW averaged over a year, with 8 or more kWh/day/kW during summer months. This amounts to 2190 kWh/yr/kW of PV. With the same lending conditions, the annual payments on the PV system will be $611/kW, which amounts to $0.28/kWh. However, if it were decided that PV-generated electricity is desirable to the extent that money is made available over a 30-year loan period at 5%, then the annual payment on the system would be $390/kW, and the cost of electricity from the PV system would drop to $0.18/kWh.

The distributed nature of PV generation is also economically and strategically attractive. Since large PV systems are generally composites of small systems, there is no particular economic advantage for large PV systems. By installing PV systems at or near the point of use, transmission and distribution loads are reduced and the corresponding losses are avoided. Large-scale distributed deployment of PV generation can reduce the need for further construction of transmission lines and the accompanying siting problems associated with the construction.

15.4.3 TECHNICAL AND NONTECHNICAL BARRIERS TO ENERGY SYSTEM DEPLOYMENT

A look at the present technical and nontechnical barriers to PV system deployment reveals the business, engineering, and political actions that are needed to encourage use of the technology. The concerns over construction of fossil or nuclear facilities have been well defined during the past few decades. Power plant and transmission line siting processes have attracted the attention of many diverse groups, all of which tend to provide reasons why either the facility should not be built at all or at least why it should not be built in a particular backyard. Whether or not all concerns expressed are valid or legitimate, the expression of these concerns causes delays in construction and leads to strained relationships between energy producers and their customers.

Because of the many problems associated with the deployment of fossil and nuclear generation and associated transmission, one might think that use of PV and wind sources would be enthusiastically supported by everyone. But this is not the case.

In particular, nontechnical barriers to PV system deployment include the cost of PV arrays; the cost of other PV system components, such as array mounts, inverters and wiring; the standardization of interconnection requirements; the training and licensing of qualified installers and inspectors; and the metering of customer-owned PV generation. Additional technical barriers facing small power producers include utility concern over the potential for low-quality power, disruption of the grid from improper interfacing, and islanding.

At the turn of the 21st century, the Institute of Electrical and Electronic Engineers (IEEE) Standard 929 (IEEE, 1999) has addressed the technical concerns of utilities and Underwriters Laboratories (UL) Subject 1741 (Underwriters Laboratories, 1997) has provided test procedures to verify compliance with IEEE 929. Since the conversion of PV dc output to ac is achieved electronically, using microcontroller-based systems, PV systems can respond within milliseconds to any grid disturbance. The remaining engineering challenges involve reducing the cost and continuing to increase the reliability of these inverters, also known as power conditioning units (PCUs). Additional engineering challenges involve the development of turn-key PV systems that will result in lower overall system costs as well as reduced installation costs. The final engineering challenge is to develop manufacturing methods to reduce the production costs and increase the conversion efficiency of PV cells.

The manufacture, marketing, installation, and maintenance of PV systems tends to be relatively labor intensive. From a business perspective, this means job opportunities in all of these sectors. From a regulatory perspective, quality control will need to be assured through training and licensing of installers and inspectors.

If PV systems are not owned by utilities, and if they supply energy to the grid, a fair price for the energy must be established. Perhaps the simplest scheme is net metering, in which the customer is paid by the utility at the same rate charged the customer by the utility. But this simple concept does not compare the time of use value of the PV energy with the wholesale cost of electricity. In some cases, the value of the PV energy may exceed the wholesale cost, while at other times, it may be less. Since PV systems tend to have high output at utility peaking times, it has been argued that net metering is particularly applicable to PV systems.

A final regulatory issue involves the standardization of utility interconnects. The Public Utility Regulatory Practices Act of 1977 (PURPA) requires that utilities purchase energy from small power producers, but it does not establish pricing or interconnect rules. It is a simple matter for a utility to discourage the use of PV by requiring exorbitant levels of liability insurance to be carried by the small power producer and to require expensive, redundant, transformers and disconnects between the producer and the utility. The uniformity of interconnection problem has already been addressed by IEEE 929, UL 1741, and the National Electrical Code (National Fire Protection Association, 1999). At the moment, however, utilities are allowed to impose additional requirements beyond these codes and standards. Regulators could prohibit such practices.

When nuclear power producers found it difficult to insure their facilities, the federal government provided the insurance. This suggests that a solution to the PV liability problem should be much simpler, such as providing underwriting of homeowner policies so insurance companies will not have any greater exposure if a homeowner chooses to install an approved, inspected PV system.

Perhaps the bottom line of the PV challenge is to provide a level playing field for competing energy sources. The field would need to account for subsidies of all forms and establish environmental costs for greenhouse gases as well as those pollutants already regulated by the Environmental Protection Agency. Such action clearly would require business, engineering, and government cooperation.

15.5 From Solarex to BP Solar: A Solar Start-Up Becomes a Solar Giant

By John J. Berger
Renewable Energy and Natural Resources Consultant
El Cerrito, California

In 1973, two Hungarian expatriate scientists, Dr. Joseph Lindmeyer and Dr. Peter Varadi, founded a company in Maryland known as Solarex Corporation to make inexpensive terrestrial solar cells. By 1989, when Siemens Solar Industries (part of the German conglomerate Siemens), purchased ARCO Solar, Solarex, which was by then next largest in size, became the biggest and oldest U.S. solar cell manufacturer. The saga of Solarex's struggle from obscurity to prominence illustrates the challenges and perils of solar energy entrepreneurship as well as the role of oil companies in the development of solar energy.

Photovoltaic (PV) solar cells transform light directly into electricity without combustion or the production of waste products, and they require no water for cooling, an important fact in hot arid lands. These sturdy and efficient electronic devices are able to produce sustainable pollution-free power on remote rooftops, on buildings connected to utility grids, or in utility-scale power plants.

Both Dr. Lindmeyer and Dr. Varadi had earlier worked on developing solar cells for space applications at the Communications Satellite Company (COMSAT). Lindmeyer was an electrical engineer and coauthor of the world's first book on semiconductors; Varadi was a physical chemist.

The pair started Solarex with the goal of making solar cells efficient and inexpensive enough for use on the ground instead of in the sky. Of course, they hoped this would also make them lots of money. They began work in a small, two-story office building in Rockville, Maryland, with very little working capital and the huge task of creating a new technology and finding markets for it, all without infringing on COMSAT's technology.

During its early years, Solarex struggled financially and made solar watches and solar calculators to survive. Their early technology development efforts were fraught with hazardous laboratory mishaps that often brought the local fire department racing to their building. Yet their processes and products proved themselves and sales increased. Then in 1979, Dr. Lindmeyer was successful in securing European investment capital and in obtaining several million dollars from Amoco oil company in return for a share in the company. Solarex was finally able to build a new $7 million manufacturing facility in Frederick, Maryland. Perhaps with the large infusions of cash, the company became less disciplined in money matters; in any case, Solarex, which had been profitable to 1979, soon was losing money, and this presently was to cause Lindmeyer and Varadi a lot of grief.

About this time, the Reagan administration, which took office in 1981, was actively demonstrating its hostility to solar energy. Solar collectors were removed from the White House; solar energy tax credits were withdrawn by Congress, which hurt Solarex's sales; and Solarex's multiyear, $9 million Department of Energy grant for the development of a new silicon refining technology was abruptly curtailed after only a year. Their new silicon refinery based on this technology soon turned into a financial and technological fiasco. Although Solarex's sales soared to a peak of $25 million in 1983, the company did not make money from 1979 until 1995, when — after 15 years of continuous red ink — Solarex's crystalline cell operations finally became profitable again.

But as if to illustrate how difficult it has been to make money in the solar cell business, the firm again went into the red in 1996 as it began investing heavily in a new $25 million thin-film solar cell factory, and Solarex has not turned a profit since 1995 as it endeavors to position itself in the thin-film arena where many solar businesspeople see large possible future cost reductions and profits.

Back in 1983, nervous about the company's mounting losses, Maryland National Bank called in Solarex's line of credit and, faced with bankruptcy, Lindmeyer and Varadi had no choice but to sell the company to Amoco that same year amid bitter recriminations.[1] For its investment, Amoco now owned a company that had not only developed a unique cast-ingot polycrystalline silicon solar cell but also had acquired a valuable amorphous silicon thin-film cell technology developed by RCA scientists.

In addition, because Solarex had absorbed Solar Power, Inc., from Exxon when Exxon exited the solar cell business, Amoco also gained Exxon's solar cell R&D. Through the 1980s and early 1990s, Amoco continued investing in Solarex and its R&D; expanding its production capacity; and finally building the new amorphous silicon plant near Williamsburg, Virginia, which began production in 1997.

During Solarex's year of profitability in 1995, Amoco succeeded in converting its Solarex holdings to a 50:50 joint venture with Enron Corporation, known as Amoco/Enron Solar. Following BP's 1998 merger with Amoco to form BP Amoco, the new conglomerate bought Enron's 50% share in the joint venture to become full owner of Solarex. The company was attractive for its outstanding polycrystalline (silicon) products and for its

---

[1]For a detailed history of Solarex and a discussion of the controversy between its founders and Amoco at the time of takeover, see Berger (1998).

patents that controlled important aspects of cutting-edge thin-film solar cell technology.

By this time, Solarex was an international company with sales of $58 million and was exploiting the RCA amorphous silicon technology to produce highly efficient multijunction amorphous solar cells in Virginia in addition to its being a leading producer of polycrystalline cells in Frederick, Maryland.

BP Amoco, which later renamed itself BP, subsequently integrated Solarex into its BP Solar subsidiary, which it founded in 1981. Today, the new firm, BP Solar, is the largest PV company in the world. The company produces 20% of the world's solar cell output, operates in more than 180 countries, and generated revenues of $179 million in 1999.

BP Solar's factories manufacture all of the world's major types of solar cell and module technologies (single-crystal, polycrystalline, thin-film, and concentrator devices), and the company is intensively involved in state-of-the-art solar cell R&D. The firm also designs and installs a wide range of solar electric products and systems for residential, commercial, and industrial uses.

Top BP executives understand the need for diversifying their revenue sources and recognize that solar energy is becoming an increasingly important part of the world's energy supply. The Solarex acquisition is part of a strategy announced in 1997 by BP Chief Executive Sir John Browne to generate a billion dollars a year in sales from solar energy by 2007.

BP's interest in solar energy fits well with the company's commitment to reducing the global risk of climate change and limiting the emission of heat-trapping gases. Unlike some other oil and coal producers and major industrial energy consumers, British Petroleum was a leader in acknowledging the risk that fossil fuel use poses to the climate and, instead of devoting itself to denying the mounting evidence of global warming, BP set a responsible corporate example for the oil industry by endorsing the concept that the world ought to begin taking some precautionary steps.

BP then voluntarily adopted a company-wide emission reduction program with the goal of reducing its own global warming gas emissions by 10% below its 1990 emissions by 2010. Given the expected growth of BP, which is seeking government approval of its 1999 merger with ARCO, the 10% reduction target will actually be as much as 40 to 50% below what business-as-usual emissions otherwise would have been.

Although PV is still a small and risky business today compared with conventional fuels, such as petroleum, coal, and gas, if current rates of PV growth continue, PV will be a very large business within 20 years, and will have a major part to play in the modernization and transformation of the world's energy system, especially in the delivery of power to rural areas without well-developed power grids, since PV can be installed where needed without transmission lines. Thus, once the installed costs of solar electric systems are reduced — through further cost-cutting and favorable regulatory policies — to levels competitive with coal, gas, and oil, PV growth will soar. During the next decades to 2050, trillions of dollars will be spent on new electrical capacity worldwide. If PV garners even a modest share of this market, the revenue will be huge.

Eventually, PV will become ubiquitous on tens of millions of rooftops throughout the world. When combined in national clean energy systems employing wind, geothermal, biomass, and the solar thermal technologies, PV offers the nation and the world an unparalleled opportunity to create a clean and sustainable energy future.

Geothermal plants tap the flow of heat from the earth's interior, brought to the surface in hot springs, geysers, and steam vents. Shown above is Coldwater Creek in Kelseyville, California. PHOTOGRAPH COURTESY OF CORBIS.

CHAPTER 16

Biomass, Geothermal Energy, and Energy Storage

16.1 Biomass Energy

Biomass is the organic matter that can be converted to useful energy forms such as heat or liquid fuels. Biomass accounts for about 12% of the world's energy supply; the largest part of this use is in developing countries. Agricultural residues, commercial wood and logging residues, animal wastes, the organic portion of municipal solid waste, and methane gas from landfills can all be used to create energy in a process that has several advantages over fossil fuels. This is a virtually limitless option because of the diversity of sources.

Biomass energy, the chemical energy stored in organic plant matter is derived from solar power via photosynthesis. Photosynthesis is about 4×10^{13} W as shown in Figure 14.1. Biomass accounts for about 3% of total U.S. energy use. As shown in Figure 16.1, photosynthesis can theoretically supply a large portion of the world's needs. The most compelling argument for the use of biomass technologies is the inherent recycling of the carbon by photosynthesis. In addition to the obvious method of burning biomass, conversion to liquid and gaseous fuels is possible, thus expanding the application possibilities.

In the context of electric power generation, the role of biomass is expected to be for repowering old units and for use in small (20- to 50-MW) new plants. Several new high-efficiency conversion technologies are either already available or under development for the utilization of biomass.

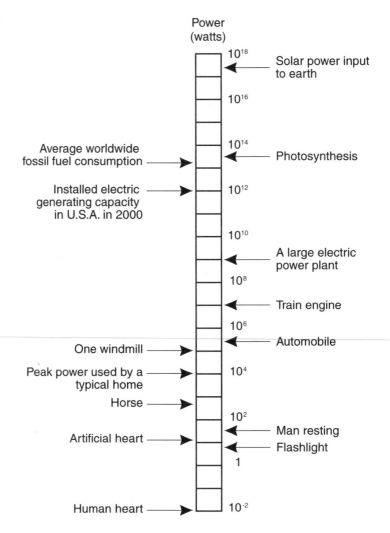

Figure 16.1 **The power of various devices and of animals and humans in watts.** The *power* of a device is the time rate of delivery of energy from the device. The unit of power in the SI system is joules per second or watts.

When biomass is grown at the rate it is used for energy, there are no net CO_2 emissions from the biomass. Air pollutant emissions in conversion to useful energy forms and energy services depend on the conversion technologies involved, except that biomass conversion is generally characterized by very low SO_2 emissions, owing to the low sulfur content of biomass. Gasification-based power-generating technologies now under development will have low emissions of all local pollutants, except in some cases NO_x emissions arising from fuel-bound nitrogen, which might require the use of emission control equipment.

Fuels, such as ethanol, can be produced from low-cost cellulosic materials (e.g., various residues in the near term and also energy crops in the longer term) via enzymatic hydrolysis. The challenge is to find cost-effective ways to convert the cellulose and hemicellulose in these feedstocks into component sugars via hydrolysis and then ferment those sugars to ethanol. However, there is little near-term prospect that ethanol derived from corn can be provided without federal subsidy, which is currently 54 cents per gallon.

U.S. production goals for ethanol derived via enzymatic hydrolysis is for 16 billion gallons annually by 2020. The development of dedicated energy crops may be key to the ultimate successes of the biopower and biofuels programs, because in the long run biomass may play major roles in the energy economy only if residue supplies are supplemented by biomass grown as dedicated energy crops.

Biomass can be only a partial answer to the global energy need because of two fundamental constraints: its high water requirements and the inherent low photosynthetic efficiency of converting solar energy into the chemical energy of plant matter. High water requirements constrain biomass production mainly to regions where rainfall is adequate to support commercial yields, whereas the low photosynthetic efficiency can lead to land-use competition with food production. In addition, there may be practical constraints relating to costs for biomass energy crops.

While it is possible to assign numbers to potential energy crop resources, it is unclear how a market for them can be created. R&D can bring the cost of biomass production down, but at the moment there is no industry active in developing energy crops as there are industries active in wind power and photovoltaics. Moreover, if energy crops are to succeed, farmers must become fuel suppliers catering to an entirely different set of customers (electric utilities) than that with which they are accustomed to dealing.

The combustion of biomass or biofuels produces air pollutants, including carbon monoxide, nitrogen oxides, and particulates such as soot and ash. The amount of pollution emitted per unit of energy generated varies widely by technology, with wood-burning stoves and fireplaces generally the worst offenders. Emissions from conventional biomass-fueled power plants are generally similar to emissions from coal-fired power plants, with the notable difference that biomass facilities produce very little sulfur dioxide and toxic metals.

Much urban waste is combustible. The potential exists for burning 100 million tons of refuse each year and using the heat to produce steam for heating or electricity generation. Another possibility is to convert the waste to alcohol-based fuels. At this time, only about 3% of the U.S. energy is derived from biomass, nearly all of it from making ethanol derived from corn or from burning wood. But many assert that the world needs to rely more on biomass in order to cut reliance on oil imports and reduce the threat of global warming.

16.2 Research, Development and Commercialization of the Kenya Ceramic Jiko (KCJ): An Example of Environment-Friendly Technology Cooperation and Transfer

By Daniel M. Kammen
Associate Professor of Energy and Society
Director, Renewable and Appropriate Energy Laboratory
Energy and Resources Group
University of California, Berkeley

16.2.1 INTRODUCTION: IMPACT AND OPPORTUNITY IN THE HOUSEHOLD ENERGY SECTOR

With roughly half of the global population cooking daily with the traditional biomass fuels of dung, crop residues, wood, and charcoal, efforts to disseminate improved efficiency cookstoves have long seemed an ideal way to address a wide range of socioeconomic and environmental goals. These include energy conservation; time saved from wood collection; expanded economic opportunities for both rural and urban families; empowerment of women; a reduction in harmful household woodsmoke exposure; reduced forest clearing and ecological alteration; and even the mitigation of global atmospheric pollution. The widespread dissemination and adoption of improved woodstoves has the potential to impact each of these economic and environmental issues, and has thus been a focal point of household development and quality of life efforts for several decades.

Source: Majid Ezzati.

Figure 16.2 **Three-stone fire for cooking (Kenya).**

The raw statistics describing the role of biomass in the energy–health–environment cycle are dramatic and instructive. Half the worldwide wood harvest of more than 3 billion tons is used as fuel. Wood and other biomass fuels comprise 40 to 60% of the total energy resource, both industrial and domestic, for many developing Asian, Latin American, and African nations. Domestic cooking accounts for over 60% of total *national* energy use in sub-Saharan Africa and exceeds 80% in several nations (Barnes *et al.*, 1994; Kammen, 1995a,b). Further, some poor families expend 20% or more of disposable income to the purchase of wood and charcoal fuels, or devote upward of 25% of total household labor to wood collecting. Reliance on biomass fuels, which are generally utilized at a low thermodynamic efficiency, thus not only entails a high opportunity cost for poor households, but also results directly in health and environmental costs as well (Figure 16.2).

The potential for improper biomass management to degrade the environment is quite dramatic. While the need for cooking fuel is only one of many demands on the biomass resource, which also include commercial timber harvesting and forest clearing or land conversion for agriculture, domestic wood consumption represents a constant need, however, and in several developing nations exceeds *one ton* per capita per year (Nkonoki and Sorensen, 1984; Goldemberg *et al.*, 1985). The environmental impact of wood use — involving industrial, agricultural, and domestic applications — of course varies widely, from areas with ecologically sustainable harvest levels, to regions where the population density and fuelwood demand alters the type of forest cover and biodiversity, and in the most extreme cases to areas of dramatic deforestation and erosion (Figure 16.3).

Figure 16.3 **Biomass combustion indoors can lead to high pollution levels, as seen emerging from this rural home.**

Biomass cooking fuels are often combusted inefficiently in a variety of traditional cooking practices, where sometimes as little as 5 to 15% of the total energy content of the fuel is utilized to heat the food (Openshaw, 1979, 1982; Barnes *et al.,* 1994; Kammen, 1995a,b). This is due to the prevalence of open "three-stone" fires and some traditional stove designs that generate large quantities of smoke and particulate matter, while at the same time directing only a small fraction of the resulting heat to the cooking pot or pots.

Health

Approximately one-half of the world's population relies on biomass — wood, crop residue, dung, and charcoal — as the primary source of domestic energy, burning two billion kilograms of biomass every day in developing countries (Reddy *et al.,* 1994; Ezzati *et al.,* 1999).

Elevated levels of indoor air pollution have serious health implications. High levels of woodsmoke exposure — often 10 or more times the recommended World Health Organization exposure limits — have been reported in emission studies throughout developing nations. This, in turn, has been and linked to acute respiratory infection (ARI), in particular pneumonia, along with a number of other ailments. The segment of the population most continuously exposed to indoor air pollution are women — who generally perform over 90% of domestic chores including cooking — and their children (Figure 16.4). The resulting pollution levels in homes and cooking huts can even exceed those measured in industrial cities, and in some cases represents the equivalent particulate dosage of smoking several *packs* of cigarettes per day (Smith, 1994). ARI is, in fact, the leading health hazard to children in developing nations and results in an estimated 4.3 million deaths per year (Smith, 1994). Among all endemic diseases, including diarrhea, ARI is the most pervasive cause of chronic illness.

Exposure to indoor air pollution, especially to suspended particulate matter, resulting from the combustion of biomass has been implicated as a causal agent of acute respiratory infections (ARI) and eye infections (c.f. Chen *et al.,* 1990; Ezzati *et al.,* 1999). Acute and chronic respiratory infections together account for over 10% of the total burden of disease in developing countries (World Bank, 1993; World Health Organization, 1999). In 1997 and 1998, the leading cause of death from infectious diseases was acute lower respiratory infections with an estimated 3.7 and 3.5 million deaths worldwide for the 2 years, respectively.

Biomass Management Issues

The implication of a continuing dependence on biomass fuels, traditional woodfuel management schemes, and traditional cooking stoves has suggested to many researchers a dramatic opportunity for development efforts: the introduction of improved cookstoves. While specific designs vary widely, all "improved" cooking stoves are intended to consume less fuel per amount of useful energy delivered and to emit less pollution to impact the health of the user and the environment generally. The dissemination and

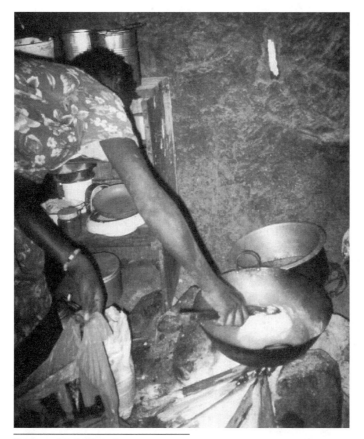

SOURCE: Majid Ezzati.

Figure 16.4 **Women in less developed nations, who generally perform more than 90% of domestic chores, including cooking, and their children are continuously exposed to indoor air pollution.**

adoption of technically and culturally appropriate stoves could reduce the need for fuelwood harvesting and human impact on forest ecosystems, while at the same mitigating human exposure to indoor air pollution. Reduction in the labor and capital required to collect or purchase cooking fuels can also provide new resources and expand the economic opportunities to women throughout the developing world. Indeed, the interdependence of so many aspects of the energy–food–health–economy development milieu hinging on the management of biomass fuels and cooking techniques has led one researcher to coin the phrase "the hearth as system central" (Smith, 1991).

16.2.2 COOKSTOVES AND THE TECHNOLOGY FOR DEVELOPMENT PARADIGM

Improved cookstove research and dissemination programs have become ubiquitous, with recent surveys identifying several hundred individual projects in over 50 nations (Barnes *et al.*, 1994). The scale of cookstove programs range from entirely local, nongovernmental advocacy and implementation groups, to national initiatives reaching over 100 million homes (Smith *et al.*, 1993), to regional efforts sponsored by multinational development agencies. The range of stove types, from massive sand and clay models to a variety of portable metal and ceramic designs (Figure 16.5), is matched by the diversity of technology transfer paradigms that have been tested and utilized: from the establishment of commercial cookstove mass-production and sales; to training programs where the eventual users, primarily women, adapt and construct stoves for their home use.

The implications of such a variety of development schemes has naturally been equally profound. The quality and efficiency of improved cookstoves has varied greatly, as has the success of individual projects at reaching the intended audience or developing a self-sustaining market or cookstove industry. The long history and wide range of cookstove projects not only defines this technology as a crucial test for ideas of the small-scale, household-oriented, paradigm of development (c.f. Boserup, 1970, and the many papers on this topic that followed) but also as a model for design and dissemination efforts with other renewable energy technologies, such as biogas digesters, solar ovens and food dryers, household photovoltaic systems, windpumps, and micro-hydro stations.

SOURCE: Majid Ezzati.

Figure 16.5 **Improved stoves in use in central Kenya.**

The evolving tools and continuing pitfalls of technology transfer and implementation projects can be critically examined through an analysis of the changing project goals, and implementation strategies for improved cookstove design and dissemination efforts. In this section, I summarize the changes in both the technology of improved cookstoves and the resulting opportunities for economic, health, and environmental conservation and social empowerment. Notable in this process are the expanding degree of communication and cooperation between project implementation and end-user groups, the growth of market-based technology transfer mechanisms, the analysis of the rural vs. urban market potential, and the design of multi-objective efforts that integrate disparate actors in the development process. It is the evolution in the understanding and interaction of these factors in the cookstove design, development, dissemination, and adoption chain that provides the clearest lessons for the role of household energy management in social change, empowerment, and development planning generally.

16.2.3 IMPROVED STOVES: DESIGN ENGINEERING AND ENERGY EFFICIENCY

Improved cookstove designs fall into two broad categories: *lorena*-type,[1] bulky and generally immobile stoves of mud, clay, sand or cement with several openings for pots and commonly a chimney over an enclosed burning box; and portable models that accommodate one or a few pots that sit atop an enclosed fuel combustion box. The *lorena* stove is just that, a "complete" stove, while the portable model is essentially a "single burner" (Figures 16.5, 16.6, and 16.7). The efficiency gains for both types of improved stoves stem from design attention paid to the same set of factors:

- *Maximize combustion* of the fuel by keeping the temperature high and ensuring the presence of sufficient oxygen.
- *Maximize radiative heat transfer* from the fire to the pot(s) by keeping the pot as close to the flame as possible.
- *Maximize convection* from the fire to the pot(s) with a stove geometry that passes as much of the hot gases over the pot(s) as is possible. Reduce drafts.
- *Maximize conduction* to the food pot(s) by using an insulating material for the stove so that heat is retained and concentrated near the pot(s).
- *Maximize user satisfaction* in terms of convenience of use (with local fuels, cooking pots and utensils), and the ability to easily prepare local dishes well.

In summary, only a stove with what one might call *robust efficiency* will consistently save fuel under conditions of actual use. The stove must be easy to use and fuel efficient when used in a variety of configurations: to boil;

---

[1]Derived from the Spanish *lodo* (mud) and *arena* (sand). A new generation of "chula" stoves, however, are lighter and more portable outgrowths of the initial *lorena* design.

slow simmer; bake; fry; and when only one opening in a large "three-pot" stove is being used; or when the stove is dirty or worn. This requirement has more often than any other been the source of failed stove programs where the potential for a stove design is evaluated under idealized laboratory conditions that is far from a reasonable approximation of the environment and practical constraints seen in real-world kitchens and cooking huts. Cookstoves are workhorses, not racehorses, and must be designed accordingly.

Early stove projects were heralded as the solutions to a tremendous array of social, economic, and environmental ills — from deforestation to the oppression of women. While improved cookstove efficiency and household energy security can lead to improvements in all of these areas, evaluations of early projects were generally disappointing.[2]

The reasons for the early failures spanned both the technical and social aspects of the stove projects. With the diversity of stove programs came a problem of quality where:

> early improved cookstoves were often designed by development workers with a great deal of zeal and enthusiasm but little technical background. Under the banner of "appropriate technology," new designs were quickly labeled "improved stoves" and construction manuals prepared, without prior serious scientific research (Krugmann, 1987).

In fact, advocates and researchers involved in many early stove projects have fallen into the trap of equating "appropriate" technology with "simple" technology. The first four design factors listed above, essentially thermodynamic criteria,[3] have proved difficult, particularly given the variability and variety of materials and real-world usage patterns.

Many early programs expected to see efficiency gains of 75% *or more* (Barnes *et al.,* 1994), often based on tremendously idealized cooking conditions never realized in the field. At the same time improved stoves were afforded too much credit technically. Claims of the inefficiency of traditional three-stone fires of 5 to 10% neglected many of the benefits of lighting, space heating, and maximized ease of use and versatility. In fact, under conditions of shielded, carefully tended use, three-stone fires have reached efficiencies of 20%. Further, many of the new stoves were expensive and difficult to use, and degraded rapidly with use. Further, construction of *lorena*-type stoves in each house (thus built one unit at a time without standardization or economies of scale) often resulted in uneven quality and thus efficiency. Overall, many new stove designs and dissemination programs failed the "robust efficiency" test.

---

[2]The demand for woodfuel leads to local shortages and often long transportation distances to urban centers, but is not a major contributor to deforestation. Agricultural land conversion and alteration as well as commercial logging are the primary causes of deforestation.

[3]For an excellent analysis of the design and evaluation of cookstoves based rigorously on the principles of heat transfer and materials science, see Baldwin (1987).

In many early projects discussion, feedback, and training programs for the *real* end-users, generally women, often did not exist or were inappropriately targeted to men or extension workers who rarely cooked. Taken in sum, the oversell of improved stoves and undersell of the reliability and versatility of traditional methods meant that the fuel savings in some of these early efforts often amounted to little or nothing.

While sometimes disappointing, the diversity of stove projects and design problems resulted in careful engineering reevaluation and redesign of improved stoves, greater end-user participation in the process, and far more rigorous and realistic measures of actual stove performance. This more realistic analysis began to yield useful efficiency and cost comparison data for second-generation stove projects.

A series of measurements of actual stove efficiencies conducted in West Africa is particularly instructive. In Table 16.1, the percent heat utilized

Table 16.1 **Efficiency Comparison of Several West African Improved and Traditional Stoves**

Stove	Description	PHU	% Increase in % PHU over:		Decrease in wood use over:	
			3-Stone	Metal	3-Stone	Metal
Three Stone	Pot supported by three stones over open fire	10.2	0	—	0	—
Metal Stove	Simple one-pot metal stove	14.5	42	—	29	—
SIM	Sand insulation placed around the simple metal stove	16.1	58	11	36	10
Sota	Clay shell stove around single pot	18.4	80	27	44	21
AIDR	Three-pot partially insulated stove, without chimney	10.9	7	—	6	—
GS Chula	Insulated two-pot lorena-type stove with chimney	15.2	49	5	32	5
Nouna	Brick and cement lorena-type stove	15.3	50	6	33	5
CATRU-B	Lorena-type, aluminum top plate and matched pots	17.2	69	19	41	15
CATRU-A	CATRU-B with improved chimney	25.9	154	79	61	43

Comparison of percent heat utilized, PHU, and wood savings for a variety of traditional and improved stoves in West Africa. The PHU measure is based on a combination of the initial boiling and sustained cooking ("simmering") phase — thus providing a proxy for realistic cooking conditions. The PHU values reported here are averaged over five or more measurements per stove. The comparison to both a traditional three-stone fire and a metal stove (Metal) is presented. The Three Stone, Metal, SIM, and Sota stoves all accommodate one pot only, while the AIDR has openings for three pots and the remaining four stoves all have two openings. The range of wood savings, as expected, range from minimal (6%) to dramatic (61%). Data sources: Baldwin (1987), Baldwin *et al.* (1985), Kammen (1995b).

(PHU) for a variety of stoves is compared. The PHU is determined by boiling and then continuing to cook a volume of water. The total fuel combusted, volume of water boiled, and water boiled off in comparable field settings provides a quantitative measure of the total (energy utilized/energy expended) ratio (Baldwin, 1987).

For the stoves listed in Table 16.1, the prices range from essentially nothing, aside from the cooking pots (three-stone fire), to $0.6 for the metal stove, $5 for the GS Chula (when home made), to for the $12 for the Nouna stove, and over $20 for the CATRU-A stove. At the prevailing wood costs these prices correspond to amortized payback times ranging from one month to about a year.[4]

With fuel savings compared to traditional stoves of as much as 60% for the expensive CATRU-A stove, to as little as 5% for some simple and inexpensive designs, embarking on stove dissemination efforts clearly should not be taken lightly. This is particularly true when one considers the financial investment by users and donors, and the long-term commitment of research, training, and support services seen in successful stove dissemination programs. One bright spot, however, has been the growing realization of the wide range of important health benefits of reducing woodsmoke exposure. A particular case study will help to focus analysis and insight into the workings and policy measures available to promote particular programs.

16.2.4 KENYA CERAMIC JIKO CASE STUDY: SUMMARY

The Kenya Ceramic Stove, or *Jiko* (KCJ), is a charcoal-burning stove that is roughly 30% efficient, and if used properly can save 20 to 50% in fuel consumption over simple "unimproved" stoves or a traditional three-stone fire (Walubengo, 1995). The KCJ was developed after study of a Thai "bucket" stove that was examined partially through a "South–South" dialog over stove characteristics and design.

The KCJ is a portable improved charcoal burning stove consisting of an hourglass-shaped metal cladding with an interior ceramic liner that is perforated to permit the ash to fall to the collection cavity at the base (Figure 16.6). A single pot is placed on the top of the stove. There are now more than 200 businesses, artisans, and micro-enterprise or informal sector manufacturers producing over 13,000 stoves each month. There are over 700,000 KCJs in use in Kenya (Walubengo, 1995). The KCJ is found in over 50% of all urban homes and roughly 16% of rural homes. Stove models adapted from the KCJ are now being disseminated in many countries across Africa, and wood-burning variants are being introduced and promoted in rural areas as well (Figure 16.7).

---

[4]Based on urban wood or charcoal costs of $0.3 to 0.4 per family per day (Baldwin *et al.*, 1985).

SOURCE: Majid Ezzati.

Figure 16.6 **The distinctive hourglass-shaped Kenya ceramic Jiko stove is about 30% efficient if used as intended.**

SOURCE: Majid Ezzati.

Figure 16.7 **The Jiko stove is found in more than 50% of all urban homes and in about 16% of rural homes in Kenya.** Here women in a workshop in rural Kenya examine and select for adaptation various stove models.

The fuel savings of the KCJ have important economic benefits to the users who in some cases devote a quarter of family income to charcoal purchases (Kammen, 1995a). The stoves can also reduce the pollution exposure of families using the stove. The WHO reports that more than two million premature deaths per year can be attributed globally to the indoor air pollution caused by household solid fuels. Reducing the harmful products of incomplete combustion produced by household stoves is an important benefit from the development of cleaner cookstoves.

16.2.5 APPROACH

The KCJ is the result of research on stove design, efficiency, and patterns of usage initiated in the 1970s and actively continued through the 1980s (Barnes *et al.,* 1994; Kammen, 1995a). A single private sector company, Jerri International, served as the initial manufacturer of the KCJ.

Since 1982 the Kenya Energy and Environment Organization (KENGO) has organized promotion and outreach efforts to encourage the use of the KCJ. A number of NGOs and national development agencies have played important roles in the evolution of the stove and the stove dissemination process, and have worked both within Kenya and across sub-Saharan Africa to promote the manufacture and sales of the KCJ through a network of informal-sector stove entrepreneurs.

A decision was made not to directly subsidize commercial stove production and dissemination. Initially stoves were expensive (~ US$15/stove), sales were slow, and quality control was a significant problem. Continued research and refinement and expanded numbers and types of manufacturers and vendors increased competition, and spurred innovations in materials used and in production methods. The KCJ can now be purchased in a variety of sizes and styles. Prices for KCJ models have decreased to roughly US$1 to $3 (Walubengo, 1995). This decrease is consistent with the "learning curve" theory of price reductions through innovations that result from experience gained in the manufacturing, distribution, marketing and sales process. Two architects of the stove program received an international award for their work, which is an important recognition for the need for research on often unheralded but important technologies.

The ceramic liner of the KCJ degrades over time and needs to be replaced. Street vendors of stoves and many of the larger stove sales outlets take "used" stoves back, discounting the purchase of a new stove. The liners of the old stove are then removed, the metal cladding is repaired, if needed, and the stove is reassembled, repainted, and resold. This process has also served to foster a wider informal-sector-based stove economy.

16.2.6 IMPACTS

The KCJ can reduce fuel use by 30 to 50%, although charcoal production itself can have significant environmental impacts, and the attractiveness of the KCJ may have increased this demand. The KCJ also reduces emissions of trace gases and particulate matter, which contributes to acute respiratory infection, the leading cause of illness in developing nations. Reported levels of emissions reductions from KCJ range up to 50% although this is a subject of ongoing research (Barnes *et al.*, 1994; Ezzati *et al.*, 1999). The KCJ and the dissemination process used in Kenya has now been widely disseminated (and adapted) across sub-Saharan Africa.

16.2.7 LESSONS LEARNED

While avoiding direct subsidies, a number of organizations provide training, outreach services, publicity, and logistical support for the local commercial industry. This "soft" subsidy can be particularly effective in facilitating the development and acceptance of a new technology without introducing the price distortions that can be associated with some forms of subsidy.

The lessons for international involvement that can be drawn from the KCJ case include:

- Support for research, both within developing nations and for research collaborations between developing nations, can lead to significant innovations in the performance and commercialization of what had been regarded by many as a simple and mature technology.
- Extended, stable, program support is invaluable, whereas short-lived, episodic funding can lead to waste and inefficiency. Significant technical, social, cultural, and economic questions must be addressed even for technologies that may appear simple.
- Support for stove programs need not take the form of direct subsidies. Partnerships between institutional groups, including NGOs and international organizations, involved in R&D, promotion, and training can support commercial producers and sellers if the mechanisms for feedback and cooperation are planned and developed.

16.3 Geothermal Energy

Over the life of the earth, thermal energy has been stored within its core. Of this energy stored within the earth, some is transmitted from the interior by means of conduction through the earth. The average rate of flow of heat to the surface has been found to be 0.063 W/m^2. For the earth's surface of $510 \times 10^{12} \text{ m}^2$, the total heat flow amounts to $32 \times 10^{12} \text{ W}$. Of this, only 1% of the total rate, or $0.3 \times 12^{12} \text{ W}$, is due to heat convection by hot springs and volcanoes.

Unfortunately, we cannot directly exploit this heat supply, but we can use local hot spots — subterranean reservoirs where the heat has been stored in the form of steam and hot water. Such reservoirs are the source of geothermal energy.

Geothermal energy is the heat energy inside the earth. It is used today in the western United States to produce electricity or heat from underground steam or hot water (hydrothermal) resources. This is the only type of geothermal resource that is currently commercially developed. Hydro-thermal resources are limited; however, in the future, hot dry rock (HDR) and other advanced resources might be tapped. Another component of geothermal technology that is also being developed is the ground source heat pump; currently more than 250,000 U.S. houses have them.

A geothermal reservoir can store enormous amounts of energy in the form of steam and hot water. Temperatures of the water or steam are as high as 700°C. Such reservoirs are most readily tapped within depths of 2,000 to 8,000 ft.

Another type of reservoir containing water at 100° to 150°C at relatively shallow depths is located in large sedimentary basins. Heat at this temperature is useful for space heating and for agricultural, mining, and other purposes.

Regions of high heat flow usually display natural hot springs, geysers, and steam vents such as in Yellowstone National Park. These features represent leakage from a geothermal reservoir.

The Geysers Power Plant in Sonoma County north of San Francisco, California, was the first geothermal plant active in the United States. The Geysers is located on the steep slopes of Cobb Mountain, an extinct volcano. Steam is obtained from fumaroles and steam vents that emit steady steam vapor — not intermittent geysers, despite the name of the site. The first generating unit began operation in 1960 and provided 11 MW. Development of the site continued and the capacity of the site was about 500 MW in 2000.

Currently the United States is a major producer of geothermal electric power, with an installed capacity of 2,700 MW. Worldwide capacity is now greater than 7,000 MW with much growth occurring in developing countries. Projections indicate that more than 10,000 MW total may be installed by 2000. In the United States, with favorable energy markets and financing, an additional 5,000 MW could be added by 2015. Although hydrothermal-based geothermal energy can make an important contribution to energy needs, wide-scale use is not possible because of the limited size of the resource and because it is confined to areas associated with recent volcanism or near the boundaries of tectonic plates, such as along the Pacific Coast.

16.4 Energy Storage

Energy storage involves the collection and retention of readily available energy for later use. Energy available at periods of low demand should be stored for times of peak demand. Often the peak power demand of an electric distribution system can be twice the minimum level of power demand. Thus, storage of hydroelectric, wind energy, or solar energy is necessary. In this section we discuss the use of batteries, pumped water, compressed air, and flywheels for energy storage. In a later section we discuss the use of hydrogen for storage.

Storage will take on added importance in the future to ensure reliable, high-quality service. It will provide for increased renewable use and system stabilization with distributed generation. Areas of importance include pumped hydro, compressed air, battery, and inertial technologies covering a wide capacity range.

Storage batteries may reach economical use for the storage of large amounts of energy within the next 20 years. Batteries have the advantage of minimal siting problems and can be placed near the anticipated load, thus reducing transmission costs. Efforts are needed to produce a battery with a specific energy of over 220 watt-hours (Wh) per kilogram (100 Wh/pound) and a specific power of over 55 W/kg (25 W/pound). In addition, a lifetime of 4 years or more and a life of 1,000 cycles is desired. Furthermore, the efficiency of storage and discharge should be relatively high — normally more than 70% in total. Unfortunately, no presently available storage batteries meet these requirements.

A compressed-air storage system uses an air compressor to place air in an underground cavity or a storage tank. When the storage energy is required, the compressed air is used to drive a turbine and an electric generator. During off-peak hours, energy is used to drive the pump and store air in the cavern or tank. Air can be stored in salt domes, mine caverns, depleted gas fields, abandoned mines, or an aquifer.

The overall efficiency of a compressed-air storage system is estimated to be 70%. During offpeak hours, electricity is fed to the motor, which drives the compressor, sending air to the storage vessel. At times of peak demand, the air is released to the turbine-generator to yield electric power to the consumer.

Compressed-air energy storage in particular is well suited for use in both increasing the value of wind power and in making possible high levels of penetration of wind power on electric grids. The technology is commercially available and has been demonstrated with storage in solution-mined caverns in bedded salt.

Hydroelectric pumped storage systems use water that is pumped up to a reservoir by electric pumps and then allowed to fall. On the eastern shore of Lake Michigan is the biggest pumped storage plant in the world. The Ludington facility takes water from Lake Michigan and stores it in a man-made reservoir 1 mile wide by 2 miles long. At times of peak electrical power demand, water is released from the reservoir and drives turbines generating 1.9 GW of electricity. Reversible turbines are used as pumps and driven turbine-generators. Water is usually pumped up at off-peak hours, usually at night and on weekends.

The flywheel, which is one of the oldest of human inventions, is an inertial storage device. The flywheel offers the potential for storing energy on a large scale to handle the peak loads of utilities and to power electric vehicles. The principle of a flywheel is that a spinning wheel stores mechanical energy in proportion to the square of the angular velocity of the wheel. As simple as they are, however, flywheels so far have been too expensive to replace traditional lead–acid batteries.

The economics of energy storage is complicated. Nevertheless, the potential for efficient storage of energy is good, and it should be developed to shift off-peak power to use at peak hours.

16.5 Energy Storage Technologies

By Mukund R. Patel
Faculty of Engineering
U.S. Merchant Marine Academy

Electrical energy is being used increasingly in our society because it is a highly ordered form of energy that can be easily converted into mechanical or thermal form with near 100% efficiency. Heat energy, on the other hand, cannot be converted into electricity with efficiency over 50%, because it is a disordered form of energy in atoms.

Electricity, however, has a disadvantage. It cannot be easily stored on a large scale. Almost all electrical energy used today is consumed instantaneously as it is generated. This poses no hardship in conventional power plants (fossil, nuclear, and hydro), where fuel consumption is varied with the instantaneous load requirement. However, a renewable power plant, such as a wind, photovoltaic, or ocean energy plant, having an uncontrollable energy source, cannot meet the instantaneous load demand all the time. Large-scale energy storage, therefore, is a desired feature to incorporate in such an intermittent nondispatchable power system for storing energy when in excess and using it when in shortage. It can significantly improve the load availability to the consumers, a key requirement for any power system, particularly for stand-alone plants.

The energy storage technologies that may be incorporated in any power system design are:

- Electrochemical battery
- High-speed flywheel
- Superconducting coil
- Compressed air.

16.5.1 ELECTROCHEMICAL BATTERY

The battery is the most widely used means of energy storage for a variety of applications. It stores energy in the electrochemical form, which is a semi-ordered form of energy having one-way conversion efficiency of 80 to 90%. The battery is made of numerous electrochemical cells connected in a series-parallel combination to obtain the desired operating voltage and current. Each cell can store a certain charge at a certain voltage, measured in ampere-hours (Ah) and volts (V), respectively. The battery rating is stated in terms of the average voltage during discharge and the amp-hour capacity it can deliver before the voltage drops below the specified limit. The product of the discharge voltage and the amp-hour capacity creates the watt-hour energy rating the battery can deliver to a load from the fully charged condition.

There are two basic types of batteries: (1) the primary battery that converts chemical energy into electrical energy in a nonreversible reaction and (2) the secondary battery, also known as the rechargeable battery, that converts energy in a reversible reaction. After a discharge, it can be recharged by injecting direct current from an independent source. The round-trip conversion efficiency is between 65 and 80%. Six major rechargeable electrochemistries are:

- Lead-acid (Pb-acid)
- Nickel-cadmium (NiCd)
- Nickel-metal hydride (NiMH)
- Zinc-air
- Lithium-ion (Li-ion)
- Lithium-polymer (Li-poly).

Their performance in terms of the specific energy (Wh/kg) and energy density (Wh/liter) are summarized and compared in Figure 16.8. Having the least cost per watt-hour delivered over the life, the lead-acid battery has

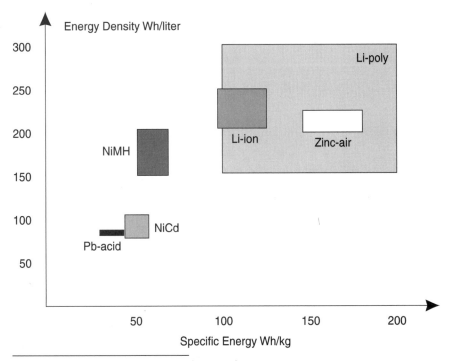

SOURCE: Reprinted with permission from Patel, 1999.

Figure 16.8 **Specific energy and energy density of various batteries.**

been the workhorse of the industry. New electrochemistries are being developed by the United States Advance Battery Consortium for a variety of applications, such as electric vehicles and utility load leveling and for renewable energy systems.

16.5.2 HIGH-SPEED FLYWHEEL

The flywheel stores kinetic energy in a rotating wheel of high inertia (Figure 16.9). This energy can be converted into electricity using an electrical generator. The flywheel energy storage is an old concept that has become commercially viable now due to recent advances made in high-strength lightweight fiber composite rotors and the magnetic bearings that operate at high speed. The round-trip conversion efficiency of a large flywheel system can approach 90%, significantly higher than the battery.

Small to medium-size flywheels have been in use for years. Considerable efforts are under way around the world to develop high-speed flywheels for large-scale energy storage. The present goal of these developments is to achieve five times the specific energy of the currently available secondary

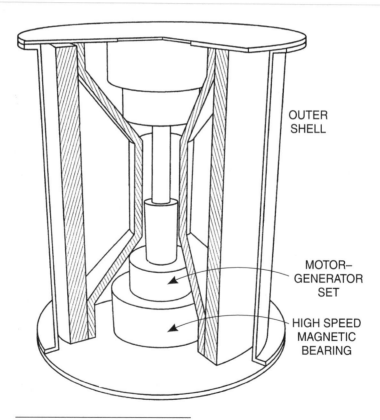

OUTER
SHELL

MOTOR–
GENERATOR
SET

HIGH SPEED
MAGNETIC
BEARING

SOURCE: Reprinted with permission from Patel, 1999.

Figure 16.9 **High-speed flywheel assembly.**

batteries with the following enabling technologies, which are already in place in their component forms:

- High-speed magnetic bearings, which eliminate friction, vibrations, and noise
- High-strength fibers having ultimate tensile strength of over one million pounds per square inch
- Advances made in designing and manufacturing fiber-epoxy composites.

The energy stored in a flywheel having the polar moment of inertia J and the angular speed of ω is given by

$$E = \frac{1}{2} J \cdot \omega^2 \text{ Joules.} \qquad (16.1)$$

The maximum energy that can be stored is limited by the mechanical stress due to the centrifugal force at high speed. In a rotor material of density ρ, the centrifugal force at radius r is given by $\rho(r\omega)^2$, which is supported by the hoop stress in the rim. The allowable stress in the material places an upper limit on the rotor tip speed. A given material has a limiting tip speed. Therefore, small rotors can run at high speed, and large rotors must run at low speed. Lightweight materials produce low centrifugal stress, allowing higher speeds. Rotor materials with high ultimate strength and low mass density result in high specific energy. Thin rim rotors have high inertia, and hence store more energy per unit of weight. For these reasons, good flywheel designs use thin rim rotor with high σ_{max}/ρ ratio to achieve high specific energy. A high E/ρ ratio is also necessary for rigidity, where E is Young's modulus of elasticity.

The flywheel energy storage system requires the following components:

- High-speed rotor attached to the shaft via compatible hub
- Bearings with good lubrication or on magnetic suspension
- Electromechanical energy converter (electrical machine)
- Power electronics to drive the motor and to condition the generator power
- Control electronics for the magnetic bearings and other functions.

Conventional bearings are used up to speeds of a few tens of thousand revolutions per minute (rpm). Speeds approaching 100,000 rpm are possible by using magnetic bearings, which support the rotor by magnetic repulsion and attraction. Mechanical contact is eliminated, thus eliminating the friction. The windage is eliminated by running the rotor in vacuum.

Electromechanical energy conversion is achieved with one machine, which works as motor for spinning up the rotor for charge, and as a generator while decelerating the rotor for discharge. Two types of electrical machines can be used, the synchronous machine with variable frequency converter or the permanent magnet brushless dc machine.

The main advantages of the flywheel energy storage over the battery are:

- High energy storage capacity per unit of weight and volume
- High depth of discharge up to 90%
- Long cycle life, which is insensitive to the depth of discharge
- High peak power capability without overheating concerns
- Easy power management, because the state of charge is simply measured by the speed
- High round-trip energy efficiency
- Flexibility in design for a given voltage and current
- Improved quality of power because the electrical machine is stiffer than the battery.

These benefits have the potential of making the flywheel the least cost energy storage alternative per watt-hour delivered over the operating life.

16.5.3 SUPERCONDUCTING COIL

In the working principle, a coil carrying electric current stores energy in the resulting magnetic field. The energy storage per unit volume is given by

$$E = \frac{1}{2}\frac{B^2}{\mu} \quad \frac{\text{Joules}}{\text{m}^3}. \tag{16.2}$$

In term of inductance, it can be expressed as

$$E = \frac{1}{2}I^2 \cdot L \text{ Joules}. \tag{16.3}$$

where $B =$ the magnetic field density produced by the coil (tesla), $\mu =$ magnetic permeability of air ($4\pi\ 10^{-7}$ Henry/m), $I =$ Current in the coil (ampere), and $L =$ Inductance of the coil (Henry).

The electrical resistance of the coil conductor is temperature dependent. In certain conductors, the resistance abruptly drops to precise zero at some critical temperature near absolute zero. The coil reaching the critical temperature is said to have attained the superconducting state having zero resistance. The coil can then carry extremely high current with correspondingly high energy storage indefinitely with no loss. The energy in the coil is viewed as "frozen."

The development efforts in the superconducting energy storage technology have started yielding promising results. Conceptual designs of large superconducting energy storage systems up to 5,000 MWh of energy for utility applications have been developed and prototypes tested.

A typical superconducting energy storage system is shown in Figure-16.10. The coil is charged by an ac-to-dc converter in the magnet power supply. Once fully charged, the converter continues providing the small voltages needed to overcome losses in the room-temperature parts of the circuit components. This keeps the constant dc current flowing (frozen) in the superconducting coil. In the storage mode, the current is circulated through the normally closed switch.

The system controller has three main functions:

- Control the solid-state isolation switch.
- Monitor the load voltage and current.
- Interface with the voltage regulator that controls the dc power flow to and from the coil.

During steady-state operation, if the system controller senses the line voltage dropping, it interprets that the system is incapable of meeting the load demand. The switch in the voltage regulator automatically opens in less than 1 ms. The current from the coil now flows into the capacitor bank until the system voltage recovers the rated level. The capacitor power is inverted into 60 or 50 Hz ac and is fed to the load. As the capacitor energy is depleted and

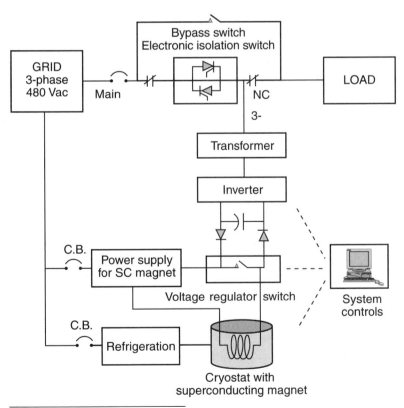

SOURCE: Reprinted with permission from Patel, 1999.

Figure 16.10 **Superconducting energy storage system.**

the bus voltage drops, the switch opens again, and the process continues to supply energy to the load continually. The system is sized to store required energy to power the load for a specified duration.

Superconducting coil energy storage has several advantages over other technologies:

- The round-trip efficiency of the charge-discharge cycle over 90%, which is higher than that attainable by any other technology
- Much longer life, up to about 30 years
- Charge and discharge times can be extremely short, making it attractive for supplying large power for short time if needed
- Has no moving parts in the main system, except in the refrigeration components.

The primary operating cost of the system results from the need to keep the coil below the critical superconducting temperature. Until now the niobium-titanium alloy has been extensively used, which has the critical temperature of about 9 K. This requires liquid helium as coolant at around 4 K. The 1986 discovery of high-temperature superconductors has accelerated industry interest in this technology. Three types of high-temperature superconducting materials are available now, all made from bismuth or yttrium-cuprate compounds. These superconductors have a critical temperature of around 100 K. Therefore, they can be cooled by liquid nitrogen, which needs orders of magnitude less refrigeration power. As a result, numerous programs around the world have started to develop commercial applications.

16.5.4 COMPRESSED AIR

The compressed–air energy storage system consists of:

- Air compressor
- Expansion turbine
- Electric motor-generator
- Overhead storage tank or an underground cavern.

In its operating principle, if P and V represent the air pressure and volume, respectively, and if the air compression from pressure P_1 to P_2 follows the gas law $PV^n = $ constant, then the work required during the compression is the energy stored in the compressed air. It is given by

$$\text{Energy stored} = \frac{n\left(P_2 V_2 - P_1 V_1\right)}{n-1},$$
(16.4)

and the temperature at the end of the compression is given by

$$\frac{T_2}{T_1} = \left(\frac{P_2}{P_1}\right)^{\frac{n-1}{n}}.$$
(16.5)

When the elevated temperature air at the end of the constant volume compression cools down, a part of the pressure is lost with the corresponding decrease in the stored energy. The energy stored is small in gas with small value of n. The isentropic value of n for air is 1.4. Under normal working conditions, n is about 1.3.

The electrical power is generated by venting the compressed air through an expansion turbine, which drives the generator. The compressed-air system may work under constant-volume or constant pressure.

In the constant volume compression, the compressed air is stored in pressure tanks, mine caverns, depleted oil or gas fields, or abandoned mines. One million cubic feet of air storage at 600 psi provides an energy storage capacity enough to supply about one-quarter million kWh_e. This system, however, has a disadvantage. The air pressure drops as the compressed air is depleted from the storage, and the electrical power output decreases with the decreasing pressure.

In the constant-pressure compression, air storage may be in an aboveground variable-volume tank or in an underground aquifer. One million cubic feet of air storage at 600 psi provides an energy storage capacity enough to supply about 0.07 million kWh_e. A variable-volume tank maintains a constant pressure by the weight on the tank cover. If an aquifer is used, the pressure remains approximately constant while the storage volume increases because of water displacement in the surrounding rock formation. During electric generation, the water displacement of the compressed air causes only a few percent decrease in the storage pressure, keeping the electrical generation rate essentially constant.

The energy storage efficiency of the compressed air-storage system is a function of a series of component efficiencies, such as the compressor efficiency, the motor-generator efficiency, heat losses, and compressed-air leakage. The overall round-trip energy efficiency of about 50% is achievable.

Once considered an inexhaustible resource, the fish in the sea are becoming scarce, leading to fishing quota systems and additional stresses on an already labor-intensive industry. The fishing boats shown here are anchored in an Alaskan harbor. PHOTOGRAPH COURTESY OF CORBIS.

CHAPTER 17

Natural Resources and Waste Systems

17.1 Material Flows

Natural resources are often extracted in one country, transformed into products in another, and consumed in a third. The result is that a significant portion of the natural resource use that supports a nation's economy often takes place outside the nation's borders. One measure of materials use is the ratio of a nation's total material requirement (TMR) to its Gross Domestic Product (GDP). A recent study shows that most industrialized nations are exhibiting a declining TMR/GDP ratio, thus, supporting the idea that economic activity is becoming less material intense (Adriaanse, 1997). One proposed goal is to decrease the TMR/GDP ratio by a factor of 10. Methods that make natural resource use more efficient or increase recycling lead to lower requirements and environmental impacts over the entire materials cycle.

Material flows account for the physical flows of national resources through extraction, production, fabrication, use and recycling, and final disposal, accounting for all losses along the way. Significant reductions in the TMR/GDP ratio will be necessary in the future.

For Germany, the United States, and the Netherlands (representative industrialized nations) the value of TMR has leveled off at about 80 metric tons per year per capita. Fossil fuel is the largest contributor to TMR.

An important issue is whether industrialized nations are becoming less material intensive as they shift from an industrial economy to a service economy. The TMR/GDP ratio for the three nations mentioned above declined from 100 in 1975 to about 70 in 1994. Improvements in technology could further reduce this ratio and if a shift from fossil fuels to solar and wind energy could be increased, this ratio can be significantly reduced. In addition recycling of materials would aid significantly in the reduction of the TMR/GDP ratio.

17.2 Water Resources

Water is a critical resource for people and agriculture. Freshwater is found in lakes, rivers, wetlands, and aquifers. The idea that the world may be heading for a water shortage is not new. There is adequate freshwater on the planet. The biggest problem with water is not its quantity, but rather its distribution. People do not necessarily live near a water source. Consider the people of the Sahara or Los Angeles. It takes dams, river diversions, and aquifer depletion to serve people far from a large source of water. Egypt, for instance, seems to be burdened with the Aswan dam, which is holding back the Nile's silt. This dam may threaten the whole of the Nile delta. Furthermore, the dam, like so many others, cannot last. How can the mud stacking up against its walls be moved? Where can it be dumped? After damming so many of the world's rivers, even enthusiastic engineers are reconsidering these megaprojects — except in China, which has some of the world's worst water supply problems.

Many areas of the world are drawing down their underground aquifers such as the steady draining of the Ogallala aquifer in America's midwest, which may not last another 50 years.

The availability of freshwater is difficult to assess completely. Table 17.1 provides an estimate of freshwater availability in the regions of the world. There is adequate water for human consumption, but the water needs of agriculture are extensive and rising. Irrigated land is critically important to world food production. Some 40% of the global harvest comes from the 17% of cropland that is irrigated. Because of limited opportunities for expanding rain-fed production, many expect that share to increase markedly in the decades ahead, in order to feed the world's growing population.

Table 17.1 **Freshwater Availability per Capita by Region**

Region	Water available (million gallons per person)
United States and Canada	5.6
Latin America	9.6
Europe	2.5
Africa	2.2
Asia	1.4
Oceania	19.0

WATER SCARCITY

Water resources may be the limiting resource. One billion people may be facing absolute water scarcity by 2025. Countries such as China and India may have to drastically reduce water use in agriculture to satisfy residential and industrial water needs.

Few products can claim the life-giving properties that water boasts — a fact that sets the water industry apart. The consequence: Resource conservation is at the top of the list of concerns, with, as a close second, impacts of wastewater on the aqueous environment.

Water as a resource stored in an aquifer must be managed to be sustainable. Like any renewable resource, groundwater can be tapped indefinitely as long as the rate of extraction does not exceed the rate of replenishment. Some aquifers are replenished slowly and are essentially nonrenewable. Others are readily recharged and can be managed to be sustainable. Even where aquifers do get replenished by rainfall, few governments have established rules and regulations to ensure that they are exploited at a sustainable rate.

Since 1900 there has been a sixfold increase in world water use. As a result, access to safe drinking water for one-fifth of the world's population is limited. Irrigation has increased 60% since 1960, with serious consequences. Perhaps one-half of the world's cities still dump raw sewage into their waters and half of the earth's populations do not have adequate water sanitation. About five million humans die each year from causes related to unsanitary water.

Many observers believe that the annual water depletion is as large as 160 billion cubic meters per year (Postel, 1996b). The vast majority of this groundwater is used to irrigate grain, the staple of the human diet. It takes about 1,000 tons of water to produce 1 ton of grain.

Improved irrigation technologies will be used to increase the efficiency of irrigation. These technologies will be used as the cost of water increases. At present, however, many governments subsidize the price of irrigation water. It may be necessary to increasingly regulate water as a scarce resource.

Two relatively feasible means for expanding water availability for agriculture are (1) treatment and reuse of wastewater and (2) reducing demand through efficiency means. Another approval is to create water markets. A water market is a means for placing a proper value on this essential resource and for providing the economic incentives to move water from applications of lower value to areas of higher value, with benefit to both. When people have a financial incentive to conserve water, they will. If farmers can sell excess capacity, they will find ways of farming that are less water intensive. Urban users are more apt to save water when they have to face higher prices when it is scarce.

Much of the projected increase in water demand will occur in developing countries, where population growth and industrial and agricultural expansion will be greatest. However, per-capita consumption continues to rise in the industrialized world as well.

Water pollution adds enormously to existing problems of local and regional water scarcity by removing large volumes of water from the available supply. Water quality in most of the developed countries has steadily improved in recent years.

The situation is far worse in many developing countries. Water scarcity has increased and human health has been significantly damaged by accelerating contamination of usable water supplies, especially in rapidly urbanizing areas.

Better management of water resources is the key to mitigating water scarcities in the future and avoiding further damage to aquatic ecosystems.

17.3 Forests

Forests have many resources that are useful for human purposes. They supply humans with lumber for building houses and other buildings, wood that can be used for fuel, pulpwood used in making paper, medicines that can cure diseases, and many other products that are worth more than $150 billion a year worldwide. Many forests are also places where some livestock is grazed, and where humans enjoy the recreational aspects of forest lands. Worldwide, about half the timber cut each year is used as fuel for heating and cooking (especially in developing nations), one-third is converted into lumber and other wood products used in building, and one-sixth is used as pulp in making a variety of paper products.

During the past 8,000 years, nearly one-half of the forests that once covered the earth have been converted to farms, pastures, and other uses. Most of the forests that are left have been heavily altered by humans, often rendered into a patchwork of smaller forested areas.

Today forests cover more than one-quarter of the world's total land area. Slightly more than half of the world's forests are in the tropics; the rest are in temperate and boreal (coniferous northern forest) zones. Seven countries hold more than 60% of the world's remaining forests. In order of forest area, they are Russia, Brazil, Canada, the United States, China, Indonesia, and the Congo.

Most forests in North America have been logged at least once and are now managed forests. Managed forests are also called tree plantations and are similar to farms. Some plantations are forests planted for pulpwood production while others are established for environmental purposes, such as carbon sequestration or watershed protection. During the 1970s to 1990s, the growing role of industrial plantations in providing a source of raw material for industry has been widely recognized and encouraged. Forest products industries are looking less to natural forests as sources of wood fiber, and more toward tree farms.

Forestry-related measures with the greatest potential to offset carbon emissions include increasing the productivity of existing forests, planting trees in new areas, and growing tree crops for biomass energy.

Activities undertaken to meet wood and fiber needs can also affect local environmental conditions. For example, they may accelerate erosion and topsoil loss by removing physically stabilizing root systems and energy-absorbing forest canopies, and by reducing the capacity of the soils of these systems to absorb rainwater. Increased runoff carries large amounts of soil into nearby waterways, decreasing the fertility of the originally forested landscape and making forest regeneration more difficult.

Opportunities for more sustainable management of forest resources are similar to those for agriculture. One broad strategy is to recognize and exploit the multiple uses and services including timber supply, wildlife habitat, recreational resources, and watershed management provided by forested land. Adaptive management strategies that seek to maximize a sustainable stream of revenues from multiple uses of the same landscape can have a lower overall environmental impact than intensive harvesting alone.

17.4 Oceans and Fishing

The world fish population lives in the oceans, lakes, and rivers of the world. The world fish catch rose to a new high of 95 million tons in 1998. This translates to about 16 kg per person per year. Constant overfishing continues to threaten the productivity and viability of ocean ecosystems. It is expected that the fish catch will level off by 2005. Many observers believe that 100 million tons could be harvested annually if fish management practices were widely and uniformly followed.

Overfishing certain fish species is driven by a flourishing international trade in fish. A very labor-intensive activity, both capture fishing and aquaculture increasingly happen in the developing world, where wages are lower. Fully 75% of fish caught in the world are from developing countries or nations in transition.

Nearly half of all fish caught today are traded internationally, with the bulk flowing from developing to wealthy nations. Consumers in the industrial world account for 85% of world fish imports by value. Several forces are pushing the world's major fish stocks toward the brink of collapse — the relentless growth in demand for seafood, for example, and the fact that there are just too many boats pursuing too few fish.

The world's fisheries are increasingly overfished, lowering catches in many regions and driving some important fish stocks to near extinction. Overfishing impinges on the diets of nearly one billion people worldwide who depend on fish as their primary source of protein, risking the sustainability of 200 million global fishing jobs.

The cod of New England and eastern Canadian coasts were once abundant, but today they are an endangered species. Worldwide, 30% of fish stocks, from orange roughy and shark to swordfish and tuna, are declining because of overexploitation — and an additional 44% is on the edge.

The obvious answer is to reduce the size of the fishing fleets, but it is very difficult to do so. The fleet in 2000 was 3.5 million fishing boats worldwide.

17.5 Cropland

As the population of the world expands, food supplies depend on expanding cultivated areas and raising land productivity. Cropland is a critical world resource. During the period from 1950 to 1990, land productivity was significantly increased. Since 1990 the rate of increase of land productivity has declined and we may need to rely on increasing land use as the world population expands.

Most of the world's cropland is used to produce grain, which when consumed directly supplies about half of human caloric intake and when consumed indirectly, in the form of livestock products, accounts for most of the remainder.

Some cropland is converted to residential and industrial use in the industrialized nations. In emerging nations, cropland conversion to housing uses may result in large losses of cropland. Losses to soil erosion and desertification are also significant. The cropland per capita available in 2000 is given for selected regions in Table 17.2. Canada, the United States, Europe, and Australia benefit from sustainable cropland per capita. Asia and Africa have less cropland per capita and may face significant conversion impacts.

Table 17.2 **Cropland Availability per Capita by Region, 2000**

Region	*Cropland (acres per person)*
Canada and United States	2
Latin America	0.9
Europe	1.1
Africa	0.7
Asia	0.4
Oceania	4.5

The amount of cropland per person worldwide is given in Table 17.3 for the period from 1970 to 2000. The cropland per capita available in 2020 may decline to 0.10 hectares per capita without suitable conservation measures.

Table 17.3 **Grain Cropland per Person Worldwide**

Year	*1970*	*1980*	*1990*	*2000*
Cropland per person (hectares per capita)	0.18	0.16	0.13	0.12

17.6 Waste Systems

Waste products become a resource as we institute systems of recycling and remanufacturing. Instead of disposal, waste streams can be a resource for the manufacture of products. About 180 million tons of nonhazardous solid waste is collected in the United States annually. This waste stream has increased 80% between 1960 and 2000 and is expected to increase another 20% during this decade. The average amount of municipal solid waste generated per person in the United States is about two to five times more than that in most other developed nations.

Communities are returning organic materials such as plant material and vegetable cuttings to soil at a growing rate. The United States and Europe are composting organic municipal waste at an increasing rate. Organic waste is a relatively untapped resource since only about 15% is recycled in the United States. Because recycled municipal and human wastes supply soils with nutrients and organic matter, recycling can reduce applications of manufactured fertilizer.

There are no known methods to increase the cultivable areas of the world, but careful management of farmlands is one key to preserving vital resources. Farmers consult with the aid of a laptop computer to manage their crop. PHOTOGRAPH COURTESY OF CORBIS.

CHAPTER 18

Agriculture, Biotechnology, and Resources

18.1 Agriculture and the Farm

Agriculture is the science, art, and business of cultivating the soil, producing crops, and raising livestock that can be used for food for humans. The ability to each year increase the food produced to feed a growing population is due to the invention of the farm and the development of the science and technology to enhance the productivity of the farm.

The first significant agricultural crops were grasses: barley, wheat, rice, and so on. These grains are cereal grains. Today more than 70% of all cultivated land is devoted to cereal grains. These crops provide the bulk of human nutrition; most of the calories we consume come from grain or from eating animals raised on feed grains.

Around 3000 B.C. the invention of the plow increased the output per acre of a farm. However, plowing often left the soil vulnerable to erosion. By 1900 A.D., technology had increased the productivity of farms and enabled a single farmer to cultivate more land. A simple model of the food production system is shown in Figure 18.1. Population increases lead to cropland decrease and the invention of new technology methods. If good technology overcomes the erosion problems and productivity increases, then food production increases. As food production increases, the food per capita increases. With improved conditions, such as health and nutrition, the birth rate decreases and the population growth rate declines.

Farm productivity in the industrialized nations such as the United States, Canada, and France increased significantly during the period from 1900 to 1970 due to improved agricultural methods and technologies. For example, the farm output per man-hour worked on a farm increased 10-fold from 1900 to 1970. Agriculture is an industry and as such is subject to our concern for a sustainable future. Since the first recorded times, mankind has been concerned with producing sufficient food to absorb losses in years of poor harvest.

By 4000 B.C., people began to divert river water to their fields, thus giving rise to irrigation practices. Irrigation transformed the productivity of the farm and surpluses could be stored and distributed (Postel, 1999a). In Egypt today, 100% of cropland is irrigated and 62% of the cropland is irrigated in Japan. Because irrigated farms typically get higher yields and can grow two or three crops per year, the spread of irrigation has been a key factor in the rise in food production. Many nations, including China, Egypt, India, Indonesia, and Pakistan, rely on irrigated land for more than half of their domestic food output.

As income levels rise worldwide, people add more beef, pork, and poultry to their diets. Therefore, the world's meat consumption may double by 2050. To provide the grains for the animals and the people new strategies will be needed. Past methods for increasing agricultural production included

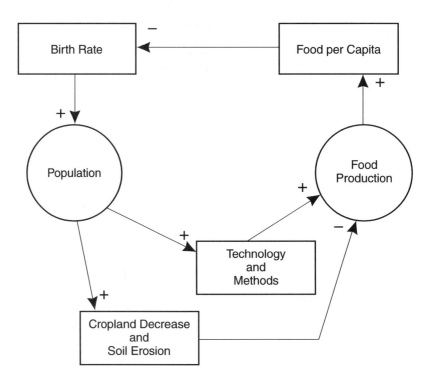

Figure 18.1 **Food production system.**

converting wildlands to croplands (extensification), using chemical fertilizers to increase yields (intensification), and using irrigation to increase yields or enable cultivation of otherwise nonarable land. Increasing food production in the future will require some combination of these strategies, each of which have risks that will need to be minimized by careful management.

18.2 Worldwide Food Supplies

Estimates of malnourished peoples vary widely, but it is probably safe to state that 10 to 20% of the population is currently undernourished. Evenly distributed, the world's production of food grains would probably provide the world's population with a basic diet. The distribution of food is uneven and many nations strive to feed their people. Most nations have increased their yields through improved technologies and practices that include energy, equipment, and water sources for irrigation; more fertilizer and pesticides; well-chosen agricultural machinery, often usable at small scale; improved storage and transport of harvested crops; and simple field instrumentation and control and better weather prediction, to guide farmers in managing their crops. More and better spread agronomic advice, now broadly available in the developed nations, and with it the spread of the best local practices, will add substantially to the yields of lands already under cultivation.

A change from near-subsistence farming by households numbering in the hundred millions to industrialized agriculture, with at least a 10-fold increase in labor productivity, may allow the developing nations to seek a high standard of living for their rural people. The task for the future is to feed the seven or eight billion people worldwide by 2060.

A sustainable agricultural system of the future could have multiple benefits for both the environment and food production. Options for increasing the sustainability of agriculture include more efficient use of resources such as water, fossil fuels, and land; maintaining a broad agrobiodiversity base; adoption of traditional, low-input agricultural technologies where appropriate; and investing in improved agricultural technologies.

Even with improved agricultural practices, undernourishment may continue. Amartya San states that (Sen, 1999a):

> A more adequate understanding of famine requires examining the channels through which food is acquired and distributed as well as studying the entitlement of different sections of society. Starvation occurs because a substantial proportion of the population loses the means of obtaining food. Such a loss can result from unemployment, from a fall in the purchasing power of wages or from a shift in the exchange rate between goods and services sold and food bought.

Poverty can be an important impediment to accessing food and only the overcoming of poverty and associated social ills can lead to widespread and equitable nourishment. Hunger is no longer considered simply a scarcity issue but rather an access issue. Reportedly, some 800 million people are undernourished.

As the world population grows over the next several decades, the present system of agricultural practices will be challenged. One vulnerable portion of the system is the potential depletion of underground aquifers as increased pumping leads to the use of groundwater for irrigation faster than nature is able to replenish the aquifers. Depletion of aquifers in India and China could lead to serious shortages of irrigation water. As we move through the 21st century, the challenge is to boost water productivity; that is, to obtain more benefit from every liter of water devoted to crop production.

In China, rising affluence has generated more demand for livestock products and the grain to feed the livestock. This rise in demand can equal the increase in demand for grain due to increases in population.

The increasing use of pesticides and herbicides may challenge the ability of the exosphere to absorb their impacts.

Vaclav Smil in a recent book, *Feeding the World,* provides one vision of how we can best feed the nine billion or so people who will likely inhabit the earth by the middle of the 21st century (Smil, 2000). He asks whether human ingenuity can produce enough food to support healthy and vigorous lives for all these people without irreparably damaging the integrity of the biosphere. He espouses neither the catastrophic view that widespread starvation is imminent nor the cornucopian view that welcomes large population increases as the source of endless human inventiveness. He attempts to show how we can make more effective use of current resources and suggests that if we increase farming efficiency, reduce waste, and transform our diets, future needs may not be as great as we anticipate. He asserts that there are no insurmountable biophysical reasons why we cannot feed humanity in the decades to come while easing the burden that modern agriculture puts on the biosphere.

18.3 The Green Revolution

The high-yield crops introduced in the period of 1960 to 1990 are part of the Green Revolution. To achieve such a revolution, new strains of plants were developed by means of breeding techniques. The **Green Revolution** can be described as the application of biological engineering to the development of edible plants. The technologies of the Green Revolution were developed on experiment stations that were favored with fertile soils, well-controlled water sources, and other factors suitable for high production. Since the 1970s, world food prices have declined in real terms by over 70%. Those who benefit most are the poor, who spend the highest proportion of their family income on food.

Plant breeders using traditional breeding techniques have largely exploited the genetic potential for increasing the share of photosynthate that goes into seed. As the rate of increase in grain yields declines, the need for new techniques emerges. Since 1950, the Green Revolution has been based on the introduction of new crop varieties and the greatly increased use of irrigation and fertilizers. The future may demonstrate diminishing returns to the new varieties and inputs of water and fertilizer.

During the period from 1980 to 2000, the world has generally experienced surplus grain crops which were not always evenly distributed throughout the world. By 2020, the world could experience a delicate balance of supply and demand. The assumptions underlying population, agricultural, and trade policies during an age of surpluses may need to be reassessed if the world moves into an age of scarcity.

The challenge is to repeat and enlarge the accomplishments of the Green Revolution by redesigning the Green Revolution system so that it is sustainable and equitable. Gordon Conway describes this redesign in his book *The Doubly Green Revolution* (Conway, 1999):

> While the first Green Revolution took as its starting point the biological challenge inherent in producing new high-yielding food crops and then looked to determine how the benefits could reach the poor, this new revolution has to reverse the chain of logic, starting with the socioeconomic demands of poor households and then seeking to identify the appropriate research priorities. Its goal is the creation of food security and sustainable livelihoods for the poor.

Conway calls for new, innovative means that will apply new technology in an environmentally sensitive way through the creation of partnerships between scientists and farmers.

18.4 Agricultural Technology

Since the second world war, agriculture has become bigger, more intensive, and more dependent on technology, both in the developed and much of the developing world. It is one of the world's largest industries, employing 1.3 billion people and producing $1.3 trillion worth of goods a year. World output of food per capita has gone up by some 25% during the past 40 years, even though land use has grown by only 10% and world population has increased by 90%. Food prices in real terms have fallen by two-fifths, so that consumers in America, for example, now spend only 14% of their household income on food.

Agriculture technology and systems use many inputs to increase the output of crops and livestock as shown in Figure 18.2. The complex, industrial-like system uses seeds and plants, land, energy, chemicals, and water as inputs. The farming processes include planting techniques, application methods for chemicals, and water irrigation methods. The output of the system is the desired plants, grains, and livestock.

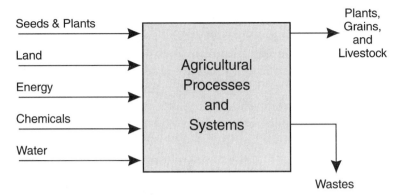

Figure 18.2 **Agricultural system.**

Synthetic fertilizers and pesticides have played a large role for decades. Chemical pesticides are a source of concern. World pesticide consumption reached about 2.8 million metric tons of chemical ingredients in 1998. Global use of pesticides creates substantial health impacts in all parts of the world, although the exact toll is difficult to identify, given both the various chemicals and types of exposure. In short, not all pesticides are equally risky, and not all people are equally at risk. Agricultural chemicals and practices can have far-reaching effects. They influence food quality, drinking water quality, air quality, and the soil on which we live and grow food.

In a sense, modern agriculture can be called "industrial farming." It is characterized by rapid technological innovation and application, large-scale farming operations, highly specialized enterprises, large capital investment, high labor efficiency, and extensive dependency on agribusiness. With the industrialization of agriculture have come increased environmental problems, including excessive topsoil erosion, water pollution, depletion of aquifers, and loss of wetland, prairie, woodland, and wildlife habitat. Also, industrialization has led to large displacement of farm families.

To counteract the problems, the sustainable agriculture movement emerged in the 1990s. The goal of this activity is to reduce the use of chemicals while increasing yields. They also seek better land and water management methods.

18.5 Biotechnology

For centuries, breeders have concentrated on modifying the traits of plants to influence their growth performance. As the limits of breeding plants are reached, scientists are turning to biotechnology and the genetic engineering of plants. For example, the use of insecticides can be reduced by using a plant that has been genetically modified to be resistant to insects.

Biotechnology includes any technique that uses living organisms to make or modify products, improve plants or animals, or develop microorganisms for specific uses. Scientists are able to isolate and transfer specific genes across organisms and can improve crops by introducing into a plant a copy of a gene for a desired trait, such as resistance to drought or disease.

Biotechnology opens up the possibility of controlling the pests and weeds that lower agricultural yields without the use of powerful pesticides and herbicides that often have negative side effects. Humans are not going to stop growing enough food to feed themselves, but they can learn to grow it in a more environmentally friendly way.

Using the techniques of genetic engineering, scientists have been able to insert new genes into plants. Thus, crop seeds and plants will be genetically engineered to tolerate dry, low-nutrient, or salty soils, which could permit degraded farmland to flourish. Other crops could produce their own pesticides thus reducing the need for chemical pesticides. The basis of these new techniques is the science of the genetic code and gene splicing. The plants are transgenic, that is, genes have been inserted into their chromosomes. By 2000, 25% of corn, 35% of soybeans, and 45% of cotton grown in the United States were genetically altered, either to make the crops resistant to weedkillers or to produce their own pesticides. Americans regularly consume genetically modified organisms (GMO). For example, bioengineering corn and soybeans are used widely as ingredients in processed food and soft drinks.

Some claim GMO crops are a direct threat to human health while others see GMO crops as necessary to feed the growing world population. Is biotechnology outpacing the scientific understanding of its risks and the development of a regulatory apparatus to supervise its use? By far the most important bioengineered trait today is herbicide tolerance, which accounts for two-thirds of all transgenic crops. It is possible that the transgenic plant could cross with a weed resulting in a pesticide-resistant superweed.

In less developed nations in Africa and the Middle East, weeds such as witchweed and broomrape significantly lower the yields of crops such as sorghum and grain legumes. Thus, it is attractive to introduce GMO seeds of sorghum and legumes that resist herbicides which control the weeds. The first Green Resolution crops depended heavily on irrigation, artificial fertilizers, and chemical pesticides. By contrast, the benefits of bioengineering are encapsulated in the seed. Pest-resistant seed corn, for example, needs no costly spraying equipment, is not very complicated to grow, and releases little toxin into the environment. Many scientists believe that genetic engineering will enable development of improved productivity of crops but equally recognize that the technology's potential hazards must not be ignored.

Through biotechnology, scientists are attempting to solve the real-world problems of sickness, hunger, and resource depletion. Almost everything we eat has been "genetically modified" by the hand of man through selective breeding, beginning thousands of years ago. What's new is the ability to select and transfer specific genes or sets of genes more precisely.

Almost all biotech seed currently sold has one ultimate purpose: allowing farmers to grow more food on less land at less cost. However, critics contend the methods used to modify crops have gone too far and safeguards are too loose, and also question the potential effects on the environment and the safety of related food products. Biotechnology and genetically enhanced crops hold great potential to reduce costs, improve the environment, and contribute to human heath goals. However, concerns over the safety of these foods and the environmental impact of genetically enhanced crops must also be addressed. Modern biotechnology has great potential for human well-being if developed and used with adequate safety measures for the environment and human health.

18.6 Aquaculture and the Future of Fisheries[1]

Michael De Alessi
Director
Center for Private Conservation

A decade ago, a fish Malthusian might have predicted the end of salmon as a food. Human ingenuity seems to have beaten nature once again.
— Fleming Meeks, 1990

While the world fish catch has stagnated in recent years, aquaculture production has grown dramatically. It is now responsible for nearly 20% of the world fish production, and is one of the world's fastest growing industries. In 1991, world aquaculture production was approximately 13 million metric tons, double what it was seven years before (FAO, 1993). By 1995 that number had jumped to over 21 million metric tons (FAO, 1997).

The reason for these increases should be obvious. Of course, some aquaculture operations have been heavily subsidised, but the most important reason for aquaculture's success has been that there is no tragedy of the commons within an aquaculture facility. A fish not harvested today will be there tomorrow, normal rates of mortality notwithstanding. Private ownership has invigorated entrepreneurs to tinker, to experiment, and to innovate. Salmon is one of the most commonly farmed species, and fish farmers have developed remarkable ways to manage their fish. Through genetic manipulations as well as environmental and dietary control, aquaculturalists increase

---

[1]This section has been excerpted with permission from the author's *Fishing for Solutions,* London: Institute of Economic Affairs, 1998.

the fat content for sushi chefs and reduce it for producers of smoked salmon. They can also increase a salmon's nutritional value, adjust its brilliant orange color, or set the flavor to bold or mild (Bittman, 1996).

One great advantage of aquaculture is the stability of supply. Farm-raised fish are often brought to market within a day of being harvested, while wild-caught fish sometimes take a week. Aquaculture facilities have fresh fish in holding tanks and can either slow or accelerate growth as they please. Markets and restaurants can count year-round on the availability of fresh fish of uniform quality and size. No wild fishery approaches that.

Another reason for the success of aquaculture is the survival rate of juveniles; only 10% of salmon fry survive in the wild, whereas in captivity the number jumps to almost 90% (Munk, 1995). In 1980, the total worldwide catch of salmon (wild and farmed) was just over 10,000 metric tons (Meeks, 1990). In 1990, farm salmon alone from Norway, Chile, Scotland, Canada, and Iceland amounted to over 220,000 metric tons. As a result, in real terms, the retail price of salmon in 1990 was about half of what it was in 1980 (Meeks, 1990). Shrimp aquaculture is booming as well, and the production of farmed shrimp should exceed the wild harvest by the year 2000 (Gujja and Finger-Stich, 1996).

Besides salmon and shrimp, catfish and tilapia are common aquaculture species. The majority of fish grown around the world is finfish (approximately 70%), followed by mollusks such as oysters and clams (24%); crustaceans, mostly shrimp (6%), make up the rest (FAO, 1993b).

Aquaculture is not without its problems. Most aquaculture (approximately two-thirds) occurs near the coast or in shallow estuaries where pollution from outside sources can cripple an aquaculture operation. In addition, intensive aquaculture in these areas can produce significant amounts of organic pollution, which can lead to reduced levels of oxygen in the water and an increase in quick-growing algae that is harmful to marine life. In some cases there is also growing concern over the antibiotics used. It is worth noting that when pollution does occur, it is generally because property rights have not been appropriately defined and/or are not readily enforceable. Government subsidies and incentives to expropriate coastal areas for aquaculture often further undermine nearshore private property rights.

Many environmentalists ignore these perverse incentives and instead simply vilify aquaculture in developing nations, particularly Thailand and Ecuador, holding it singly responsible for vast amounts of coastal habitat destruction. They compare shrimp farming to slash-and-burn agriculture and have even gone so far as to hold mock trials for poor Thai shrimp farmers in New York, accusing them of "despoiling their country's coastal wetlands by raising shrimp" (Mydans, 1996). In one sense they are right. According to United Nations estimates, in Thailand only 40,000 acres of mangrove forest remain, down from nearly a million acres just 30 years ago (Mydans, 1996). Shrimp farming has certainly been a significant factor in this decline. Indeed, abandoned ponds can "saturate the surrounding soil with salt and pollute the land and water with a chemical sludge made up of fertiliser and antibiotics as well as larvicides, shrimp feed and waste" (Mydans, 1996). But the root cause of this problem is a lack of secure property rights in marine resources, which is the result of government intervention, *not* an inherent feature of aquaculturists (who are merely operating within the incentive structure defined by the extant institutions).

In Thailand, aquaculture is heavily subsidized and in many cases farms are built in areas that were previously managed much more sustainably by a system of customary tenure (C. Bailey, 1988). In Malaysia, the Land Acquisition Act was amended in 1991 to allow the state to grab land for any reason deemed beneficial to economic development, including the construction of fish ponds (Murray, 1995). Similarly in Ecuador, bribes, corrupt government partnerships, and land grabs are common because "by law, coastal beaches, salt water marshes, and everything else below the high tide line is a national patrimony" (Southgate, 1992). Not only shrimp farms but city slums regularly invade these areas, even in national ecological preserves (Southgate, 1992).

Alfredo Quarto, a director of the Mangrove Action Project, has pointed out that the main reason why shrimp farmers choose to clear mangrove forests is that they are usually government owned (Weber, 1996). In other words, government-sanctioned open access and expropriation of common property rights are what is really to blame for coastal habitat destruction in places like Thailand.

Nearshore aquaculture problems can often be solved by moving operations offshore, where water circulation is better and risks from pollution, both exogenous and endogenous, are limited. Offshore aquaculture is now beginning to move beyond the experimental stage. The engineering problems of raising fish far from protected shores are substantial. Nevertheless, offshore net pens and cages are increasingly appearing off the coasts of places like Norway and Ireland.

One of the greatest setbacks for offshore aquaculture is the question of tenure. For this reason one of the more promising avenues involves utilizing decommissioned oil rigs. These rigs offer a fixed platform over ground that has already been leased for oil exploration. Most oil leases specify that they are only for oil production, which complicates the matter, but using existing structures is still promising — they would certainly be far cheaper than starting from scratch.

Self-contained, indoor aquaculture facilities are another relatively new development, but one with tremendous potential. Aquafuture, a firm in Massachusetts, has already had some success raising striped bass in a closed tank system (Herring, 1994). The process uses much less water and feed than conventional fish farms, produces fewer wastes that are easily converted to fertilizer, and by changing the water temperature fish can be grown to market size either faster or slower than in the wild depending on the current market. The enclosed environment is also more sanitary, so Aquafuture's mortality rate is half the industry average.

Conclusions

The crash of so many once-plentiful marine fisheries is a clear indication of institutional failure. But there is hope for the future. Evidence indicates that fish stocks are highly resilient and are likely to recover rapidly if given the chance (Myers *et al.,* 1995). To give them that chance, government should consider turning over the management of fisheries to private interests. As this section has attempted to demonstrate, private property, owned individually or in common, offers the best hope of creating incentives for conservation, stimulating the production of new technologies, and protecting the marine environment. Although they may stumble occasionally, mistakes by one individual or group will be more than offset by the successes of others. In the end, the environment, the health of the fish stocks, their harvesters and the fish-eating public will be better off.

From satellite technology used to monitor the oceans to artificial reefs that attract and nurture fish populations to fish farming, technologies exist that could facilitate private ownership in the oceans. Indeed, the mere prospect of ownership in the oceans is already fomenting innovative thinking. One such example is Ocean Farming Inc., a firm in Virginia that hopes to take advantage of limitations on productivity in the oceans that have been created by the limited availability of naturally occurring iron by fertilizing huge swathes of the ocean with a patented iron delivery system (Markels, 1995). Founder Michael Markels (1995) is well aware that for this kind of venture to succeed, "Some kind of [private] property rights need to be created so that those who bear the costs of [enhancing the oceans] can reap the rewards."

The issue of liability is also crucial. The problems of open access and government control pervade not only the ways in which fish are harvested, but also the quality of the water everywhere. Making those responsible for pollution liable for their actions is potentially one of the greatest attributes of any private ownership régime.

The FAO recently announced that world fish production in 1996 was 112.3 million metric tons, another substantial increase. Fish farming contributed most of the growth, but not all. The FAO also estimated that number could increase by 20 million tons if underdeveloped resources were exploited and bycatch reduced. The potential for tremendous gain is there. All that is required to begin tapping it and to begin restoring those resources that have suffered is to unleash the vast potential of private initiatives to conserve, protect, and enhance the marine environment.

18.7 International Ecosystem Assessment[2]

By Edward Ayensu, Daniel van R. Claasen, Mark Collins, Andrew Dearing, Louise Fresco, Madhav Gadgil, Habiba Gitay, Gisbert Glaser, Calestous Juma, John Krebs, Roberto Lenton, Jane Lubchenco, Jeffrey A. McNeely, Harold A. Mooney, Per Pinstrup-Andersen, Mario Ramos, Peter Raven, Walter V. Reid, Cristian Samper, José Sarukhán, Peter Schei, José Galízia Tundisi, Robert T. Watson, Xu Guanhua, and A. H. Zakri

Despite technological developments, we are still intimately connected to our environment. Our lives depend on ecosystem goods such as food, timber, genetic resources, and medicines. Ecosystems also provide services including water purification, flood control, coastline stabilization, carbon sequestration, waste treatment, biodiversity conservation, soil generation, disease regulation, maintenance of air quality, and aesthetic and cultural benefits (Watson *et al.*, 1998; Daily, 1997). We know too little of the current state and future prospects of these goods and services: a system of international assessment is urgently needed. Without such a system, development will not be sustainable.

18.7.1 MAKING ENDS MEET

Historically, changes in technology and land use helped to reduce harmful social and economic consequences of imbalances between the supply and demand for ecosystem goods and services. For example, between 1967 and 1982, 0.24% per year growth in the extent of agricultural lands combined with a 2.2% per year increase in cereal yields led to net increases in per capita food availability, despite a 32% increase in world population (Pinstrup-Anderson *et al.*, 1997). Similarly, declining production of fish and timber in natural ecosystems has been partially offset by increased production through aquaculture and plantations (although often with significant ill effects such as increased water pollution and loss of biological diversity) (Naylor *et al.*, 1998).

These changes in land use and technology have had profound impacts on natural ecosystems. About 40 to 50% of land on the Earth has been irreversibly transformed (through change in land cover) or degraded by human actions (Vitousek *et al.*, 1997). For example, more than 60% of the world's

[2]This section has been reprinted with permission from *Science,* Vol. 286, Oct. 22, 1999, pp. 685–686. Copyright 1999 American Association for the Advancement of Science. The authors are members of a Steering Committee exploring the merits of launching a Millennium Assessment of the World's Ecosystems. Walter Reid is the coordinator of the project.

major fisheries will not be able to recover from overfishing without restorative actions (Granger and Garcia, 1996). Natural forests continue to disappear at a rate of some 14 million hectares each year (Roberts, 1998).

The magnitude of human impacts on ecosystems, combined with growing human population and consumption, means that the challenge of meeting human demands will grow (Figure 18.3). Models based on the United Nations' intermediate population projection suggest that an additional one-third of global land cover will be transformed over the next 100 years (Walker *et al.*, 1999). By 2020, world demand for rice, wheat, and maize is projected to increase by ~40% and livestock production by more than 60% (Pinstrup-Anderson *et al.*, 1997). Humans currently appropriate 54% of accessible freshwater runoff, and by 2025, demand is projected to increase to more than 70% of runoff (Postel *et al.*, 1996). Demand for wood is projected to double over the next 50 years (Watson *et al.*, 1998).

These growing demand can no longer be met by tapping unexploited resources, and trade-offs among goods and services have become the rule. A nation can increase food supply by converting a forest to agriculture but, in so doing, decreases the supply of goods that may be of equal or greater

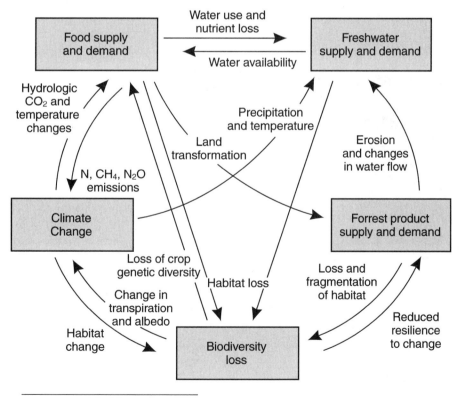

SOURCE: Modified from Watson *et al.*, 1998.

Figure 18.3 **Linkages among various ecosystem goods and services (food, water, biodiversity, forest products) and other driving forces (climate change).**

importance such as clean water, timber, biodiversity, or flood control. Finally, projected climate change may well exacerbate the problem of balancing supply and demand, particularly in developing countries where adaptation will be constrained by financial and other resources. Although no one questions that these are significant changes, we need to develop ways to quantify their impacts.

18.7.2 THE INTEGRATED APPROACH

Sectoral approaches to management — focused on agriculture, forestry, or water supply — made sense when trade-offs among goods and services were modest or unimportant. They are insufficient today, when ecosystem management must meet conflicting goals and take into account the interlinkages among environmental problems (see diagram). For this reason an integrated, or "multiple functions," approach to analysis of ecosystems must be adopted.

Reactive management was inevitable when ecological knowledge was insufficient to allow more reliable predictions. Today, given the pace of global change, human welfare is utterly dependent on forward-looking, adaptive, and informed management decisions.

An integrated, predictive, and adaptive approach to ecosystem management requires three basic types of information.

First, reliable site-specific baseline information on ecosystems (including the amount, economic value, and condition of the goods and services produced) must be more widely available. In particular, information on the output and value of nonmarketed ecosystem goods and services has rarely been available historically, despite evidence that these economic values may be significant to management decisions (Arrow *et al.*, 1995), nor is information available on the capacity of the ecosystem to maintain production of particular goods and services.

Second, knowledge of how the production of goods and services in specific ecosystems will respond to biophysical changes must be made available to public and private sectors. Ecosystem management will ultimately require quantitative answers to such questions as (i) How do ecosystems differ in their response to elevated nitrogen, carbon dioxide, and sulfur concentrations, and how will this affect the goods and services they produce? (ii) How do ecosystems differ in the manner in which land cover change affects the local hydrological cycle, including amounts of precipitation and the timing and amount of runoff? (iii) How do changes in biological diversity affect the supply and resilience of various goods and services produced by different ecosystems? (iv) What thresholds are likely to exist in different ecosystems, and to what types of changes will those ecosystems be most sensitive?

Better forecasting tools also enable exploration of potential "win–win" opportunities for ecosystem management, such as managing land cover to maximize biodiversity conservation, watershed protection, and carbon sequestration simultaneously.

Third, integrated regional models that incorporate biophysical, economic, and technological change must be developed to provide policy-makers with better understanding of the consequences of different management options. A key element of the development of these models will be the need to ensure coherence between data collected at various scales, so that global models can be informed by regional and local data and can be downscaled for regional analyses.

18.7.3 ASSESSMENT DESIGN

Other major international science assessments, such as the Global Biodiversity Assessment and the assessments of the Intergovernmental Panel on Climate Change, have been conducted over 3- to 4-year periods, with budgets of $5 million to $20 million, and with important contributions of time and expertise from the research community. A worldwide ecosystem assessment conducted with a similar scale of effort could significantly aid national and international decisionmaking. Ideally, such an assessment would be repeated at 5- or 10-year intervals to facilitate monitoring of ecosystem changes, progress in response to those changes, and to incorporate new research findings. Such a process would galvanize international attention around the importance of ecosystems for human development and the consequences of actions that we might take, or fail to take, to ensure effective management of these systems.

An international assessment could be either fully independent of governments or established through an arrangement among governments with a formal link to one or more international bodies, such as U.N. conventions. A system of strict peer review could maintain the scientific independence of its findings. Experience with past assessments suggests that, in order to succeed, assessors must ensure that their product is (i) demand driven — with the choice of issues guided by the decision-makers who will use its findings; (ii) inclusive — involving natural and social scientists from all relevant sectors and organizations and representing all geographic regions; (iii) peer reviewed and independent of political and economic influence on its findings; and (iv) relevant to a wide range of public and private sector stakeholders.

A global ecosystem assessment would also need to build on and not duplicate various international activities, including research programs, such as the Diversitas Programme; monitoring activities, such as the Global Terrestrial Observing System; data sets held by national governments and international institutions, such as the Food and Agricultural Organization (FAO) and the World Conservation Monitoring Centre; recent assessments of issues, such as food production and biodiversity (Heywood *et al.*, 1995), and several other ongoing assessments, such as the FAO Global Forest Resources Assessment 2000 and the Global International Waters Assessment. Without the information from these related activities, an integrated assessment of world ecosystems would be impossible, but these activities alone are insufficient to meet the needs we have identified.

Because ecosystems are differentiated in space and time, sound management requires careful local planning and action. An international ecosystem assessment must ultimately be complemented by, and informed by, detailed local monitoring and assessment. Local and regional assessments alone are insufficient, however, because some processes are global and because local goods, services, matter, and energy are often transferred across regions. The worldwide assessment should thus help to catalyze the establishment of appropriate monitoring and assessment institutions from highly centralized processes at a global level to highly decentralized processes at a local level.

Both the challenge of effectively managing earth's ecosystems and the consequences of failure will increase during the 21st century (Lubchenco, 1998). Decisions taken by local communities, national governments, and the private sector over coming decades will determine how much biodiversity will survive for future generations and whether the supply of food, clean water, timber, and aesthetic and cultural benefits provided by ecosystems will enhance or diminish human prospects. The scientific community must mobilize its knowledge of these biological systems in a manner that can heighten awareness, provide information, build local and national capacity, and inform policy changes that will help communities, businesses, nations, and international institutions better manage Earth's living systems. We believe that the time is right — at the turn of the millennium — to undertake the first global assessment of the condition and future prospects of global ecosystems.

18.8 Agriculture, Technology, and Natural Resources

By David Pimentel[*] and Marcia Pimentel[†]
[*]*Professor of Ecology and Agricultural Sciences*
College of Agriculture and Life Sciences
Cornell University
[†]*Senior Lecturer in Nutrition*
Division of Nutritional Sciences
Cornell University

More than three billion humans are currently malnourished worldwide (WHO, 1996). This is the largest number and proportion of hungry humans ever recorded in history! Based on current rates of increase, the world population is projected to double to approximately 12 billion in about 50 years (PRB, 1998). At a time when the world population is expanding at a rate of 1.4%/year, adding more than a quarter million people daily, providing adequate food becomes an increasingly severe problem for technology.

Reports from the FAO of the United Nations and the U.S. Department of Agriculture, as well as numerous other international organizations, further confirm the serious nature of the global food supply problem (NAS, 1994). For example, the per-capita availability of world cereal grains, which make

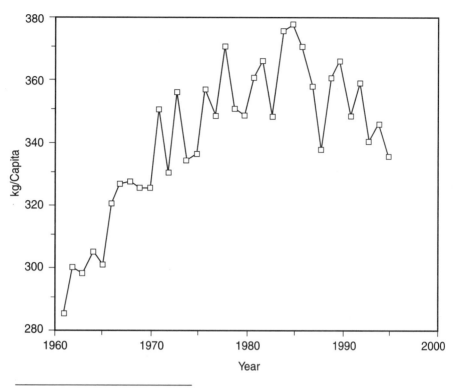

SOURCE: Adapted from Harris, 1996; WATI database system compiled by the Economic Research Division, USDA, and FAO production yearbooks.

Figure 18.4 **World cereal grain production per capita.**

up 80 to 90% of the world's food supply, has been declining for the past 17 years (Figure 18.4). These shortages are reflected in the rapidly growing number of humans who are malnourished worldwide (WHO, 1996).

Thus, as the world population continues to expand, more pressure than ever before is being placed on technology and on all the basic resources that are essential for food production and a sustainable world. Unfortunately, while the human population grows exponentially, food production can only increase linearly. Furthermore, degradation of land, water, energy, and biological resources vital to agriculture continues unabated (Pimentel *et al.,* 1999).

18.8.1 AGRICULTURAL RESOURCES

More than 99% of the world's food supply comes from cropland, while less than 1% comes from oceans and other aquatic habitats (FAO, 1991). The continued production of an adequate food supply is directly dependent on the availability of ample quantities of cropland, freshwater, energy, biodiversity, and the appropriate technologies to sustainably and efficiently utilize these resources. Obviously, as the human population grows, the requirements for these resources and related technologies will escalate. Even if cropland, water, and energy resources are never completely depleted, their supply will decline significantly *on a per-capita basis* because they must be divided among more and more people.

Land

Throughout the world, fertile cropland is being lost from production at an alarming rate. True for all cropland, it is illustrated by the diminishing amount of land now devoted to cereal grains (Figure 18.5). Soil erosion by wind and water, as well as by overuse, is responsible for the loss of about 30% of the world's cropland during the past 40 years (WRI, 1994; Pimentel *et al.*, 1995). Once soil is lost, it takes 500 years or more to form a mere 25 mm of fertile soil. For crop production, at least 150 mm of fertile soil is required.

Most replacement for eroded and unproductive agricultural land is coming from cleared forest land and marginal land. The need for more cropland accounts for more than 60 to 90% of the world's deforestation (Myers, 1994; Haerdter *et al.*, 1997). Despite such land replacement strategies, per-capita world cropland is declining and now stands at only 0.27 ha per capita, or about half of the 0.5 ha per capita considered the minimum for the production of a diverse diet similar to that of the United States and Europe (Lal and Stewart, 1990). China now has only 0.08 ha per capita, or about 15% of the accepted minimum.

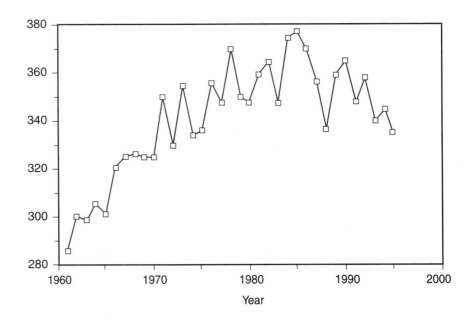

SOURCE: Adapted from Harris, 1996; WATI database system compiled by the Economic Research Division, USDA, and FAO production yearbooks.

Figure 18.5 **World cropland in cereal grain production (x 1,000 ha).**

Water

Rainfall, and its collection in rivers, lakes, and vast underground aquifers, provides the water needed by humans for their survival and diverse activities. Freshwater is critical for all vegetation, including crops. All plants use and transpire massive amounts of water during the growing season. For example, a hectare of corn, producing about 8,000 kg/ha, will transpire more than 6 million liters of water during one growing season (Pimentel *et al.*, 1996). This means that more than 10 million liters of water must reach each hectare of corn during the season. In total, agricultural production consumes more freshwater than any other human activity. Specifically, about 70% of the world's freshwater supply is consumed, or used up, by agriculture, and thus is unavailable for other uses (Postel, 1999).

Water resources are under great stress as densely populated cities, states, and countries increase their withdrawal of water from rivers, lakes, and aquifers every year. For example, by the time the Colorado River reaches Mexico, it has almost disappeared before it finally trickles into the Gulf of California (Sheridan, 1983). Also, the great Ogallala aquifer in the central United States is suffering an overdraft rate that is about 140% above its recharge rate (Gleick, 1993). Water shortages in the United States and elsewhere in the world already are reflected in a per-capita decline in crop irrigation during the past 20 years (Postel, 1992).

To compound the water problem, about 40% of the world population lives in regions that directly compete for shared water resources (Gleick, 1993). In China, for example, more than 300 cities already are short of water, and these shortages are intensifying as Chinese urban areas expand (WRI, 1994). Competition for water resources among individuals, industries, and regions both within and between countries is growing throughout the world community (Gleick, 1993).

Along with the quantity of water available, water purity is also important. Diseases associated with impure water and unsanitary systems rob people of their health, nutrients, and livelihood. These problems are most serious in developing countries, where about 90% of diseases can be traced to a lack of pure water (WHO, 1992). Worldwide, about four billion cases of disease are contracted from impure water and approximately six million deaths are caused by water-borne disease each year (Pimentel *et al.*, 1996). Furthermore, when a person is stricken with diarrhea, malaria, or another serious disease, from 5 to 20% of an individual's food intake is used by the body to offset the stress of the disease, diminishing the benefits of his/her food (Pimentel *et al.*, 1996).

Disease and malnutrition problems appear to be particularly serious in the developing world, where poverty and poor sanitation is endemic (Shetty and Shetty, 1993). And the number of people living in urban areas is doubling every 10 to 20 years, creating other environmental problems, including shortages of water and the lack of sanitation, increased air pollution, and increased food shortages. For these reasons, the potential for the spread of disease is great in urban areas (*Science,* 1995).

Energy

Energy from many sources, especially fossil energy, is a prime resource used in food production. About 70% of the fossil energy used each year throughout the world is consumed by populations living in developed countries (BP, 1999). Of this, about 17% is expended in the production, processing, and packaging of food products (Pimentel and Pimentel, 1996). In particular, the intensive farming technologies characteristic of developed countries rely on massive amounts of fossil energy for fertilizers, pesticides, irrigation, and for machines that substitute for human labor. In contrast, developing countries use fossil energy primarily for labor, fertilizers, and irrigation to help maintain yields, rather than to reduce human labor inputs (Giampietro and Pimentel, 1993).

Because fossil energy is a finite resource, its depletion accelerates as populations expand and their food requirements increase. The U.S. Department of Energy warns that our country will exhaust all of its oil reserves within the next 15 years (Youngquist, 1997). Consider that at present, the United States is importing more than 60% of its oil. To sustain its energy-based activities, U.S. oil imports will have to increase in future decades, further worsening the U.S. energy trade imbalance. The cost of fuel will also increase. The impact of price increases is already a serious problem for developing countries, where the relatively high price of imported fossil fuels makes it difficult, if not impossible, for poor farmers to power irrigation technology as they try to sustain needed harvests.

Worldwide, per-capita supplies of fossil energy show a significant decline, and this trend can be expected to continue. Furthermore, the current decline in per-capita use of fossil energy, caused by the decline in oil supplies and increasing prices, is generating direct competition between developed and developing countries for fossil energy resources.

Biodiversity

A productive and sustainable agricultural system, and indeed the quality of human life, also depends on maintaining the integrity of the natural biodiversity that exists on earth. Though often small in size, diverse invertebrate and microbe species serve as natural enemies to control pests, help degrade wastes, improve soil quality, fix nitrogen for plants, pollinate crops and other vegetation, and provide numerous other vital services for humans and their environment (Pimentel *et al.,* 1996). Consider that in New York State on a bright sunny day in July, wild and other bees pollinate an estimated 1,000,000 million blossoms that are essential for the production of fruits and vegetables and other plants. Humans have no technology that can substitute for this task and many of the other contributions provided by the estimated 10 million species that inhabit the earth.

Food Distribution

Assumptions are made by some that market mechanisms and international trade are effective insurance against future food shortages. However, when the biological and physical limits of domestic food production are reached by all nations, food importation will no longer be a viable option for all countries, because at that point, food importation for countries like the rich can only be sustained by starvation of the poor. In the final analysis, existing biological and physical resource constraints will regulate and limit all food production systems.

These concerns about the future are supported by two observations. First, most of the 188 nations of the world now are dependent on food imports. Most of these imports are cereal grain surpluses produced only in those countries that now have relatively low population densities, where intensive agriculture is practiced and where surpluses are common. For instance, the United States, Canada, Australia, and Argentina provide about 80% of the cereal exports on the world market (WRI, 1992). This situation is expected to change in the next 75 years when U.S. population doubles (USBC, 1998). Then, based on this projection, instead of exporting cereals and other food resources, these foods will have to be retained domestically to feed 540 million hungry Americans (Pimentel and Pimentel, 1996). The United States, along with other exporting countries, will cease to be able to export food.

In the future, when the four major exporting countries retain surpluses for domestic use, Egypt, Jordan, and countless other countries in Africa and Asia will be without the food imports that are basic to their survival. China, which now imports many tons of food, illustrates the severity of this problem. If, as Brown (1995) predicts, China's population increases by 500 million beyond their present 1.2 billion, and their soil erosion continues unabated, it will need to import 200 to 400 million tons of food grains each year starting in 2050. This minimal quantity is equal to more than the current grain exports of *all* the exporter nations mentioned earlier (USBC, 1998). Based on realistic trends, sufficient food supplies probably will not be available for import by China or any other nation on the international market by 2050 (Brown, 1995).

18.8.2 TECHNOLOGY AND SUSTAINABILITY

Over time, technology has been instrumental in increasing industrial and agricultural production, improving transportation and communications, advancing human health care, and, overall, improving many aspects of human life. However, much of its success is based on the availability of the natural resources of the earth.

In no area is this more evident than in agricultural production. No known or future technology will be able to double the amount of the world's arable land. Granted, technologically produced fertilizers are effective in enhancing the fertility of eroded croplands, but their production relies on the diminishing supplies of fossil fuels.

In addition, increases in the size and speed of fishing vessels has not resulted in increases in per-capita fish catch (Pimentel and Pimentel, 1996). It is actually to the contrary. For example, in regions like eastern Canada, where overfishing has become so severe that about 80,000 fisherman have no fish to catch, and the entire industry has been lost (W. Rees, University of British Columbia, personal communication, 1996).

Consider also the world supplies of freshwater that are available must be shared by more individuals, and for increased agriculture and industry. No available technology can double the flow of the Colorado River; the shrinking groundwater resources in vast aquifers cannot be refilled by human technology. Rainfall is the only supplier of water.

Certainly, improved technology will help increase food production. Technology can help with the more efficient management and use of resources, but it cannot produce an unlimited flow of those vital natural resources that are the raw material for sustained agricultural production. And what of the available technology: Where is it and why has it not been employed, now that cereal grain production per capita has been declining for the past 17 years and continues to decline (Figure 18.4)?

Biotechnology has the potential to make some advances in agriculture, provided its genetic transfer ability is wisely used. However, the biotechnology developed more than 20 years ago still has not stemmed the decline in per capita food production during the past 17 years (Figure 18.4). Currently, about 40% of the research effort in biotechnology is devoted to the development of herbicide resistance in crops (Paoletti and Pimentel, 1996). This technology does not increase crop yields, but it does increase the use of chemical herbicides and the related pollution of the environment.

What of a Sustainable World?

Strategies for global food security and sustainability must be based first and foremost on the conservation and careful management of cropland, freshwater, energy, and all biological resources required for food production. Our stewardship of world resources will have to change. The basic needs of all people must be brought into balance with the availability of life-sustaining natural resources worldwide. The conservation of these resources will require the coordinated efforts of all individuals and all nations. Once these finite resources are exhausted, they cannot be replaced by human technology. In addition, more efficient and environmentally sound agricultural technologies must be developed and put into practice to support the continued productivity and sustainability of agriculture (Pimentel and Pimentel, 1996).

Unfortunately, none of these conservation measures will be sufficient to ensure adequate food supplies and sustainability for future generations unless the growth in the human population is simultaneously curtailed.

Several studies have confirmed that to enjoy a relatively high standard of living, the optimum human population should be less than 200 million for the U.S. and less than 2 billion for the world (Pimentel *et al.*, 1999). This harsh projection assumes that from now until such an optimum population is achieved, *all* currently available strategies and technologies for the conservation of soil, water, energy, and biological resources are successfully implemented and an ecologically sound, productive environment is maintained. The lives and livelihood of future generations depend on what the present generation is willing to do now.

There is no question that reducing population numbers over the next 100 years to approximately two billion, so that everyone on earth can enjoy a relatively high standard of living, will infringe on our freedom to reproduce. However, this freedom to reproduce can infringe on our freedoms from malnourishment, hunger, diseases, pollution, and poverty. In addition, we lose our freedom to enjoy a quality environment and a bountiful nature.

The data of the World Health Organization and other world specialists concerning the number of people who are currently malnourished and diseased confirms that nature already is putting pressure on the quality of human life. If humans don't limit their numbers, *nature will!*

During the past two decades, recycled aluminum has become a major element of overall and domestic aluminum supply. The cost benefits of recycling include savings of 95 percent in energy usage and 90 percent in capital and labor. Aluminum can be recycled again and again without loss of its original properties. FROM IMCO RECYCLING. PHOTOGRAPH COURTESY OF CORBIS.

CHAPTER 19

Materials and Manufacturing

19.1 The Manufacturing Process

Businesses are seeking ways to prosper while reducing their use of materials. This process can be called **dematerialization** or simply "doing more with less material." Dematerialization is based on a closed-loop system in which materials, once used, are returned to the system for reuse. In such a system, materials and energy sources are continually "cycled" within the economy, rather than consumed and disposed of. Waste is redefined as by-products that have no useful application anywhere in the system. New products will evolve from or consume available waste streams, and processes are developed to produce usable waste. For instance, recycling packaging materials or burning waste products to recover their energy is preferable to putting materials into landfills.

The manufacturing process of industrial companies can be represented by the model shown in Figure 19.1. Energy and materials are inputs to the process and the desired output is the product to be sold. Any process has undesired outputs, called **wastes**. From this waste stream, we can separate products that can be remanufactured ①. Also, we can separate parts of the waste that can be recycled back into the original manufacturing process ② or other manufacturing processes ③. Finally, if waste materials cannot be used, they are subject to storage or disposal ④.

Six factors that can be considered for reduction in the manufacturing process are shown in Table 19.1. In the next section we consider the reduction of material intensity and energy intensity.

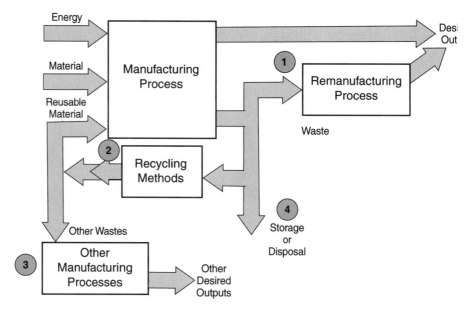

Figure 19.1 **The manufacturing process.**

Table 19.1 **Six Factors for Reduction in the Manufacturing Process**

Reduction of:
1. Material intensity
2. Energy intensity
3. Amount of waste not recycled or reused
4. Amount of depleting resources used
5. Service intensity
6. Health and environmental risk

19.2 Material Intensity and Energy Intensity

Materials and energy are the resource inputs to manufacturing processes as well as throughout the supply chain of a company. The materials and energy supply chain is shown in Figure 19.2. The goal is to reduce the total materials intensity and energy intensity of a product. The materials intensity of a product may be measured in the total materials used to produce one product unit. Similarly, the energy intensity is the energy used to produce one product unit. As an example, consider the production of a plastic. The output of the production is measured in pounds (or kg) of plastic. The materials input, primarily petroleum, is measured in pounds (or kg) of petroleum. Then, the materials intensity for this plastic would be measured as a ratio with the units kg/kg. Similarly, we could measure the energy intensity as the ratio of energy, in joules, per kilogram of plastic produced.

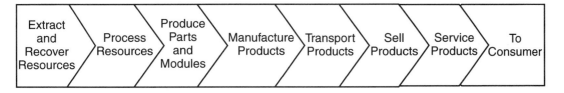

Figure 19.2 **The materials and energy supply chain.**

An example of a lower intensity product is Procter & Gamble's Ultra detergent. This product uses only one-half a cup of powder for washing compared to the traditional one-cup detergents. Procter & Gamble claims reduced soap and packaging as well as halving of the energy used (Fussler, 1996).

The design of an automobile is a good example for reduced materials and energy intensity. By the use of plastics and aluminum, an automobile may be lighter and gain better fuel consumption. Reduced materials intensity and energy intensity may be obtained by means of improved design. (For a review of design for the environment methods see Section 5.3.)

The chemical industry is an excellent example of an industry that has developed improved methods that reduce energy and materials intensity. By means of chemical engineering advances chemical companies in the United States have increased yields with reduced intensities during the past several decades. Dupont, for example, a decade ago had a materials yield of 75%. That is, every pound of raw material yielded only three-quarter pound for products such as Lycra. That yield exceeded 90% in 2000. The goal is to reduce the materials intensity while increasing the value of the product.

An example of an industry that may dematerialize and reduce energy intensity is the paper industry. In 1997, the world produced 299 million tons of paper, over six times the amount produced in 1950. Paper is used for many different purposes. Today packaging claims about 48% of all paper use. Printing and writing papers accounts for 30%, newsprint another 12%, and sanitary and household papers, about 6%. Perhaps paper consumption will decline as better means are developed for reading electronic documents and books. The supply chain for producing paper products (see Figure 19.2) affords many possibilities for reducing energy and material intensities. From forest to harvest to processing, the chain is ripe with opportunity.

The design of a product can depend on "design for the environment" (DFE), which requires that environmental objectives and constraints be inserted into process and product design, and materials and technology choices.

The focus is on the design stage because, for many products, that is where most, if not all, of their life cycle environmental impacts are explicitly or implicitly established. Traditionally, electronics design has been based on a correct-by-verification approach, in which the environmental ramifications of a product (from manufacturing through disposition) are not considered until the product design is completed. DFE, in contrast, takes place early in a product's design phase as part of the engineering process to ensure that the environmental consequences of a product's life cycle are understood before manufacturing decisions are committed.

It is estimated that a majority of the environmental impacts generated by product manufacture, use, and disposal are locked in by the initial design. The design of a product and its component selection control many environmental impacts associated with manufacturing. Product design also establishes the ease with which a product can be refurbished or disassembled for parts or materials reclamation after consumer use. DFE tools and methodologies offer a means to address such concerns at the design stage.

While every organization in the supply chain bears some responsibility for effecting environmental change, producers have the greatest responsibility because they have the most ability to reduce the adverse effects of their products on the environment. In the past, however, environmental factors entered into product design with low priority. At the same time, retailers and consumers seldom took the environment into account in their choice, use, and ultimate disposal of products. The goal is to extend responsibility up and down the product chain so that everyone involved has an incentive to minimize the environmental impacts of energy and resource use. This approach could lead to meaningful source reduction, extensive recycling of materials, and significant product innovation.

Consider the 40 million personal computers (PC) manufactured for the U.S. market in 2000. Each PC contains at least one printed-circuit board, the components of which include lead-based surface finishes that are attached with lead solder. The electronics industry uses some 8,000 metric tons for solder applications. In 1998 only 10% of the PCs taken out of service were recycled or refurbished. Most of the rest went into waste dumps.

Lead is a heavy metal. Like other heavy metals, including cadmium and mercury, it is toxic; in humans, it affects the nervous system, blood circulation, and kidneys. Placed in landfills or incinerated, it can leach into the groundwater, where it can get into the food chain. In recent years, lead use has been restricted. The electronics industry is attempting to reduce or eliminate lead solder from a majority of its products by 2001.

Similarly, another goal is to reduce the energy requirements in the manufacturing process and along the supply chain. The goal is to design energy efficient products that can be manufactured with energy efficient methods. The designs for new electronics, appliances, machines, and autos provide examples of ways to reduce the energy intensity of products. For example, about 65% of the electric energy in an industrial country is consumed by constant-speed motor drives, about 80% driving fans and pumps. One estimate is that new and improved designs could improve the efficiency of lightly loaded pumps and fans by 30%. These new designs could save up to 20% of the nation's electric energy consumption (Kaplan, 2000).

19.3 Recycling and Reusing Waste Products

Natural systems recycle or reuse waste products and industrial systems can be designed to do so. The aim is to make our wastes into useful products or raw materials for other systems as shown in Figure 19.1. We can remanufacture, recycle, convert to inputs, or store and dispose of waste. The highest goal is to use waste for its best and most valuable use. Thus, remanufacturing or recycling are usually the best use since the part or module being reused is already fitted for use in our manufacturing process.

Remanufacturing is the process of reconditioning a part or module so that it is suitable for use in a new product. A subassembly or module is taken apart, cleaned, and reconditioned so that it can be used again.

Voluntary take-back of products is one form of product stewardship that supports remanufacturing and reuse. The five largest manufacturers of rechargeable nickel-cadmium batteries formed an organization to handle the collection and recycling of such batteries throughout the United States. This corporation plans to collect more than 85% of the batteries for recycling by 2001.

An aggressive initiative is Xerox Corporation's asset recycle management program, which is aimed at attaining "waste-free" factories and developing waste-free products. Under this program, for instance, the company achieved a 60% return rate for copier cartridges in 1998. These cartridges were then sent to dismantling centers where the component parts were sorted, cleaned, and repaired to meet the standards for new parts.

Some of the best opportunities for reducing the overall environmental impact of manufactured goods lie upstream with suppliers. Consequently, involving the entire supply chain in the design of reusable products is becoming more and more prevalent in industry. Federal Express has announced a goal to all its suppliers of using fully recycled paperboard for its envelopes.

The European Commission (EC) is moving toward a directive that stipulates that the EC's member states must "take the necessary measures to ensure that producers set up systems so that last holders and distributors can return end-of-life electrical and electronic equipment." Manufacturers are also held responsible for financing the collection system, which is to apply to all products, new and old.

The disposition process for a product at the end of its initial useful life is to collect it, disassemble it, and process its recyclable materials and parts and remake or upgrade older parts. Many engineers assert that it is cheaper to collect and update older machines than it is to build new ones. The design process can specify products that will be naturally upgradable after its normal life.

It is possible to sell the waste material from a company's product stream to other industries for use in their products. The Internet is providing new e-commerce marketplaces where a company can sell its wastes. Almost every industry seems to have at least one Web site dedicated to buying, selling, bartering, or exchanging surplus goods and waste materials.

Table 19.2 **Recycling and Reuse Methods**

- Establish recovery and recycling programs for all major materials, including oils, paper, metals, glass, and plastics.
- Develop an integrated waste management program that makes judicious use of reuse, recycling, incineration, composting, and landfilling.
- Develop a remanufacturing and upgrade program for most products.
- Find new markets for wastes in other industries.

Table 19.3 **Issues for Reuse and Recycling of Products**

- What portion of a product is economically feasible to recycle or reuse?
- What is the most profitable/least costly way to retire a product?
- How long would it take to fully or partially disassemble the product for recycling?
- What are the economical trade-offs between reuse, remanufacturing, recycling, and disposal?
- How much of it will end up in the landfill?
- How can the recovery process itself generate the highest possible return on investment?
- What would happen if material prices or disposal fees radically changed?

The four methods of reducing wastes are summarized in Table 19.2. Issues that should be considered in the development of a recycling and reuse program are provided in Table 19.3.

19.4 Resource Conservation

It is important to reduce the amount of depleting resources used. Fossil fuels, such as petroleum, may run out eventually. Products that rely on renewable and biodegradable materials are more sustainable. Examples of industries that practice resource conservation include fishing, forestry, and agriculture.

Natural cycles involve the production, transport and breakdown of an enormous mass of materials. Borrowing materials from natural resources and returning them at the end of their use (through biodegradation) does not affect the overall balance of nature. This can reduce overall environmental impacts since synthetic materials involve large amounts of fuels and create nonbiodegradable emissions and wastes.

Plastic's combination of toughness and resilience is a drawback as well as a benefit. Some plastics are so durable that may remain unchanged in landfills for thousands of years. New plastics made from agricultural products such as corn or beets may be biodegradable. Cargill and Dow Corporations plan to jointly build a plant in Nebraska to make plastics from corn or wheat grains. The new plastic, called polylactide, is versatile and strong enough to compete with plastics now used for clothing, carpets, and food containers.

Another possibility is to derive chemicals from plants. These are called **agrochemicals** and are generally biodegradable and less toxic than chemicals derived from petroleum. Currently the price of agrochemicals is higher, but they may become competitive in the near future.

19.5 Service Extension

It is possible to reduce the service intensity of a product by extending its serviceable life. Service extension of a product is obtained by designing products with (1) increased durability, (2) improved repairability, (3) multifunctionality, and (4) shared use.

Long-lasting or durable products postpone the environmental impacts that result from product replacement. Durable products are the alternative to throw-away (disposable) products. For nonconsumable goods like cars, refrigerators, washing machines, or computers, it is important that products be designed for easy refurbishing and upgrading. Truly durable products are likely to be more expensive than throwaways. For some items, the upfront cost could be higher. However, the life-cycle costs may be less for a durable product.

It is important to ensure that a long-lived, durable product does not translate into an obsolete product. This calls for a modular approach that allows for easy updating of modules. Ultimately, durable products would lend themselves to remanufacturing. For products subject to high wear and tear, such as carpets, remanufacturing is crucial; they should be easy to take apart so as many of the components can be reused as possible.

19.6 Health and Environment Risk

It is important to reduce health and environmental risks. Such risks include long-term health effects from toxic materials use, emissions to the atmosphere, and accident risks. Effluents and wastes containing toxic chemicals and emissions are potential hazards.

For production processes, cleaner production includes conserving raw materials and energy, eliminating toxic raw materials, and reducing the quantity and quality of all emissions and wastes before they leave a process.

For products, the strategy focuses on reducing impacts along the entire life cycle of the product, from raw material extraction to the ultimate disposal of the product. Examples are electronics manufacturers who seek to reduce the uses of toxic materials in their processes. Emissions must be reduced to reduce risk for people. The focus of concern over environmental hazards to human health has traditionally been in relation to chemical and microbial contamination of air, water, soil, or food.

19.7 The Interface Corporation Case

Interface, Inc., is the world's largest producer of free-laying carpet tiles for office, commercial, and health care facilities. Their sales were about $1.3 billion in 2000. The Atlanta-based company is lead by its CEO, Ray Anderson. In the words of Ray Anderson in Interface's annual report for 1999:

> At Interface, we seek to become the first sustainable corporation in the world, and, following that, the first restorative company. It means creating the technologies of the future — kinder, gentler technologies that emulate nature's systems. I believe that's where we will find the right model. . . .

The goals of Interface are those listed in Table 19.1 and 19.2. They seek a model of operations that leads to zero waste with every material recycled or reused. They started in 1995 with waste minimization projects. They also shifted from new materials to 100% recycled fiber made from PET (polyethylene terephthalate), which comes from post-consumer soda bottles. Recycling materials also reduces energy usage.

As a first step, the company focused on the elimination of toxic effluents at its 26 manufacturing plants. The next goal is to reduce the energy intensity of its products by shifting to recycled materials. New nylon carpet is made from discarded carpet.

The company adopts the idea that the value to the customer is a service that includes flooring systems with desirable attributes. To transform a durable carpet into a service they introduced the Evergreen Lease. They lease the carpet to the building owner. As carpet tiles wear out and are replaced, the old ones are broken down and remanufactured into new tiles as part of the lease fee. The customer does not pay an installation cost, only a monthly fee for constantly fresh-looking and functional carpeting.

The Interface, Inc., model (see *www.ifsia.com*) is based on the use of natural materials that flow through its processes as well as compostable materials as the only discarded waste. Solar energy is the desired energy input and recycled modular products are part of the total service offered.

High-speed trains at Gare de Lyon, France. PHOTOGRAPH BY CARL PURCELL, COURTESY OF CORBIS.

CHAPTER 20

Transportation Systems

20.1 Sustainable Transportation Systems

Transportation has played a major role in shaping the industrialized world, influencing the location of economic activity, the form and size of cities, and the style and pace of life. The mobility and access provided by transportation have been instrumental to economic and social development worldwide and throughout history. Transportation systems foster economic growth by facilitating trade, permitting access to resources, and enabling greater economies of scale and specialization. They also expand cultural and social connections, increase employment and educational opportunities, and offer more options for where to live.

What is perhaps most important about the U.S. transportation system is the unmatched scale and the extent to which one mode of travel, the automobile, has become so integrated into the daily lives and activities of Americans, influencing where and how people reside, work, shop, and socialize. Americans drive their cars about one hour each day; both as drivers and as passengers, they travel by car more than 15,000 miles each year.

The global population of automobiles and trucks was about 510 million in 2000 and is expected to reach one billion motor vehicles by 2050. For example, China increased its motor vehicle population from 613,000 in 1970 to 13 million in 2000.

The transportation system integrates people, modes, and land uses. Industrialized nations have relatively fully developed systems so that it is not possible to just build our way out of congestion and other problems. Throughput of the system must be managed and new technologies introduced to mitigate congestion and pollution emissions.

A sustainable transportation system is one that satisfies the economic, environmental, and social equity components in an integrated manner. The economic component is concerned with the infrastructure funding, organization, costs, and scale. The environmental component includes issues of how transportation investments and mode options influence travel and land use patterns and how these in turn influence energy consumption, emissions, air and water quality, and habitats. The social component emphasizes adequate access to transportation services by all segments of society.

Integration of the system includes linking various modes together into a robust, economic system. The system is concerned with transporting both people and freight. The primary modes of transportation are auto, train, and air travel. Auto and air travel have grown steadily during the past decades and can expect to do so in the future.

The various forms of transportation that have been developed over time are called **modes**. The distinctions among modes include uses, right of way, propulsion type, fuel source, support, fixed vs. variable route, and control. The success of various transportation modes depends on available technology and socioeconomic conditions at any particular time, as well as on geographic factors. As technology or socioeconomic conditions change, new transportation modes appear, develop, and may later decline as more effective competitors appear. For many centuries water transportation was considerably cheaper than overland transportation. Access to waterways was quite influential in the location of economic activities and cities. Technological developments have so drastically improved the relative effectiveness of air transportation that within a short period (1950 to 1965) aircraft almost totally replaced ships for transporting passengers across oceans. It is also notable that as economic prosperity grows, personal transportation tends to shift from the walking mode to bicycles, motorcycles, and then automobiles.

Common measures of performance of a transportation system are listed in Table 20.1. Users choose a transportation mode to suit their unique per-

Table 20.1 **Transportation Performance Measures**

- Capacity = Maximum throughput
- Utilization rate = Fraction of time in use
- Load factor = Fraction of payload used
- Energy efficiency = Energy used per output-mile
- Pollution emissions = Pollutant per hour
- Cost or price per output-mile
- Safety
- Reliability
- Availability or access

NOTE: Output is persons or volume of goods (transported).

sonal needs. Thus, when transportation planners advocate a new or alternate system they must be concerned with the possibility that the new vehicles or system will not meet people's needs.

People use transportation modes that fit their need for mobility and reliability. The choice of an automobile is logical, but traffic congestion can often result in a slow speed of travel. In congested London today, average automobile speed is similar to that of the horse-drawn carriage of a century ago.

Many advocates call for less use of automobiles with a required shift to rail mass transit systems. Others call for increased reliance on telecommuting and teleconferencing to reduce the need for travel. During the past 20 years, the share of vehicle-miles traveled under urban conditions has grown from 50% to 66%, and it is expected to reach 73% in 2010. Today, most car journeys are short trips of under 20 miles and the car is normally carrying only one or two people. This means there is a rising need for a city car in the developed countries.

Government policies to change travel choices include many methods. Table 20.2 summarizes policy instruments for modifying travel mode selection. For instance, some cities restrict available parking so that the number of automobiles that can access the city are limited. The actual package of policies used may not gain the sought-after result.

Table 20.2 **Transportation Policy Measures**

- Gasoline and diesel fuel taxes
- Vehicle taxes and rebates
- Fuel economy standards for vehicle type
- Tax credits for manufacture of efficient vehicles
- Transportation control measures — restricted lanes, high parking charges
- Control of land use patterns

The original attraction of the automobile was its promise of unlimited mobility, something it could deliver when societies were largely rural. But in an urbanized world, there is an inherent incompatibility between the automobile and the city, leading to air pollution and traffic congestion.

Often the only alternative to the automobile in urban settings is a state-of-the-art rail passenger-transport system augmented by other forms of public transportation and bicycles. Whether the goal is mobility, clean air, the protection of cropland, limits on congestion, or stabilization of climate, the automobile will be combined with other means of mass transit and bicycles. Many countries encourage the use of bicycles in urban areas. Table 20.3 shows the number of automobiles and bicycles manufactured in selected years.

Table 20.3 **Autos and Bicycles Manufactured Worldwide**

Vehicle	1969 (millions)	1980 (millions)	1995 (millions)
Automobiles	23	29	36
Bicycles	25	62	109

20.2 Air Transportation

Air transportation is relatively recent, having become practical for transporting mail and passengers in the 1920s. Its growth was paced primarily by technological developments in propulsion, aerodynamics, materials, structures, and control systems. These developments have improved its speed, load capacity, energy efficiency, labor productivity, reliability, and safety to the point where it now dominates long-distance mass transportation of passengers overland and practically monopolizes it over oceans. Airliners have put ocean passenger liners out of business because they are much faster and also, remarkably, more fuel and labor efficient. For cargoes that are perishable, high in value, or urgently needed, air transportation is preferred over long distances. There are approximately 10,000 commercial jet aircraft in use in the world.

Since the first commercial jet aircraft were introduced in the 1950s, the energy efficiency of air transport has improved significantly. Aircraft energy efficiency can be improved primarily by increasing propulsion efficiency, reducing aerodynamic drag, and reducing air-frame and engine weight. Of these, propulsion efficiency offers the greatest opportunities for improvement.

Exhaust emission controls are a key environmental issue. During the past 20 years, Boeing has reduced emissions per aircraft by 50% and unburned hydrocarbons by 70%. However, aircraft account for only 1.4% of the total carbon monoxide emissions. As air travel grows in popularity, it will be difficult to reduce the total emissions from aircraft without introducing new jet engine technology.

20.3 Railroad Transportation

The main advantages of railroad technology are low frictional resistance and automatic lateral guidance. The low friction reduces energy and power requirements but limits braking and hill–climbing abilities. The lateral guidance provided by wheel flanges allows railroad vehicles to be grouped into very long trains, yielding economies of scale and, with adequate control systems, high capacities per track. The potential energy efficiency and labor productivity of railroads is considerably higher than for highway modes.

Substantial traffic is required to cover the relatively high fixed costs of railroad track. Moreover, U.S. railroads, which are privately owned, must pay property taxes on their tracks, unlike their highway competitors. By 1920 highway developments had rendered low-traffic railroad branch lines noncompetitive in the United States.

Diesel electric locomotives with a power of up to 5000 hp haul most trains in the United States. Electric locomotion is widespread in other countries, especially those with low petroleum reserves. High-speed passenger trains have been developed intensively in Japan, France, Great Britain, Italy, Germany, and Sweden. The most advanced are the latest French TGV versions, with cruising speeds of 186 mph and double-deck cars. At such high speeds, trains can climb long, steep grades. Even higher speeds are being tested in experimental railroad and magnetic levitation trains.

Freight railroads have probably accomplished the most in fuel conservation and, in addition, are the least polluting mode of freight transport. Railroads move many more ton-miles of freight than other principal freight modes (trucks, waterways, and pipelines), carrying 38.3% of the total. Railroads tend to carry dense, heavy loads on a low-cost basis.

The efficiencies of railroads can be increased by using more efficient locomotives, improved rail lubrication, better dispatch control, improved track quality, and increased use of multimodal freight systems.

High-speed trains are popular with the traveling public, carrying one-eighth of the railway traffic in western Europe. And they are set for expansion. In France alone, nearly 2,000 km of high-speed track will be open by 2001.

Environmentally, high-speed trains are attractive. With a combination of speed and comfort, they draw people away from cars and airliners, while polluting substantially less than either. Studies by French Railways show that a plane uses four times more energy than a TGV per passenger-kilometer, and a car 2.5 times more than a TGV.

20.4 Public Transportation Systems

Public transportation is the term for ground passenger transportation modes available to the general public. It connotes public availability rather than ownership. Traditional public transportation modes have fixed routes and fixed schedules and include most bus and rail transit services. Nontraditional modes include taxis, carpools and van pools, rented cars, and dial-a-ride services.

The main purposes of public transportation services, especially traditional mass transportation services in developed countries, are to provide mobility for persons without automobiles; to improve the efficiency of transportation in urban areas; to reduce congestion effects, pollution, accidents, and other negative impacts of automobiles; and to foster preferred urban development patterns.

Traditional services such as bus and rail transit networks are quite sensitive to demand density. Higher densities support higher service frequencies. Compared to automobile users, bus or rail transit users must spend extra time in access to and from stations and in waiting at stations. Direct routes are much less likely to be available, and one or more transfers may be required. Thus, mass transit services tend to be slower than automobiles unless exclusive rights-of-way can favor them. Such exclusive rights-of-way can be quite expensive. Even when unhindered by traffic, average speeds may be limited by frequent stops and allowable acceleration limits for standing passengers. Prices are usually lower for mass transit, especially if parking for automobiles is scarce and expensive.

The effectiveness of a public transportation system depends on many factors, including demand distribution and density, network configuration, routing and scheduling of vehicles, fleet management, personnel management, pricing policies, and service reliability. Demand and economic viability of services also depend on how good and uncongested the road system is for automobile users.

Designers can choose from a great variety of options for propulsion, support, guidance and control, vehicle configurations, facility designs, construction methods, and operating concepts. Improved information and control technology can significantly improve public transportation systems. It may foster increased automation and a trend toward more personalized service.

20.5 Water Transportation

Water transportation can be classified into (1) marine transportation across seas and (2) inland waterway transportation. Inland waterways consist mostly of rivers, which may be substantially altered to aid transportation. Lakes and artificial canals may also be a part of inland waterways. The waterway share of U.S. freight transportation has increased substantially in recent years. This is largely attributable to extensive improvements to the inland waterway system undertaken by the responsible agency, the U.S. Army Corps of Engineers.

The main advantage of both inland waterway and marine transportation is low cost. The main disadvantage is relatively low speed. Provided that sufficiently deep water is available, ships and barges can be built in much larger sizes than ground vehicles. Ship costs increase less than proportionally with ship size, for ship construction, crew, and fuel. Furthermore, energy efficiency is very good at low speeds.

20.6 Motor Vehicles

Motor vehicles are an important transportation mode for most industrialized nations. Motor vehicles include automobiles and trucks. The automobile has three main advantages: flexibility, personal comfort, and low marginal cost. The auto is more flexible than any alternative in terms of getting from point A to point B at the time you need to go. The traveler does not have to get off a bus or a train and wait for another bus or train to come. She can change homes or jobs and not worry about whether the transportation system can get her between the two new points.

Second, the auto is a personalized vehicle that the owner can custom design for his or her needs. Third, the auto has a large difference between fixed and marginal costs. Once the car is purchased and insured, the marginal cost is low. Public transit systems, which must cover average costs, cannot compete with autos, which must only cover marginal costs, even if the average costs of those public systems are below the average costs of the automobile.

The growth of the use of the automobile in the United States is illustrated in Table 20.4. The availability of autos is so attractive that there is almost one auto for every adult person in the United States. The average U.S. car travels about 15,000 miles each year carrying an average of 1.3 persons.

Table 20.4 **Growth of the Use of Automobiles in the United States**

Year	Passenger cars	People	Persons per auto
1947	30,718,800	144,126,000	4.69
1959	59,561,700	177,830,000	2.99
1972	96,397,000	208,837,000	2.17
1998	202,734,000	270,312,000	1.33

Transportation of all types already accounts for more than one-quarter of the world's commercial energy use. That makes the rapid increase in the global transport sector, particularly the world's motor vehicle fleet, a real concern. Motor vehicles, cars, trucks, buses, and scooters account for nearly 80% of all transport-related energy.

Motor vehicles have brought enormous social and economic benefits. They have enabled flexibility in terms of where people live and work, the rapid and timely distribution of manufactured goods, and ready access to a variety of services and leisure activities. However, the widespread use of motor vehicles has also created real environmental and economic costs, which have ballooned as vehicle numbers have risen sharply in the past few decades. Vehicles are major sources of urban air pollution and greenhouse gas emissions. They also represent an important threat to the economic security of many nations because of the need to import oil to fuel them.

Currently, the transport sector consumes about one-half of the world's oil production as motor fuel.

Americans purchase about 16 million motor vehicles each year. There are about 510 million motor vehicles in use worldwide (see Table 20.5). In 1999, 8.7 million autos and 8.3 million light trucks were sold in the United States.

Table 20.5 **Motor Vehicles in Use Worldwide**

Year	1970	1980	1990	2000
Autos (millions)	190	310	430	510

Interest in reducing emissions and the energy used by automobiles is high. Thus, advocacy of pollution control equipment and better fuel economy are the goals of government. For the first time, by 2009 in the United States, minivans, pickup trucks, and sports utility vehicles will have to meet the same emission standards as automobiles.

20.7 Heavy-Duty Trucks

Heavy trucks account for 30% of the ton-miles of freight hauled in the United States and about 30 billion gallons of fuel consumption annually. Heavy-duty trucks are classified as those weighing over 8,500 lb. About 11 million heavy trucks are active in the United States today.

The share of freight movement by mode is given in Table 20.6. Pipelines and marine modes are used heavily for petroleum and chemical products, as well as natural gas. Railroads typically specialize in high-density bulk commodities, such as coal, metallic ores, agricultural products, lumber, and chemical products, and in long-haul operations. Truck freight can be broadly characterized as manufactured products, low-density products, high-value-added products, or products being shipped to locations with no rail service.

Table 20.6 **Freight Movement by Mode in the United States in 1995**

Mode	Truck	Railroad	Water	Air	Oil pipeline
Share %	30	38	12	1	19

In the United States trucks have steadily increased their share of the freight transportation market, mostly at the expense of railroads. They can usually provide more flexible, direct, and responsive service than railroads, but at higher unit cost. They are intermediate between rail and air transportation in both cost and service quality.

The typical fuel economy for a heavy-duty truck is about 7 miles per gallon. Heavy trucks using diesel engines are significant contributors to air pollution. The hazardous mixture that comprises diesel exhaust contains hundreds of different chemical compounds that wreak havoc on air quality, playing a role in ozone formation, particulate matter, regional haze, and acid rain.

The future continues to be seen as one of a growing economy, with increases in the already high levels of freight shipments. The shipment industry is now moving to "guaranteed" delivery times as competition remains intense among trucking firms. Railroads can be considered as possible competition, but this is unlikely due to the fact that on-time performance for the railroad deliveries is only about 80% to 85%.

20.8 Improved Motor Vehicles

The improvement of motor vehicles can address improved fuel efficiency and emission performance. One possibility is to reduce the weight of the vehicles by replacing the steel with carbon-fiber composites and plastics. These composites are more costly but may be competitively priced with steel in a few years.

Another possible route is to improve the internal combustion engine and reduce emissions while increasing fuel economy. Another option is to switch to a diesel engine. The diesel engine may shift to a new generation of electronically controlled, high-pressure fuel-injection systems. This promises to make diesel as quiet and smooth as gasoline, while offering much greater fuel economy.

Many designers are working on smaller cars that weigh less and provide improved fuel economy. DaimlerChrysler has built a small two-seat car that is marketed in Europe.

Another approach is to switch to alternative fuels that are less polluting. Motor vehicle fuels that provide an alternative to conventional gasoline or diesel include ethanol, methanol, natural gas, propane, electricity, and biodiesel. At present, the price difference between gasoline and alternative fuels is so great that crude oil prices could double without prompting an increase in consumer purchases of alternative fuel vehicles. In 1998, the U.S. Department of Energy estimated that about 3% of gasoline highway usage was replaced by ethanol and methanol.

New additives such as oxygenates are making gasoline burn more cleanly and efficiently. If aerodynamic drag could be reduced and tire and road losses also reduced, autos could perform better. For example, a lighter car with a smaller diesel engine, lower aerodynamic drag, and two seats might result in a one-third improvement in both pollution and fuel economy.

Most surveys show that the auto buyer desires durability, reliability, a well-made vehicle with a good engine and transmission that is easy to maintain and available at a reasonable price relative to alternatives. Any alternative vehicle will need to satisfy all of these criteria.

20.9 Intelligent Transportation Systems

Computers, communications systems, and vehicle technologies provide the basis for intelligent transportation systems (ITSs). These systems apply new technologies to address the many challenges of surface transportation: safety, productivity, environmental concerns, and mobility in the face of increasing congestion. ITS applications include, among others, on-board computer mapping devices to help drivers find unknown locations, robotic systems that automatically control braking and acceleration, beacons for stranded vehicles, sophisticated traffic control centers, and electronic toll collectors that do not require stopping. Most ITS initiatives emphasize highway and vehicle improvements, though some focus on public transportation.

The long-term goal of the move toward ITSs is intended to enhance the capacity, efficiency, and safety of the federally supported highway system and the states' efforts to attain mandated air quality goals. The use of ITSs should reduce societal, economic, and environmental costs associated with traffic congestion.

Advocates of ITS state that collision prevention systems could reduce accidents by 70% or by as much as 90% on fully automated highways. Fully automated highways are those that incorporate all available automatic control devices for the vehicle and the highway system itself.

Congestion avoidance is another potential benefit of ITS. Fully automated highways could have flows three times those of conventional highways.

Evolutionary processes are envisaged whereby automobiles will be progressively automated to improve their safety in conventional driving circumstances. This electronic intelligence would first warn of dangerous situations then take the action needed to avoid collisions. Relatively soon, there may be enough vehicles enhanced by intelligent technologies to justify the conversion of existing lanes and construction of new facilities for the ITS system.

Critics allege that ITS could be prohibitively expensive and take decades to fully implement. The ITS potential benefits will only fully accrue as every car on the road has the necessary automation equipment.

The Honda Insight gasoline-electric hybrid automobile can achieve up to 70 miles per gallon on highways and travels up to 75 miles per hour. PHOTOGRAPH COURTESY OF AMERICAN HONDA MOTOR CO., INC.

CHAPTER 21

Electric and Hybrid Vehicles

21.1 The Electric Vehicle

In the beginning of the automotive age, electric and steam-powered vehicles were the dominant forms of private transportation. In 1900, for example, steam-powered vehicles accounted for 40% of the market, electrics 38%, and gasoline-powered vehicles only 22%.

Yet, by the end of World War I, just 18 years later, the electric vehicle industry was almost dead, largely because gasoline-powered vehicle technology had advanced faster than electric vehicle technology. Charles Kettering's invention of the electric starter overcame the need to hand-crank the car, and advances in carburation and ignition significantly improved driveability.

Compared with the rapid progress being made in gasoline-powered vehicle technology in the early part of the century, electric vehicle technology proceeded slowly. Little could be done to lower the cost or improve the performance of the most expensive part of the electric vehicle — the batteries.

Electric vehicles have received new attention recently because the California Air Resources Board (CARB) developed a plan that would require 10% of all vehicles sold in California by the year 2003 to be zero-emission vehicles (ZEV) and other states are considering similar mandates. The only type of vehicle currently available that can meet the ZEV regulation is an electric vehicle (EV) powered by a bank of batteries. These batteries are limited, at present, to lead-acid or nickel-metal hydride batteries. The basic EV consists of a bank of batteries that supply power, via a controller, to an electric motor that drives the wheels.

The total cost of operating an electric vehicle consists of the purchase price and operating expenses. Electric vehicles are not yet in mass production and the cost of producing vehicles available in 2000 was more than that of equivalent gasoline-powered vehicles. In addition to the cost of electric power required to charge the batteries, operating costs also include battery replacement. A typical battery pack for an electric vehicle costs between $2,000 and $8,000 and must be replaced after approximately 500 charge and discharge cycles for lead-acid batteries and 1,000 to 2,000 cycles for nickel-based batteries. Battery development is focusing on increasing charge cycles, reducing recharge time, and eliminating maintenance by using sealed units. The main problem with deploying batteries in vehicles is their low power-to-weight ratio, called specific power, which is between 100 and 200 W/kg with currently available technology. The driving range of currently available electric vehicles is between 50 and 100 miles.

The primary problem with the electric car is the batteries for storage of the electric energy. The batteries of the General Motors EV1 weigh a total of 533 kg. They store only 31 Wh/kg compared to gasoline with 11,840 Wh/kg. The result is 1000 lb. of batteries and a car with a range, between charges, of only 70 miles. Furthermore, its batteries take several hours to recharge. A new battery, or other storage device, with a high energy storage per kilogram, a long life, and low cost is needed. Conventional lead-acid batteries are too heavy, short-lived, expensive, and limited in capacity.

Although electric vehicles produce no tail pipe emission, air pollution is generated by the power plants that provide the electricity to charge the batteries, and environmental and health problems may also be associated with the manufacture and disposal of batteries. Thus, the batteries are the key to improved performance of an electric vehicle.

The three main characteristics of a rechargeable battery — energy performance, power performance, and lifetime (both in actual time and in charge–discharge cycles) — are inextricably linked. If we increase one, then one or both of the others must decrease. Increase the size of a current collector in the battery to boost power density, for example, and there will be less room for active electrode materials, which will decrease energy density. The comparative performance of three types of batteries is given in Table 21.1.

Table 21.1 **Battery Performance**

Battery type	Specific energy (Wh/kg)	Energy density (Wh/L)	Specific power (W/kg)	Life (cycles)	Cost ($/kWh)
Lead-acid	40	70	150	500	100
Nickel-metal hydride	60	175	200	1000	300
Lithium-ion	90	200	900	NA	NA

NOTE: NA, not available.

Electric cars available in 2000 have little attraction for consumers because of their high cost and limited range. Honda and General Motors ceased production of their EV in 2000. The industry shifted toward the development and production of hybrid powered vehicles, as discussed in Section 21.3.

21.2 The Soft and Hard Elements of Technological Development: The Electric Vehicle Case

By Alan Pilkington
Senior Lecturer in Production and Operations Management
School of Management, Royal Holloway
University of London

21.2.1 RESHAPING TRANSPORT TECHNOLOGY

The technology needed for a viable battery EV has been available for longer than that required for petrol vehicles. However, issues of range — largely dependent on vehicle weight, battery power density, and charging times — and direct comparisons with the performance of internal combustion engine (ICE) powered vehicles have meant that electric vehicles have featured only occasionally in the market since the turn of the century. Recently though, zero-emission regulations developed in response to environmental concerns have generated a new impetus for firms to invest in technology and to try and resolve some of the technological problems present in the EV. But even so the development of the electric vehicle has not proceeded along a smooth path.

Successful EV development does not rest exclusively with technological innovation, but also on advances in supporting areas. For example, governments have felt the need to manipulate market demand through legislation and tax incentives despite public resistance from car makers and the oil companies. Similarly, the technologies that could foster the evolution of viable EVs have received financial support from public funds. Despite these initiatives, few technologies seem to have reached the stage whereby they can be applied to commercially successful products. This section explores the issues of technological change and environmental innovation that emerge in the EV case as they relate to the development of sustainable technologies.

21.2.2 THE REBIRTH OF THE ELECTRIC VEHICLE

For the past 100 years development by oil and automotive companies has focused on ICE technology and so reciprocating engine design has been developed to a fine art. Huge investments, on the order of tens of billions of U.S. dollars, have generated a great deal of knowledge about the control, performance, and manufacture of the ICE in all its derivatives. As a result the development of alternative transportation technology has been largely ignored, but the introduction of tough emission legislation has seen renewed interest. The California Air Resource Board's zero-emission vehicle (ZEV) mandate, first introduced in 1990, required that a rising percentage of new cars should produce no pollution when used. This effectively legislated for the introduction of EVs owing to limitations in other technologies. Despite a major revision in 1996 as a result of legal opposition from the car makers, the mandate has been a major factor in the development of the EVs entering production in the late 1990s.

Because of the nature of sustainable and environmentally focused innovation, existing models of technological development can prove limited in trying to help us understand the processes at work. Noori and Radford (1990) proposed a framework that is useful in explaining the EV case because it recognizes the value of separating *technological-hard* and *knowledge based-soft* elements. In the EV case software includes the complex surrogate market supported by tax regimes, shifts in environmental awareness in society, the development of a suitable infrastructure, and government-sponsored pilot programs; while hardware factors include the technological advances needed for success in the areas of batteries, charging systems, and other aspects of product engineering.

Despite the recent alignment of the hard and soft factors and the resulting advances in EV technology, there remain a number of questions concerning the future viability of an EV industry. Relatively short ranges afflict many EVs on the road, particularly those made by converting existing ICE vehicles to battery, while another concern surrounds the time required to charge the batteries. However, the average motorist travels less than 50 miles per day and most vehicles spend many hours parked each night. These frequently cited barriers are the result of trying to compare EVs with existing vehicles. This type of problem is one that is common in introducing sustainable technology as it generally has to offer improved performance over that which it replaces. Perhaps a more significant problem for EVs is that the generation of the electrical power needed for the battery produces pollution. However, the benefits of concentrating pollution at the generating station, away from the streets of populated areas, depends on the makeup of the electricity generating network. In areas that have low emission technologies such as nuclear, hydroelectric, or natural gas, even existing EV technology would make a significant difference to pollution. Despite these concerns, the consensus appears to be that there are many niches within the capability of existing EV technology (MacKenzie, 1994), but the critical question of how big those markets might be remains.

Market Development

There is a wide range of views on the size of the EV market. For example, a survey commissioned by the U.K. government concluded that specialist markets are the only ones open for the successful use of EVs, resulting in little or no growth above present levels, even with the technologies currently under development. This contradicts a *Financial Times* report which predicted the market for new EVs to exceed 630,000 worldwide in 2000 (Harrop, 1995). However, most commentators fall in the middle ground and agree that the initiatives currently in place will provide sufficient markets to encourage the future development of EV technology.

Most of the studies that report a limited market for EVs are based on estimates of present market structures. Here the market analysis is characterized by trying to identify consumers with very "green" lifestyles who own more than two vehicles. The result is a group of "affluent greens" believed to be willing to live with the extra cost and limited ranges of existing EVs (Bunch *et al.*, 1993). Under these conditions the number of potential sales is limited.

A more considered study of market demand for EVs, which discards the market segmentation approach above, has produced more encouraging results. Kurani, Turrentine, and Sperling (1994) developed a "reflexive survey" method in which 454 people were asked to keep travel diaries of trips. The respondents were then asked to see if they could arrange their weekly transportation needs around certain combinations of ICE and limited-range vehicles (EVs). The price–performance trade-off decisions of the respondents identified that there were a significant number of "hybrid households" who, once they had been forced to consider the implications and costs involved, found that they could accommodate an EV without sacrifices to their existing lifestyles (Kurani *et al.*, 1994). The work suggests that "hybrid households" represent a significant market for EVs over and above that identified in segmentation-based surveys and that the EV could quickly establish itself as a significant sector of the market. These results, with the change in buying behavior of consumers following exposure to the technology, suggest that the EV market is likely to be susceptible to enhancement through the manipulation of soft factors by education and demonstration programs.

Technological Development

It has often been argued that before the mass production of EVs can begin, existing technology needs refining to improve usability and efficiency. It is in this hard-technological development that the ZEV mandate may have had the greatest impact, providing public funding and an incentive for private investment. Despite advances in motor controllers and the general design of lightweight vehicles, many questions surrounding EV introduction still center on the batteries. Table 21.2 shows existing and potential battery performance compared to the criteria established by the auto-industry group, the U.S. Advanced Battery Consortium (USABC). Again, as with estimates of the potential EV market, there are a number of conflicting results in the performance predictions. These are a result of a range of investigator interests, with each pushing their own agendas through the manipulation of test conditions.

Despite the continuing debate surrounding the lack of a suitable battery, the ZEV mandate has prompted significant investment by the big car makers. In the United States all the "big three" car producers (GM, Ford, and

Table 21.2 **Performance Status of Electric Vehicle Batteries**

Battery type	Specific energy (Wh/kg)	Energy density (Wh/L)	Specific power (W/kg)	Cycle life (cycles @ 80% discharge)	Projected cost (US$/kWh)	Energy efficiency (%)
USABC criteria	80–100	135	150–200	600	150	75
Lead-acid	18–56	50–82	67–138	450–1000	70–100	65–80
Nickel-cadmium	33–70	60–115	100–200	1500–2000	300	65–75
Sodium-sulfur	80–140	76–120	90–130	250–600	100+	60–90
Nickel-metal hydride	80	n/a	200	1000	200	90

SOURCE: C. O. Quandt, "Manufacturing the Electric Vehicle: A Window of Technological Opportunity for Southern California," *Environment and Planning A,* 27, 1995, p. 842.

Chrysler) have been involved in pilot schemes and development programs, with GM perhaps having the upper hand as a result of its EV1 (Shnayerson, 1996). Ford and Chrysler have chosen to produce stripped-down versions of existing vehicles, called gliders, for conversion to battery power by outside specialists. The Japanese firms facing restrictions from the CARB ZEV mandate — Honda, Toyota, Nissan, and Mazda — have been at the forefront of developing EVs destined for use in Japanese pilot schemes as well as California. Similarly, European manufacturers have also produced a range of vehicles testing alternative technology with one eye on the U.S. market and another on regulations and initiatives at home (Quandt, 1995). All of these vehicles show that functioning EVs can be produced, but they have required much support from public bodies to integrate the soft and hard elements.

21.2.3 BRINGING TOGETHER HARDWARE AND SOFTWARE

As the leading place for regulatory initiatives, it is no surprise that the greatest density of EV technology programs is found in the United States. A range of different instruments is being used to identify and further the preparation of EV technology. The primary involvement of the U.S. government is through the U.S. Department of Energy and the Environmental Protection Agency, with similar departments in many states. There are also a number of consortia combining government and industry to develop an EV infrastructure, primarily through local pilot programs in firms with large fleets of vehicles.

Another important public–private-funded player in the development of the technology needed for the EV is the USABC. The USABC includes the U.S. Department of Energy and the U.S. Council for Automotive Research (the big car makers and parts suppliers), and aims to develop the advanced battery technology needed to provide increased range and improved performance for electric and hybrid vehicles. The USABC has established programs for advanced technology development and set targets for battery performance (see Table 21.2). However, it is not looking to support the development of already available near-term technologies such as lead–acid and nickel-cadmium power cells.

The configuration of players in the development of EV technology cannot be summarized here because the picture is one of complex interaction between firms, suppliers, and national and local government. However, the result is essentially that governmental resources are being applied to developing zero-emission technology in partnership with the existing car makers. There is some debate as to whether such programs might suffer inertia and resistance from the involvement of the car industry, but given the expense and complex nature of designing, making, and selling vehicles in large numbers, it is unlikely that EVs can have any significant impact on the market without their involvement.

21.2.4 CONCLUSIONS

This section has used the case of the EV to highlight some of the difficulties faced in introducing sustainable technologies. The clearest view of the status of the EV and its technological trajectory arises when a combined *software–hardware* view is adopted. Software covers the social and market aspects, and in the case of the EV, includes the complex surrogate market of tax regimes as well as the shifts in social system resulting from increased environmental awareness. This also includes the development of a supporting infrastructure and the emergence of new networks of technology and parts suppliers. These soft factors have only generated progress when in conjunction with hardware aspects such as the technological advances needed for success in batteries, charging systems, and other areas of the product. However, in this case advancement is not a linear or mutually exclusive process because both areas progress together in close unison and with heavy linkage. Technological developments spur market advances at some times while at others socially derived changes lead to increased hardware innovation. Until recently one or more of these aspects has been diluted and EV development remained faltering, but now there is increasing evidence of the alignment of hard and soft, with many manufacturers now introducing EVs into the market. One striking issue is the problem sustainable technology faces through comparison with existing systems. Often environmental technology must offer improved performance over the technology it replaces to gain acceptance. In the case of the EV, repeated comparison with the ICE has certainly slowed the pace of technology development.

The case also highlights some of the industrial and organizational issues of introducing sustainable technology. The EV has seen much resistance from the car makers and oil companies, where the size, complexity, and resulting inertia in the industries have resisted the threat of upheaval inherent in introducing the EV. It is necessary to be aware that the development of technological systems is assessed not just in terms of benefits or profit, but also in terms of the change in systems and organization. In this case the influence and resistance of the oil and car industries have been sufficient to delay regulations and, hence, the development of new technology. This delay may be just a number of years, or it may prevent the technology from ever becoming a commercial success; the key will probably rest with the environmental wishes of the consumer.

21.3 The Hybrid Vehicle

In view of the shortcomings of battery-powered electric vehicles, efforts were undertaken to develop an automotive hybrid vehicle. A hybrid vehicle (HEV) employs a small internal combustion engine and an electric generator and battery to supplement battery power to the electric motor drive, thereby increasing the vehicle's range and reducing the size of the engine and the battery pack. Although hybrid vehicles are technically not "zero emission," they do reduce emission appreciably and overcome some of the shortcomings of battery-powered pure electric cars.

The range limitations of the electric vehicle and its potential cost have driven many companies to develop a hybrid electric vehicle combining an internal combustion engine with an electric motor system as a more optimal and more viable prospect.

One configuration of a hybrid, called the parallel system, consists of a small gasoline engine linked to a compact direct-current electric motor. A controller directs their interaction. During hard driving, such as accelerating or climbing a hill, both of them send power to the wheels. The battery is called on to supply power to the electric motor in that case. At cruising speeds, the gas engine goes it alone. While slowing or rolling downhill, the electric motor becomes a generator that sends electricity back to the batteries. Hybrids are always ready to go because the gas engine assists with the recharging duties. The hybrid car drive system is shown in Figure 21.1. This is called a **parallel drive system** since the wheels can be powered by both the electric motor and the IC engine when both are needed. When rolling downhill the controller switches to the electric generator which charges the batteries.

Honda has a parallel system hybrid car called the Insight available in 2000 at a price of about $20,000. The two-passenger Honda Insight uses nickel-metal hydride batteries, a three-cylinder 1.0-liter gasoline engine, and a 7-hp electric motor drive. The hybrid car's engine consists of a small gasoline engine linked to a compact, direct-current electric motor. A computer directs their interaction. During hard driving, both of them send power to the wheel.

A series arrangement uses a configuration so that only the electric drive is connected to the drive axle as shown in Figure 21.2.

A hybrid vehicle configuration that drives the wheels of a vehicle only electrically is called a series arrangement. When the wheels can be powered both electrically and mechanically, it is called a parallel arrangement. Both

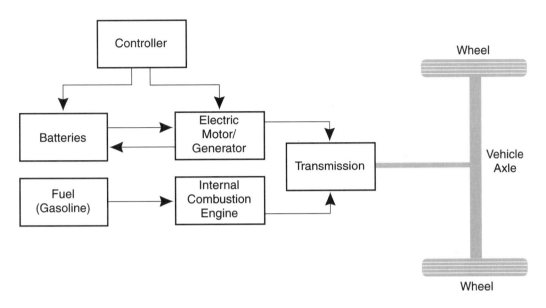

Figure 21.1 **The hybrid parallel drive vehicle.**

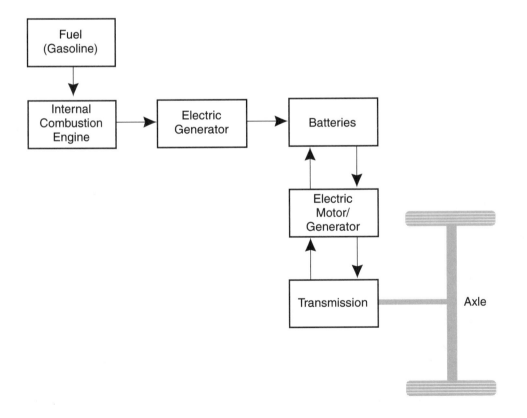

Figure 21.2 **The hybrid series drive vehicle.**

parallel and series configurations can be used for city and highway driving. The parallel system may require more on–off cycles of the gasoline engine for city driving but will be more efficient in general. Both series and parallel vehicle configurations can operate in a zero tailpipe emission mode when the batteries are charged. When the batteries are drained to a certain level, the engine can turn on to recharge them.

The advantages and best conditions of the parallel and series arrangements are summarized in Table 21.3.

Table 21.3 **Hybrid Vehicle Characteristics**

Configuration	*Advantages*	*Best suited for:*
Parallel	More power for a given motor size since both the engine and the motor can supply power.	Longer trips and highway driving
	Greater continuous operation efficiency because the engine is directly coupled to the drive shaft.	
Series	Since the engine drives a generator, it never idles and can operate at a speed that gives optimum performance.	Start-and-stop city driving
	A smaller engine can be used, which reduces pollution and improves mileage.	

The Toyota Prius uses a parallel-series arrangement that attempts to gain the advantages of both systems. The Prius parallel-series design extracts propulsion power from the gasoline engine and the battery separately (in series) and together (in parallel). That is, its parallel-series design uses a separate, engine-driven generator to charge the battery or power the motor (series operation), but it also combines engine and motor power to directly drive the vehicle (parallel operation). Toyota describes it as a parallel-series hybrid, but strictly speaking, Prius is a parallel hybrid.

When the Prius starts from rest, or is moving slowly, the engine shuts down and only the electric drive powers the wheels by drawing its power from the batteries. In normal cruise, the controller divides the engine output between the axle-wheels and the generator. The generator supplies surplus power to the batteries. During full acceleration, as on a hill, the engine gets a boost from the battery-generator. The Honda and the Toyota hybrid vehicles can average 45 mpg in highway cruising.

Fuel cells provided a cost-effective solution for the New York Police Department when a power upgrade was required for the Central Park station. This PC25 200kW system runs on natural gas, produces heat as well as electricity for the building, and was installed with limited disruption to the surrounding landscape. PHOTOGRAPH COURTESY OF INTERNATIONAL FUEL CELLS.

CHAPTER 22

Fuel Cells and the Hydrogen Economy

22.1 Hydrogen as an Energy Carrier

In spite of the many advantages of electricity at the stage of energy utilization, its generation and transmission creates many difficult problems. Electricity is difficult and costly to store for periods of peak demand and costly to transmit to the point of final use. Furthermore, the generation of electricity by the conversion of solar, wind, geothermal, nuclear, or thermal energy is a low-efficiency process. Electricity is a carrier of energy that has been converted from its original form in fossil fuels or the sun, for example, to electricity, which can be transmitted over distances to cities or towns where it is used.

Other means for storing and transporting energy can be utilized during the 2000s. The qualities of a suitable **energy carrier** are summarized in Table 22.1. A suitable energy carrier is efficiently obtained from the original energy source, efficiently transported and distributed at a low cost, readily stored, readily used in industry and transportation, and is safe to handle.

In the United States, we have made a large financial commitment to the generation, transmission, and distribution of electric power. Electric power is relatively safe to handle and can be generated and transmitted to the customers. However, electricity cannot be *stored* inexpensively. Furthermore, electric power cannot be readily used for transportation purposes without more extensive efforts toward development of an electric auto system. Electricity, however, is used efficiently for electric streetcars and rapid-transit trains.

Table 22.1 Characteristics of a Desirable Energy Carrier

- High efficiency of conversion of original energy source to the carrier energy form (fuel).
- Availability of a method of transporting and distributing the carrier fuel with low attendant energy losses and at low cost per unit distance.
- Availability of a method of storing the carrier fuel for relatively long periods of time and at low cost.
- Availability of several methods of utilizing the carrier fuel for transportation and industry.
- Safety in handling and storage.

What is required for an energy carrier is a liquid or gaseous carrier, which can be obtained from abundant resources. This carrier should be capable of readily restoring the energy at the final consumption state, without introducing significant environmental pollution. The desired process is shown in Figure 22.1, and its goal is for a minimum of environmental pollution to be generated at each conversion stage.

Hydrogen has the necessary properties and can fulfill the role of an energy carrier that can be derived from an abundant source, water. It can be substituted for petroleum and coal in almost all industrial processes which require a reducing agent, such as in steel manufacturing and other metallurgical operations. Further, hydrogen can easily be converted to a variety of fuel forms such as methanol, ammonia, and hydrazine. The use of hydrogen as a fuel would allow the industrial establishment to retain its present structure.

Hydrogen as an energy carrier can be favorably compared to electricity. In Table 22.2 we compare the projected characteristics of electricity and hydrogen which are expected to be true by 2020. Hydrogen is more efficiently converted to the carrier, transported readily and more inexpensively, stored less expensively, and readily used in industry and transportation, compared to electricity. Hydrogen does have some inherent safety hazards compared to electricity, which is used safely now.

The cost of transmission of an energy carrier becomes a very important characteristic as the new sources of energy become wind, solar geothermal, and other renewable resources.

Hydrogen can be readily derived from the primary source by the decomposition of water. Hydrogen conversion and its end-use cycle are shown in

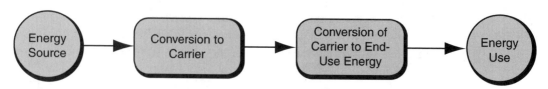

Figure 22.1 Schematic diagram of the desired energy carrier system.

Table 22.2 **Projected Comparison of the Characteristics of Electricity and Hydrogen**

Characteristic	Electricity	Hydrogen
Efficiency of conversion to the carrier	35–40%	35–50%
Transportation and distribution	More costly	Less costly
Storage	Costly	Low cost
Methods of using the carrier	Good in industry; poor in transportation	Good in industry and transportation
Safety	Proven safe	Some hazards

Figure 22.2. Energy is converted to hydrogen by the thermal or electrical decomposition of water, stored, and transported to its end use. Hydrogen is readily used as a combustible fuel, in combination with oxygen or air. The end product of the combustion is water, which recycles to the oceans and lakes. A small amount of nitrogen oxide (NO) is formed from the air entering the combustion, but it can be controlled. Otherwise, the hydrogen flame is free of pollutants.

Hydrogen can be transmitted and distributed in liquid or gaseous form. In gaseous form, hydrogen can be transmitted in much the same way that natural gas is today — by means of pipelines. The movement of a gas by pipeline is one of the cheapest methods of energy transmission, and a gas delivery system is located underground where it is inconspicuous and safe. With a power source such as the wind or solar energy, it is important to store the energy for use at times other than when the source is available. Hydrogen, as an energy carrier, can be stored in its gaseous form in an underground space. Also hydrogen can be stored in liquid form at cryogenic temperatures. Finally, hydrogen can be stored as a metal hydride. The environmental pollution from hydrogen would be much less than that from gasoline. Hydrogen can be used as a fuel for road transport in the liquid or gaseous form or as a metal hydride.

The safe handling of hydrogen presents a challenge to technology to work out the methods necessary to utilize hydrogen for commercial, residential, and transportation uses. Hydrogen has a relatively low ignition temperature, and a spark can ignite a leaking tank. However, in open air and well-ventilated places, leaks or spills diffuse so rapidly (hydrogen being the lightest of all elements) that the risks of ignition or spreading flames are actually less than those for gasoline. As a consequence, hydrogen explosions are a rarity.

Two processes are regarded as competitors for the method used to decompose water to obtain hydrogen. The first method, electrolysis, uses electric power generated by wind, geothermal, solar, or nuclear energy to decompose the water into hydrogen and oxygen. The second method uses high-temperature heat to directly decompose water in a process called **thermochemical decomposition**.

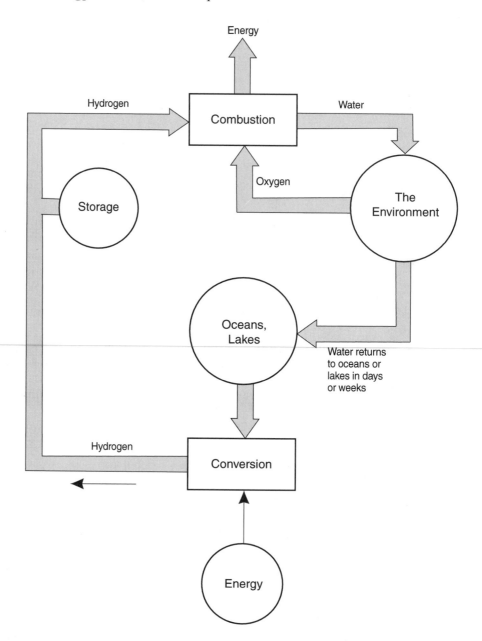

Figure 22.2 **The hydrogen conversion and end-use cycle.**

Electrolysis requires the step of electric production prior to the production of the hydrogen, and up to two-thirds of the original energy is lost. Nevertheless, the process is presently utilized and could be expanded and used with remote solar or wind generators, for example.

An alternative to electrolysis is the closed-cycle thermochemical methods of directly dissociating water into hydrogen and oxygen. This process will be more efficient overall than the electrolytic process, since it does not require that the original energy be used to generate electricity (at an efficiency of 35%) and then the electricity can be used to generate hydrogen. The thermochemical methods use the original energy, in the form of heat, to directly separate the hydrogen and oxygen.

The key to the practical use of hydrogen is efficiency. Even if the cost of the electricity used to produce hydrogen were only U.S. 1–2¢ per kilowatt hour, which may be possible in well-sited facilities built to produce power as well as hydrogen, the delivered cost of the fuel would still be more than two or three times as much as U.S. consumers pay for natural gas. Hydrogen lends itself to highly efficient applications, and as a result, the actual delivered cost of energy services to customers could be lower than they are today.

22.2 Fuel Cells

Hydrogen, like electricity, is a high-quality but high-cost energy carrier. Its adoption by the market depends on the availability of technologies and/or policies that put a high market value on hydrogen. One such enabling technology for hydrogen is the low temperature fuel cell (FC), which has wide market opportunities in both transportation and stationary combined heat and power applications. Currently the most promising low-temperature FC is the proton-exchange membrane (PEM) FC, a focus of industrial R&D efforts. The PEM fuel cell offers the potential of low cost in mass production and power densities high enough even for demanding applications such as the automobile.

In 1839 William R. Grove, a British physicist, demonstrated that the electrochemical union of hydrogen and oxygen generates electricity. Fuel cells based on this concept, however, remained little more than laboratory curiosities for more than a century, until the 1960s, when the National Aeronautics and Space Administration began deploying lightweight expensive versions of the devices as power sources for spacecraft. Today the technology, which promises clean, efficient, and quiet operation, is being developed for a host of applications, including cellular phones, laptop computers, automobiles, and power supplies.

The fuel cell is an electrochemical device that generates electricity by chemical reaction without altering the electrodes or the electrolyte materials. This distinguishes the fuel cell from electrochemical batteries. The concept of the fuel cell is the reverse of the electrolysis of water, in which the hydrogen and oxygen are combined to produce electricity and water. The fuel cell is a static device that converts the chemical energy directly into electrical energy.

A fuel cell is a simple device, consisting of two electrodes (an anode and cathode) that sandwich an electrolyte (a specialized polymer or other mate-

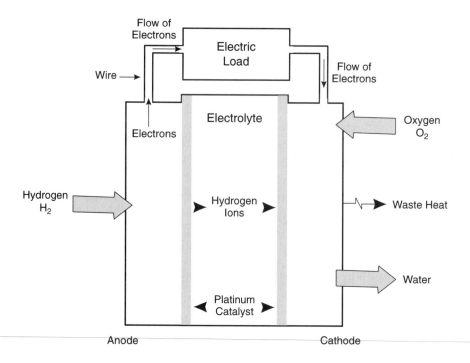

Figure 22.3 **The fuel cell.**

rial that allows ions to pass but blocks electrons). A fuel containing hydrogen flows to the anode (see Figure 22.3), where the hydrogen electrons are freed, leaving positively charged ions. The electrons travel through an external circuit while the ions diffuse through the electrolyte. At the cathode, the electrons combine with the hydrogen ions and oxygen to form water, a by-product. To speed the reaction, a catalyst such as platinum is frequently used. Fuel cells and batteries are similar in that both rely on electrochemistry, but the reactants in a fuel cell are the hydrogen fuel and oxidizer, whereas in a battery they are the materials used in the electrodes.

Several industrial firms have a goal of supplying a fuel cell and batteries in a system that can supply all the electricity needs of a 2,500-ft^2 house. Fuel cells for homes and office buildings could have an efficiency of 50% or better.

Others are working toward the development of a hybrid system consisting of a fuel cell and a gas turbine that uses the hot exhaust from the fuel cell to generate additional electricity. This system could generate 220 kW of power, enough for 200 homes.

22.3 Fuel Cell Powered Vehicles: Big Business, Fast Cars, and Clean Air

By Richard Counts and Anthony Eggert
Graduate Research Assistants
Institute of Transportation Studies
University of California, Davis

As we entered the new millennium, gasoline, adjusted for inflation, was cheaper in the United States than it had ever been since World War II, even cheaper than it was before the 1973 oil embargo. At the same time, today's cars and trucks get better mileage and emit less pollution than at any time in history. There is more good news: Vehicles have become more reliable, comfortable, longer lasting, and better handling. In fact, our vehicles are not just a means of transportation anymore — with sport utility vehicles (SUVs) they now enable more active lifestyles and increased access to the great outdoors. People love their vehicles today as much as ever; they are cheaper to operate than they have ever been, and our consumers like what they drive.

With this picture of today's vehicle consumers, there appear to be no vehicle technology problems that need to be addressed, let alone a need for a vehicle technology revolution. We don't exactly hear the average car buyer in the developed nations make comments like, "I wish there was new, revolutionary vehicle technology out there to replace my internal combustion engine. I just don't feel like my complete automotive needs are being satisfied!" Certainly, some environmentalists would be happier if this was the consumer reality, but it's not.

Yet before we completely determine that there is no need or demand for vehicle technology change, is this really the complete picture of automotive needs and wants? If the only focus of our inquiry is how people feel about their actual vehicles, then the case seems closed and static. Yet there is definitely more to the picture. To the same extent that America's love affair with the automobile is as strong as ever, Americans are realizing the environmental impacts of their vehicles. As a result, there has recently emerged a near ubiquitous concern about the quality of the air we breathe and its health effects on ourselves and, in particular, on our children. The public is becoming increasingly aware and concerned about not only the *need,* but also our *ability,* to balance our environmental interests with those of private industry and economic growth.

So what is the complete picture of our "automotive needs"? We have people who love, need, and want to drive. Yet these same people also want to clean up the air and do what they can to protect the environment, provided that only if doing so will be good for business, the economy, and our personal lives and expenses. In other words, we generally want to protect the environment, but we don't want to have to make any personal sacrifices to do it.

What are the consequences of these dual values for the automotive industry? In short, people aren't buying vehicles today that would be good for air quality or the environment. And if we look at the current "green" vehicle products that are available, who can blame them? Zero-emission vehicles like battery electric cars take hours to recharge, cost over $50,000, and most have a range of 100 miles or less. Conventional vehicles with high gas mileage tend to be underpowered, undersized, and undesirable for the image conscious. With these kinds of unacceptable options, people may want to do the right thing for the air and the environment, but the current auto market simply isn't providing them with any reasonable product options.

This is the picture of the present. Now, let's look at a vision of the future. The year is 2010 and on this exciting day you and your spouse are going to buy two new vehicles. One of you wants a large, 4WD SUV for access to skiing, hiking, and mountain biking. The other wants a high-powered, sports car convertible — all the pleasures that a pair of vehicles could provide. On your way to the car dealership, you are both thinking about the fun you will have in your new cars, the places you will go, and the looks on people's faces when they see you drive by in style. You are both as concerned about the environment as the next person, but these purchases are about enjoying your hard-earned money — you're not trying to save the world. After an hour of browsing, it comes down to one decision. The dealer offers you a simple choice — for the same price, options, performance and convenience of operation, did you want those vehicles to have the standard, polluting engines or would you like them to be zero-emission vehicles? And, by the way, the "engine" on the zero-emission choice is quieter and will require less maintenance. Well, putting it that way, if we can have the same, or even a better car, *and* save the world, then our automotive and environmental needs could actually be complementary instead of conflicting.

The best part of this vision of the future is that the technology to make this story a reality is already here — it's called a fuel cell powered vehicle (FCV). The FCV is the technological solution that will allow people to meet both their automotive and environmental needs all at once — no conflict of values, no extra cost, no inconvenience, and with all the fun, glamor, and clean skies you can imagine. What is more, not only is this new, too-good-to-be-true technology here today, but major auto makers like Mercedes, Chrysler, Ford, GM and Toyota all plan to start selling FCVs by 2004. Fuel cell powered vehicles are not science fiction; they are big business, and their introduction is just around the corner.

22.3.1 WHAT IS A FUEL CELL?

While fuel cells were first invented long ago in the 1800s, most of the key technological developments that have made them a practical automotive reality have been accomplished within the last decade.[1] But what is a fuel cell? It's not a gasoline or diesel engine, battery, turbine, windmill, solar panel, or nuclear power. Rather, fuel cells are a means of producing power

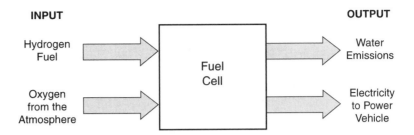

Figure 22.4 **Hydrogen fuel cell.**

that is completely different from any other kind of technology with which most people are familiar. Quite simply, a fuel cell combines hydrogen and oxygen in a controlled electrochemical reaction that does not involve combustion or burning, and the only products of this reaction are electricity and pure water. The electricity can then be used to power an electric motor to drive any type or size of vehicle. And you can literally drink the car's emissions, which are pure water (see Figure 22.4).

At the same time that fuel cells can offer the benefit of zero emissions for the environment and human health, vehicle consumers may very well prefer them over today's cars for several reasons. To begin with, because fuel cells have very few moving parts, they require very little maintenance and can provide us with more reliable, longer lasting vehicles that are amazingly quiet in operation. Further, unlike other "green" vehicle technologies to date, all of these benefits can be combined with whatever amount of power is desired for any vehicle application, from high-performance sports cars to full-size industrial trucks and buses. In all, fuel cells can power any type of vehicle with several distinguishing characteristics compared to conventional vehicles: FCVs use hydrogen as a fuel source; they don't produce any pollution; they require very little maintenance; and they can provide powerful yet quiet performance for any vehicle application.

Prior to the 1980s, the primary use of fuel cells had been to provide electric power and drinking water on U.S. spacecraft. But in the last few years, fuel cells have been brought down to earth from the space program and onto our streets. This has been accomplished primarily by the work in the last decade that has made fuel cells more reliable, compact, lighter, and less expensive to build, thus making them perfectly suited to drive our vehicles in a convenient and affordable package.

22.3.2 THE FUEL CELL VEHICLE INDUSTRY AND THE HYDROGEN ECONOMY

Fuel cells are, however, more than just a new vehicle technology. They are the technological key to an entirely new transportation and energy infrastructure and economy. Today's transportation economy is based on vehicles that run on fossil fuels such as gasoline, which can be purchased at every street corner filling station. This present energy system not only produces harmful emissions from the tailpipes of its many vehicles, but the process of refining crude oil into usable vehicle fuels like gasoline and diesel also produces vast quantities of pollutants as well.

By contrast, the vision of a fuel cell based energy economy would instead replace today's energy storage and fuel mediums of crude oil, gas, and diesel with hydrogen. As the most abundant material in the universe, hydrogen does not pose the scarcity problems of oil and gas, and when hydrogen is used as a power source for vehicles via fuel cells, the only exhaust is water. Applying an electrical current to water can, in turn, produce hydrogen by a well-known process called **electrolysis**, which separates water into its two chemical components — oxygen and hydrogen. The hydrogen fuel economy is thus impressively simple and clean: Electricity can be used to make hydrogen from water, and fuel cells are employed to turn hydrogen back into electricity and water. It is a complete energy cycle that is based on a resource that covers two-thirds of the earth's surface — water.

The major changes that this new hydrogen economy will entail for our current energy system will undoubtedly require the cooperation of fuel cell makers, auto manufacturers, oil and energy companies, and governments. Toward that end, all of these key players have already begun to form new partnerships, joint ventures, and companies to bring about this new, clean world order. One of the first such strategic partnerships was formed in 1996 when DaimlerChrysler (then Mercedes Benz) and Ford jointly invested close to a billion dollars in the Canadian company Ballard Power Systems, which is considered one of world leaders in fuel cell technology development (*Financial Times,* 1998). Ballard, Daimler, and Ford also established two joint ventures to design, build, and market the specialized components other than fuel cells that FCVs require: The joint venture Ecostar was formed to develop and build the electric drivetrain components like motors and power electronics. The joint venture Xcellsis was established to design, build, and integrate the various, specialized components such as the fuel processors that are used in some FCVs to convert liquid fuels, like methanol, into hydrogen. All of these joint ventures were created to mutually develop and commercially market FCVs by 2004 (Figures 22.5 and 22.6).

More recently, in 1999 the California Fuel Cell Partnership was formed. Membership includes the U.S. Department of Energy, the California Air Resources Board, the California Energy Commission, Ballard Power Systems, DaimlerChrysler, Ford, VW, Honda, Texaco, Arco, Shell, and others. The partnership was formed for the purpose of bringing together government regulatory agencies, automobile companies, and fuel suppliers to demonstrate fuel cell vehicles under real day-to-day driving conditions and to identify fuel infrastructure issues in order to prepare the California market for this new technology.[2] In another showing of industrial cooperation, auto giants GM and Toyota announced in 1999 their agreement to form a cooperative venture between their fuel cell research efforts. In sum, major auto and oil companies and government agencies have been entering into

Figure 22.5 **DaimlerChrysler's fuel cell powered NECAR 4.**

Figure 22.6 **Ford's fuel cell powered P2000.**

large-scale investments and cooperative ventures with the aim of turning the promising technology of fuel cell vehicles into a highly profitable, global business.

22.3.3 TECHNOLOGY AND INFRASTRUCTURE CHALLENGES FOR FUEL CELL VEHICLES

While the fuel cell business future is bright, there are admittedly several key technological and infrastructure challenges that must be addressed in order to make this clean hydrogen energy economy a reality.

First, a means of producing and distributing hydrogen needs to be developed so that its price will be cost competitive with the conventional fuels like gasoline and diesel which hydrogen is intended to replace. Currently, the most cost-effective method of producing hydrogen for commercial purposes is to chemically refine it from the cheap and abundant resource of natural gas. Refining natural gas thus offers a relatively affordable source of hydrogen for the near term. However, unlike the pollution-free vision of hydrogen being produced from clean sources of electricity, the refining of natural gas does produce some pollution and greenhouse gas emissions. Even so, the total emissions from a natural gas–to–hydrogen FCV energy system would be significantly less than our current system based on fossil fuels and conventional vehicles (Lipman and Delucchi, 1996).

Second, in order to offer the greatest total reduction in emissions from our present fossil fuel energy system, hydrogen would have to be made via electrolysis from electricity that was produced from a clean energy generation method like wind, solar, or hydroelectric dams. Farther on down the road, there is also talk that electricity could also be produced for this purpose by the radiation-free method of nuclear power called fusion. For now, however, the only clean and cost-effective means of producing hydrogen via electrolysis is to use cheap, off-peak hydroelectric power as Ballard is currently doing to fuel its public fuel cell buses in Vancouver, British Columbia. For the time being, the other technologies that can be used to provide clean electrical power like solar and wind are too expensive to produce cost competitive hydrogen fuel; but as the cost of these technologies continues to decline, this may change.

Surely, however, any plan to transform our current energy system based on oil into one based on hydrogen is doomed to fail unless it has the support of the current oil giants. Fortunately, several of the world's major oil companies are already teaming up with auto makers and government agencies to determine how best to produce and distribute hydrogen for this new world order that will be based on hydrogen and the technology of fuel cells. For example, Shell, one of the world's largest oil companies, recently formed a new division called Shell Hydrogen whose sole corporate mission is to provide hydrogen for fuel cell vehicles. Other oil giants like BP/Amoco have made public statements that they believe hydrogen will be the fuel of the future, and they want to be a part of it. Lastly, as discussed above, Texaco,

Shell, and Arco have all joined the California Fuel Cell Partnership in an effort to commercialize hydrogen as a transportation fuel.

Third, hydrogen must be stored on vehicles in a safe and reliable manner. Many private companies and government agencies around the world are currently conducting extensive R&D on this issue. While not all of the technological challenges of hydrogen storage have yet been solved, experts in the industry are confident that hydrogen will one day be as safe to store on vehicles as gasoline.[3]

Fourth, since FCVs operate on hydrogen as their fuel, a new refueling infrastructure needs be installed to allow FCV operators to refuel their vehicles with comparable convenience to the current gas station infrastructure.

Finally, while industry and government work to determine how hydrogen can best be stored and provided for refueling, an interim solution under consideration might be to use conventional liquid fuels like gasoline or alcohol for FCVs. By this means, a fuel processor can be installed on a FCV that chemically refines either gasoline or alcohol to produce the hydrogen needed to power a fuel cell. An advantage of these liquid fuels is that they are easier to store with our current technology than is gaseous hydrogen. Also, gasoline and alcohol are more readily available today for refueling than is hydrogen. However, the disadvantage of using these liquid fuels is that the process of refining them on board the FCVs will likely produce some form of pollution, depending on the technology that is used for that purpose. Even so, the emissions from this onboard refining process could be much less than those of conventional gasoline or diesel vehicles.

22.3.4 MARKET DEVELOPMENT AND POLICY CHALLENGES FOR FUEL CELL VEHICLES

When most companies want to market a new technology, they can do something conventional, like run a media advertising campaign to inform their customers about their new product. But given that FCVs are not exactly a conventional product, bringing them to market will require some unique tactics. The first order of business is that the FCV industry needs to establish a customer demand for its products. In this challenge, the industry is faced with a chicken and egg marketing problem: Fuel cell vehicles will not be competitively priced until people buy them in mass quantities, but the public isn't going to buy them in large numbers until the necessary refueling stations are as convenient as our current gas stations. And to further frustrate matters, no one is going to install these needed refueling stations until there are enough FCVs on the road to make the new stations profitable.

Central Fleets as the Market Beginning

Due to the above chicken-and-egg refueling infrastructure challenges, in the short term, the only practical use of fuel cell powered vehicles will likely be in centrally refueled fleets. Many municipalities and large companies currently use centrally refueled fleets for their operations, and many of these fleets have been converted to run on cleaner alternative fuels such as compressed natural gas (CNG). These same fleets are likely to be the fuel cell industry's first customers since compressed hydrogen fueled FCVs can be centrally refueled in much the same way as CNG fleets. Hydrogen FCVs, however, offer the added benefit over CNG of better performance and zero emissions. Ballard and Xcellsis have already begun the development of this initial fleet niche market by entering into business ventures with the public bus transit agencies of the cities of Chicago, Illinois, and Vancouver, British Columbia (see Figures 22.7 and 22.8). Through these ventures, full-size fuel cell powered busses are now successfully operating in both of these cities to provide real day-to-day public bus transport. Ballard has thus been able to share its fuel cell bus development and testing costs with these two public transit operators, and in so doing, it has cultivated them as its first commercial customers. In fact, Chicago eventually plans to replace all of its conventional buses with Ballard fuel cell buses.

Ballard and Xcellsis's product and customer development programs with Chicago and Vancouver thus fit very neatly into the FCV industry's product development plan based on initial fleet niche markets. What is more, there is now a growing trend at the federal, state, and local level to require both publicly and privately owned fleets to reduce their emissions with the use of alternative fuels. This combination of fleet regulation trends and marketing development strategies like Ballard's could thus create a fleet customer demand for the fuel cell industry's early products.

Figure 22.7 **Chicago Transit Authority's Ballard fuel cell powered bus.**

Figure 22.8 **Ballard fuel cell powered bus in Vancouver, British Columbia.**

Making the Jump to Privately Operated FCVs

Since the problem of establishing a new refueling infrastructure doesn't apply to centrally fueled fleets, and fleets are already on a trend toward cleaner alternative fuels and technologies, the introduction of fuel cell vehicles into this market seems like a natural progression. However, making the critical jump to marketing FCVs to individual car owners is a much more challenging prospect. This marketing venture is therefore seen as a longer term goal. Part of the reason for this is that auto consumers are not likely to purchase "cleaner" cars like FCVs if there isn't a convenient refueling infrastructure in place to service them. And with those kinds of convenience problems, automakers are not likely to force FCVs on consumers if they're not demanding them. It would thus seem that the natural market forces of supply and demand are poised directly against the development of the FCV market. However, there are two key fronts upon which these market forces can be changed, both to facilitate consumer demand for cleaner vehicles and to encourage automakers to supply them.

Public Demand for Cleaner Vehicles

As a general economic principle, a stable market like the auto industry where the public is happy with the current product can only be changed if one of two things happens: The public demands it or industry really wants it.[4] As for industry, with the healthy sales and profits that the auto companies are currently enjoying, there is no compelling reason for them to change their conventional vehicle technology. Unless, that is, the public demands a new, cleaner technology from them. In the case of centrally fueled fleets, it was no coincidence or act of industry altruism that led to the widespread introduction of cleaner, alternative fuel fleets. Rather, two key forces acted to shape this change: First, laws that required the transition to cleaner fleets were passed with the aim of meeting the public's demand for cleaner air. At the same time, this same public concern for air quality gave central fleet operators a public relations incentive to do their part to clean up their operations. In sum, both of these factors boil down to one common, first cause — public demand for cleaner air.

As a result, if additional clean air laws and regulations are going to get passed to expand the fuel cell market beyond its initial fleet applications, the public is going to have to demand that its lawmakers initiate these changes. Such public demand for cleaner air and vehicle technologies, in turn, can probably best be ignited with public outreach and educational efforts designed to inform the public about the human health and environmental hazards that our current vehicle technology is wreaking upon us.

Educating the public about air quality and environmental issues is thus one of the key strategies to the development of the retail FCV market. Such public outreach will make the public increasingly aware of the harm that unnecessary vehicle pollution causes to the things they care about, like their lungs, their children, and the environment. Further, if people learn in the same education breath that FCVs can not only solve these pollution problems but also provide us with better cars, an indignant public so informed could demand sweeping regulatory reform to usher in the new era of zero emission fuel cell transportation.

Economic Market Incentives from Global Warming Regulations

While the public is most likely to be galvanized by the human health aspects of pollution as a means of developing the FCV market, another potential market driver for the fuel cell industry is the reduction of greenhouse gases. Global warming used to be associated, at worst, with environmental extremists, and more moderately, it was often considered unproven by private industries. The tide, however, has recently turned in the environmentalists' favor. Just as the tobacco industry recently had to admit that its products were harmful to health, even the world's oil giants like Shell, Arco, and BP are now making unprecedented public admissions that global warming is a real problem and that we need to start doing something serious about it, today (*Economist,* 1990). Like the situation with cleaner central fleets, this critical shift by industry toward the reduction of greenhouse gases was driven not by business altruism, but by public demand and subsequent regulations. The Kyoto Protocol, an international treaty on greenhouse gas reductions, has played a key role in this transformation. Under the terms of the Protocol's CO_2 trading credit schemes, big industries are beginning to shift millions and soon billions of dollars worldwide into subsidizing business activities that produce less CO_2 and penalizing those that produce more (*Economist,* 1990). Potentially, the FCV industry may be able to use the proceeds from these trading schemes to subsidize its early products until economies of scale enable FCVs to be cost competitive with conventional vehicles on their own.

Ending the ICE Age — Being "Green" Isn't Enough

If we assume that the public will continue to demand ever cleaner air and lower polluting vehicles, and that the resulting regulations and economic forces will force the auto industry to produce increasingly cleaner cars, will this necessarily result in the proliferation of fuel cell technology to meet these environmental and health demands? In short, probably not. For every million dollars that is currently being spent by the major automakers and fuel cell makers to develop fuel cells into a cost competitive, zero-emission technology to replace the internal combustion engine (ICE), billions are spent each year on efforts to clean up ICEs to prolong their profitable presence in the market. Honda has even developed a new ICE that runs so clean on normal gasoline that the exhaust that comes out of its tailpipe can be cleaner than the air it takes in. What's more, automakers are always making advancement in catalyst technologies which can, in theory, reduce most of a car's emissions to very low levels, possibly even close to zero. And while no car maker has been able to produce an ICE that can do all of the above and also refrain from producing greenhouse gases like CO_2, there is talk that hydrogen powered ICEs could possibly achieve this incredible goal as well. What all this means is that perhaps we can eliminate vehicle sources of pollution and greenhouse gases by simply improving the conventional technology of ICE's in combination with an alternative fuel like hydrogen. So if ICEs might one day be just as "green" as fuel cells, what use do we really have for this new technology?

Dr. Ferdinand Panik, the senior project manager for DaimlerChrysler's Fuel Cell Project, is well aware that being "green" isn't going to be enough to transform the ICE age into the FCV age. As a solution to this problem, he offers a lesson from history: "The Stone Age didn't end because we ran out of stones — we found something better!"[5] In the same way, he and other FCV industry advocates argue that FCVs don't just offer consumers "greener" cars. Rather, FCVs will be better performing, quieter, more reliable, longer lasting, and they will be cheaper and easier to maintain than even the best zero-emission ICE. Just as the steam engine replaced the horse and buggy and oil replaced coal, auto consumers will *prefer* FCVs because they are a superior product, even with all environmental concerns aside.

To illustrate how the ICE age would end in this way, lets step back into our vision of the future in 2010 when you were closing the deal on those two new dream cars that you were buying. In line with the current promises that some elite automakers like BMW are making that future cars will have zero-emission ICEs that run on hydrogen fuel, let's assume that by the year 2010, many people are either driving hydrogen-powered ICE cars, or hydrogen-powered FCV's. Either way, both types of cars fill up with hydrogen at the local filling station right next to the phasing-out gasoline cars of yesteryear. The car salesperson asks if you want your hydrogen-powered convertible and SUV to be ICE or fuel cell powered, and you ask her to explain the difference. "Well," she explains, "both cars have zero emissions, so they're both great for the environment. But as an electrically powered car, the FCV is quieter, it provides noticeably better acceleration, and it gets twice the fuel economy so its cheaper to operate. What's more, since the FCVs primary 'engine' has fewer moving parts, it is expected to last the life of vehicle with very little maintenance and is thus warranted for as long as you own the vehicle." After that, perhaps your only remaining question would be to ask why they still make those old ICEs.

Further Reading

Appleby, A. J., and F. R. Foulkes. *Fuel Cell Handbook.* New York: Van Nostrand Reinhold, 1989.

Ballard Power Systems, *www.ballard.com.*

International Fuel Cells, *www.internationalfuelcells.com.*

Koppel, Tom. *Powering the Future — The Ballard Fuel Cell and the Race to Change the World.* New York: John Wiley, 1999.

Kordesch, Karl, and Gunter Simader. *Fuel Cells and Their Applications.* Germany: VCH, 1996.

Mark, Jason. *Zeroing Out Pollution: The Promise of Fuel Cell Vehicles.* Union of Concerned Scientists, 1996; available at *www.ucsusa.org.*

Motavalli, Jim. *Forward Drive — The Race to Build "Clean" Cars for the Future.* Sierra Club, 2000.

Ogden, Joan. "Developing an Infrastructure for Hydrogen Vehicles; a Southern California Case Study." *International Journal of Hydrogen Energy,* **24,** 1999, pp. 709–730.

Shell Oil Company. *Renewable Energy.* 1999; available at *http://www.shell.com/ shellreport/issues/issues4a.html.*

Sperling, Daniel. *Future Drive.* Washington, DC: Island Press, 1995.

U.S. Department of Energy, Energy Efficiency and Renewable Energy NETWORK, available at *www.eren.doe.gov/transportation.*

World Fuel Cell Council Web page, *http://members.aol.com/fuelcells/.*

Endnotes

[1] Interview with Ballard Power Systems founder Geoffrey Ballard, October 1999, Institute of Transportation Studies, University of California, Davis.

[2] See the California Fuel Cell Partnership Web page, available at *www.driving thefuture.org.*

[3] *Fuel Cells — Green Power,* Los Alamos National Laboratory, available at *www. education.lanl.gov/resources/fuelcells.*

[4]Conference statement of Linda Lance, White House Council on Environmental Quality, at the Transportation Energy and Environmental Policy for the 21st Century Conference, Asilomar Conference Center, Monterey, California, August 24–27, 1999. Conference summary available at *http://www.ott.doe.gov/facts/asilomar99.htm.*

[5]"The Fuel Cell — A Powertrain Stretched between the IC Engine and Alternative Forms of Energy," presented by Dr. Ferdinand Panik at the Transportation Energy and Environmental Policy for the 21st Century Conference, Asilomar Conference Center, Monterey, California, August 24–27, 1999. Conference summary available at *http://www.ott.doe.gov/facts/asilomar99.htm.*

22.4 Fuels for Fuel Cell Vehicles

BY JOAN M. OGDEN, THOMAS G. KREUTZ, AND MARGARET M. STEINBUGLER
Research Scientists
Center for Energy and Environmental Studies
Princeton University

Fuel cell electric vehicles offer the promise of high fuel economy, zero or near-zero tailpipe emissions, good performance, and competitive costs in mass production. Because of their desirable environmental characteristics, many analysts see fuel cell vehicles as a key technology in a future energy system with low emissions of pollutants and greenhouse gases. Most major automotive manufacturers are developing fuel cell vehicle prototypes, and several have announced plans for commercial introduction of fuel cell automobiles in the 2003–2005 time frame.

Fuel cells being developed for use in automobiles require hydrogen as a fuel. Hydrogen can either be stored directly or produced onboard the vehicle by "reforming" a more easily handled liquid fuel such as gasoline or methanol to make a hydrogen-rich gas to run the fuel cell. The vehicle design is simpler with direct hydrogen storage (fuel processors add complexity and cost, and decrease efficiency), but hydrogen requires developing a more complex refueling infrastructure than gasoline (which has unique advantages because the infrastructure already exists) or methanol (a liquid fuel more readily compatible with the existing gasoline system).

Here we compare, with respect to vehicle characteristics and infrastructure requirements, three possible fuel options for use with fuel cell vehicles (see Figure 22.9):

- Compressed hydrogen gas
- Methanol with onboard steam reforming
- Gasoline with onboard partial oxidation

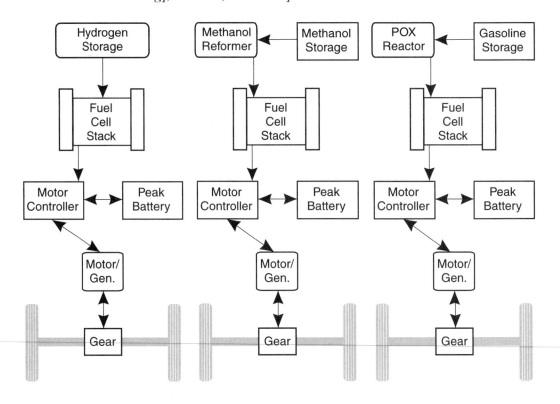

Figure 22.9 **Possible Fuel Cell Vehicle Configurations**

22.4.1 MODELING ALTERNATIVE FUEL CELL VEHICLES

To estimate the performance, fuel economy, and cost of alternative fuel cell vehicle designs, researchers at Princeton University's Center for Energy and Environmental Studies have developed computer models of proton exchange membrane fuel cell (PEMFC) vehicles (Steinbugler, 1996; Ogden *et al.,* 1998, 1999). Table 22.3 summarizes the results from the model, for streamlined, lightweight, four- or five-passenger midsize fuel cell automobiles fueled with hydrogen, methanol, or gasoline. Each vehicle is designed to satisfy identical performance criteria. Hydrogen fuel cell vehicles are projected to be simpler in design, 10–20% lighter, 50% more energy efficient, and $500–1,200 less expensive per car than methanol or gasoline fuel cell vehicles (assuming mass produced cost targets are reached fuel cell technology).

22.4.2 REFUELING INFRASTRUCTURE REQUIREMENTS

We have also compared the feasibility and cost of building a new fuel infra-structure (production, transmission, and distribution systems) for hydrogen and methanol.

Table 22.3 **Model Results: Comparison of Alternative Fuel Cell Vehicle Designs**

Fuel storage/ H_2 generation system	Vehicle mass (kg)	Peak power (kW) (FC/battery)	Combined 55% FUDS 45% FHDS (mpeg) (range, mi)		Fuel storage	Annual energy use (GJ/yr)	First cost of vehicle as compared to H_2 FCV FCV— H_2 FCV ($)
Direct H_2	1170	77.5 (34.4/43.1)	106	425	3.75 kg H_2 gas @5000 psi	13.7 (40 kscf H_2)	—
Methanol steam reformer	1287	83.7 (37.0/46.7)	69	460	13 gal MeOH	21.1 (306 gal MeOH)	$500–600
Gasoline POX	1395	89.4 (39.4/50.0)	71	940	13 gal gasoline	20.5 (157 gal gasoline)	$850–1,200

NOTE: FC, fuel cell; FCV, fuel cell vehicle.

Development of Refueling Infrastructure for Hydrogen Vehicles

Various near-term possibilities exist for producing and delivering gaseous hydrogen transportation fuel and employing commercial technologies, which are widely used in the chemical industry. These include (see Figure 22.10):

(a) Hydrogen produced from natural gas in a large, centralized steam reforming plant, and truck delivered as a liquid to refueling stations

(b) Hydrogen produced in a large, centralized steam reforming plant, and delivered via local, small-diameter hydrogen gas pipeline to refueling stations

(c) Hydrogen from chemical industry sources (e.g., excess capacity in refineries which have recently upgraded their hydrogen production capacity), with pipeline delivery to a refueling station

(d) Hydrogen produced at the refueling station via small-scale steam reforming of natural gas (in either a conventional steam reformer or an advanced steam reformer of the type developed as part of fuel cell cogeneration systems)

(e) Hydrogen produced via small-scale water electrolysis at the refueling station

In the longer term, other methods of hydrogen production might be used including gasification of biomass, coal, or municipal solid waste, or electrol-

CENTRALIZED REFORMING

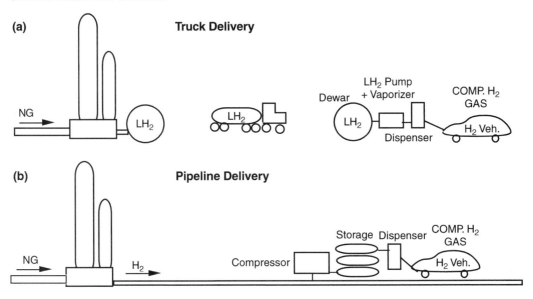

(a)   **Truck Delivery**

(b)   **Pipeline Delivery**

CHEMICAL BY-PRODUCT HYDROGEN

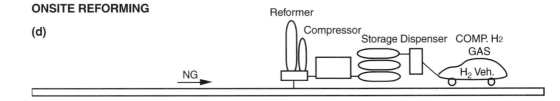

(c)

ONSITE REFORMING

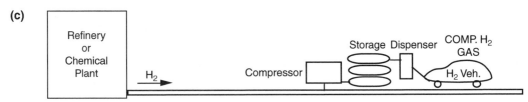

(d)

ONSITE ELECTROLYSIS

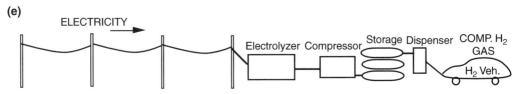

(e)

Figure 22.10 **Near-term options for producing and delivering hydrogen transportation fuel.**

ysis powered by off-peak hydropower, wind, solar, or nuclear power (Figure 22.11). Sequestration of by-product CO_2 (for example, in deep aquifers or depleted gas wells) might be done to reduce greenhouse emissions from hydrogen derived from hydrocarbons (Williams, 1996).

H_2 via BIOMASS, COAL or MSW GASIFICATION

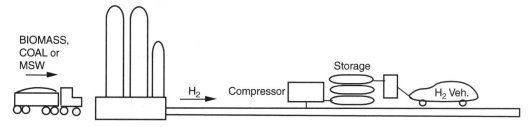

SOLAR or WIND ELECROLYTIC HYDROGEN

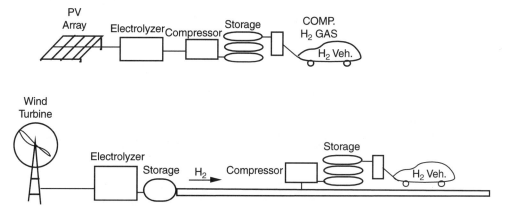

H_2 FROM HYDROCARBONS w/CO_2 SEQUESTRATION

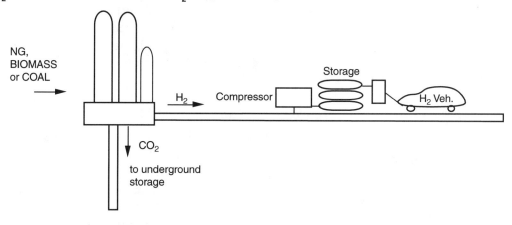

Figure 22.11 **Long-term options for producing hydrogen transportation fuel.**

We estimate the capital cost of building a hydrogen refueling infrastructure for the near-term options described above (Ogden, 1999a, 1999b) to be in the range of $310–620 per fuel cell car served by the system. The delivered cost of hydrogen transportation fuel is estimated to be $1.6–5.2/gallon gasoline equivalent, depending on the refueling station size and hydrogen supply technology. Although this is several times the current untaxed price of gasoline, the fuel cell vehicle's higher efficiency yields a fuel cost per kilometer comparable to that for today's gasoline cars. Onsite small-scale steam reforming is attractive as having both a relatively low capital cost (for advanced fuel cell type reformers), and a low delivered fuel cost. [Another comprehensive study on this topic undertaken by Directed Technologies, Inc. (1997) for Ford Motor Company found similar results.]

Developing a Refueling Infrastructure for Methanol Fuel Cell Vehicles

Methanol is made at large scale today from natural gas and is used for production of MTBE, formaldehyde, and other chemicals. Initially, to serve up to about one to two million fuel cell cars worldwide, it might be possible to use the existing methanol infrastructure, without building new methanol production capacity. The only capital costs would be conversion of gasoline refueling stations, marine terminals, and delivery trucks to methanol (U.S. Department of Energy, 1990). At these low market penetrations of methanol fuel cell vehicles, infrastructure capital costs will be small (probably less than $50/car). However, once demand for methanol fuel exceeded today's excess production capacity, new methanol plants would be needed, bringing the methanol infrastructure capital costs to levels similar to those for hydrogen, about $340–800/car (Ogden *et al.,* 1999).

Cost of Infrastructure for Gasoline Fuel Cell Vehicles

We assume that there is no extra capital cost for developing a gasoline infrastructure for fuel cell vehicles. This may be an oversimplification. For example, if a new type of gasoline (e.g., very low sulfur) is needed for gasoline reformer fuel cell vehicles, this would entail extra costs at the refinery. The costs of maintaining (and gradually replacing existing equipment with new) or expanding the existing gasoline infrastructure to meet projected future demand are not considered.

A Comparison of Hydrogen, Methanol, and Gasoline

Defining "infrastructure" to mean all the equipment (both on and off the vehicle) required to bring hydrogen to the fuel cell, it is clear that hydrogen, methanol, and gasoline fuel cell vehicles all entail infrastructure capital costs. For gasoline vehicles, these costs are entirely for onboard fuel processing, assuming that no extra off-vehicle gasoline infrastructure costs are incurred for using fuel cells. For methanol, there are "infrastructure" costs both on the vehicle (for fuel processors) and off the vehicle (for fuel production, dis-

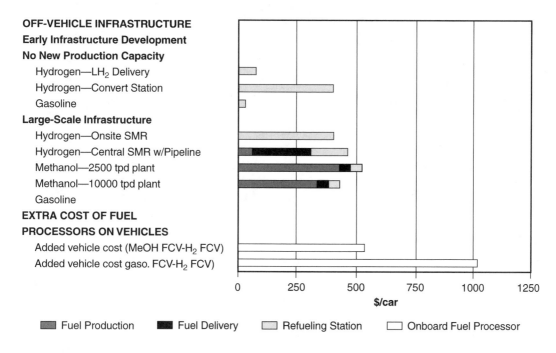

OFF-VEHICLE INFRASTRUCTURE
Early Infrastructure Development
No New Production Capacity
  Hydrogen—LH$_2$ Delivery
  Hydrogen—Convert Station
  Gasoline
Large-Scale Infrastructure
  Hydrogen—Onsite SMR
  Hydrogen—Central SMR w/Pipeline
  Methanol—2500 tpd plant
  Methanol—10000 tpd plant
  Gasoline
EXTRA COST OF FUEL
PROCESSORS ON VEHICLES
  Added vehicle cost (MeOH FCV-H$_2$ FCV)
  Added vehicle cost gaso. FCV-H$_2$ FCV)

0   250   500   750   1000   1250
$/car

Fuel Production   Fuel Delivery   Refueling Station   Onboard Fuel Processor

Figure 22.12 **Comparison of incremental capital costs for alternative fuel cell vehicles (compared to H$_2$ fuel cell vehicles) and refueling infrastructures (compared to gasoline).**

tribution, and refueling systems). In the case of hydrogen, there is no "onboard" infrastructure cost (e.g., no fuel processor) and all the infrastructure capital cost is paid by the fuel producer (and passed along to the consumer as a higher fuel cost).

In Figure 22.12 we combine our estimates of the cost of alternative fuel cell vehicles and an off-board refueling infrastructure. To within the accuracy of our cost projections, it appears that the total capital cost for infrastructure on and off the vehicle would be roughly comparable for methanol and gasoline fuel cell vehicles, with hydrogen costing somewhat less. In the longer term (once new production capacity is needed), hydrogen appears to be the least costly alternative.

22.4.3 FUEL STRATEGIES FOR FUEL CELL AUTOMOBILES

Scenario 1: *Gasoline fuel cell cars → hydrogen fuel cell cars.* Introduction of fuel cell cars with onboard gasoline fuel processors would allow the rapid introduction of large numbers of fuel cell automobiles to general consumers without changes in the refueling infrastructure, decreasing the cost of fuel cells via mass production to competitive levels. In the longer term, the more favorable economics of hydrogen for fuel cells plus energy supply and environmental concerns would motivate a switch from gasoline. A widespread hydrogen refueling infrastructure would be implemented for hydrogen fuel cell cars, allowing diverse primary sources to be used, and enabling significant reductions of greenhouse gas emissions.

Scenario 2: *Hydrogen moves from centrally refueled fleets to general automotive markets.* In this scenario, hydrogen is implemented first in centrally refueled fleet vehicles (such as buses and vans), and later moves to general automotive markets in response to a societal decision to move toward a zero-emission transportation system. Figure 22.12 suggests that this strategy might involve the lowest overall capital outlay, counting both vehicle costs and infrastructure costs.

Scenario 3: *Methanol fuel cell vehicles are introduced for fleet applications, moving to general automotive markets. Eventually a switch to hydrogen is implemented.* The advantages are that methanol is easier to store and handle than hydrogen, and easier to reform than gasoline. The disadvantages are that methanol faces the same "chicken and egg" problem as hydrogen in reaching beyond fleets to general automotive markets, and that the methanol vehicle will be more costly than one with hydrogen. Another serious disadvantage of this scenario is that the fuel infrastructure would have to be changed twice (once from gasoline to methanol, and then from methanol to hydrogen).

In the near term (2000–2005 or so), it seems probable that hydrogen or methanol will be the fuels of choice for fuel cell vehicles, and that fleets will be the main application. The optimum fuel strategy for introducing fuel cell vehicles to general automotive markets is uncertain, awaiting the results of demonstrations of alternative fuel cell vehicle types. (Technical issues remain to be resolved for gasoline fuel processors and cost issues for onboard hydrogen storage.) Ideally, this choice should be made to give fuel cells the best chance of reaching general mass markets, paving the way for economically competitive mass produced fuel cell vehicles and long-term use of hydrogen. In the long term, hydrogen is the fuel of choice because of its advantages over methanol and gasoline in terms of vehicle cost, complexity and fuel economy, and environmental and energy supply benefits. The total capital cost of implementing fuel cell vehicles appears to be lowest for hydrogen, as well.

Portland's strong commitment to improving quality of life (QOL) standards includes public transportation by an electric light rail system. The transit system has helped to revitalize neighborhoods, build a stronger and more livable downtown, and to connect the city and surrounding countryside, while reducing air pollution by eliminating 52 million car trips each year. PHOTOGRAPH COURTESY OF TRI-MET, PORTLAND, OREGON.

CHAPTER 23

Corporate Performance Indicators

23.1 Measurement and Indices

To capably discuss and manage the sustainability of business, the economy, the environment, and society we need sound performance measures. Government, business, and society operate in a complex environment that requires solid, reliable, and accurate information.

Good information enables decision makers to make good decisions and measure the actual results of their actions. The first step is to identify the important variables. If some of the important variables are environmental, for example, it would be sensible to distinguish between those that measured change in some stock of critical natural capital from those that measured change in some noncritical resource. Similarly, if some of the important dimensions are economic, it would be sensible to distinguish between those that represented factors essential to the continued survival of the sector or nation concerned from those that could be conceded against some reciprocal concession.

Benchmark indices can, with accurate information, provide information for analysis of the options available to us. A partial list of global benchmarks is provided in Table 23.1. This list could be modified for a nation or a region. Our goal is to provide measures of natural capital, social capital, and economic capital for a designated region.

For example, natural capital includes the entire environmental patrimony of a country. For any given country certain resources may not be included.

Table 23.1 **Global Benchmarks**

- World domestic product
- Temperature of the atmosphere
- Population
- Fertility rates
- Wetland losses and gains
- Arable cropland
- Production of cereals
- Fish catch
- Emissions of carbon dioxide
- Crude oil production
- Motor vehicles in service
- Total consumption of energy
- Air and water quality
- Atmospheric concentration of ozone

The elements included in the natural capital estimates considered here include agricultural land, pasture lands, forests, protected areas, metals and minerals, and coal, oil, and natural gas.

There are three types of indicators. The first type, environmental condition indicators, measures the actual conditions of the environment. An example is the global atmosphere temperature. The second type of indicator is an input measurement such as the amount of fossil fuel consumed annually. The third type of indicator is a measurement of recycling and mitigation measures.

Unlike measures of financial performance, environmental and social indicators do not lend themselves to a common unit. Rather, they are recorded in such disparate units as pounds of waste generated, liters of water used, Btus of energy consumed, or fertility rates in number of births.

There is much interest in extending the U.S. national income and product accounts to include assets and production activities associated with natural resources and the environment. Environmental and natural-resource accounts would provide useful data on resource trends and help governments, businesses, and individuals better plan their economic activities and investments.

Accounting for social, environmental, and economic activity is useful. Some observers assert that while we have doubled global wealth during the last 25 years, we have lost a significant portion of the world's natural capital. Measures of social and natural capital are important for assessment of the quality of life and proper policy decisions.

A recent report by several centers has attempted to show that it is possible to construct a single index measuring environmental sustainability (Global Leaders, 2000). Their analysis suggests that decisions of how vigorously to pursue environmental sustainability and of how vigorously to pursue economic growth are in fact two separate choices. These results are consistent with the hypothesis that high levels of environmental protection are compatible with, or possibly even encourage, high levels of economic growth, though they do not prove it. (See *www.yale.edu/envirocenter* for further details.) The index consists of five components. The components describe the current environmental systems; stresses to those systems; the vulnerability of human populations to environmental disturbances and disasters; the social

Table 23.2 **Structure of the Environmental Sustainability Index**

Component	Factor	Sample variables
1. Environmental systems	Urban air quality	Urban NO_2 concentration
	Water quantity	Groundwater resources per capita
	Water quality	Nitrogen, nitrate, and nitrite concentration
	Biodiversity	Percentage of known plant species threatened
	Land	Severity of human-induced soil degradation
2. Environmental stresses	Air pollution	SO_2 emissions per land area
	Water pollution and consumption	Groundwater withdrawals as a percent of annual recharge
	Ecosystem stress	Deforestation
	Waste production and consumption pressure	Percentage of households with garbage collection
3. Human vulnerability to environmental impacts	Population	Growth rate 1995–2000
	Basic sustenance	Percentage of urban population with access to safe drinking water
	Public health	Prevalence of infectious diseases
	Disasters exposure	Deaths from natural disasters
4. Social and institutional capacity	Science and technical capacity	Research & development scientists and engineers per million population
	Capacity for rigorous policy debate	Civil liberties
	Environmental regulation and management	Transparency and stability of environmental regulations
	Eco-efficiency	Availability of sustainable development information at the national level
	Public choice failures	Fossil fuel subsidies as a percentage of GDP
5. Global stewardship	Contribution to international cooperation	Number of memberships in environmental intergovernmental organizations
	Impact on global commons	Carbon-dioxide emissions

and institutional capacity to respond to environmental problems; and global stewardship, or the degree to which an economy behaves responsibly with respect to other economies. Table 23.2 lists the five components, the factors considered for each component, and a sample variable for each factor.

The pilot study equally weighted the variables or factors. Using that method, the results for eight of the largest nations is given in Table 23.3. The index has a highest value possible of 100. Two smaller nations, Norway and Switzerland, have the highest indices, attaining 76 and 75, respectively. The methodology for this sustainability index could be used for states, provinces, or regions.

An example of a regional index is the Silicon Valley Environmental Index (see *www.mapcruzin.com/svep*). This environmental index includes such factors as energy consumed, auto-miles, trash, and habitat threats. They are tracked over time and 12 of the 23 indicators show a negative trend. One report issued in 2000 indicates that population growth has overcome environmental progress. For example, while each resident throws out 30% less trash than in 1987, the total amount of garbage being dumped has jumped sharply in the 5 years through 1999. The same pattern holds for water and energy use. Individuals may not be consuming any more than they did a decade ago, but total consumption is rising due to the country's population growth which is up nearly a third since 1980.

Table 23.3 **Results of the Pilot Sustainability Index for Eight Large Nations**

Nation	Component values[a]					Sustainability index
	1	2	3	4	5	
Brazil	67	74	71	33	59	59
China	62	73	83	20	46	54
France	69	80	99	54	65	71
Germany	60	74	99	56	63	68
Japan	75	56	97	51	64	66
Russia	77	78	95	22	50	62
United Kingdom	69	69	99	58	71	70
United States	74	69	96	51	55	68

[a]*Components:* 1, environmental systems; 2, environmental stresses; 3, human vulnerability; 4, social and institutional capacity; 5, global stewardship.

23.2 **Measures of the Environment**

Measures of the environment are important to progress toward sustainable practices. Figure 23.1 provides a model of a closed ecological system for a region or nation. The material input is called virgin resources (V). The fossil fuel input is FF and the renewable energy input is R. The waste flowing out of the system is W_O and the waste stored within the system is W_S. The extraction process provides outputs: energy, E; materials, M; and waste, W_E. The commercial and government processes (the economy) provide output product and services, P, and waste, W_P. The consumption processes, provide three outputs: (1) services and satisfaction of needs, S; (2) recyclable and

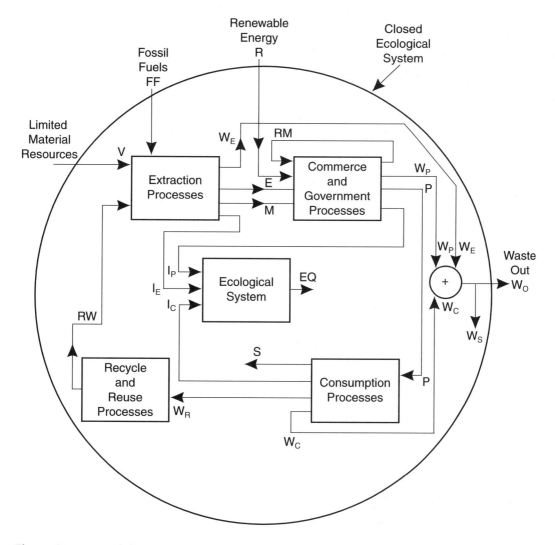

Figure 23.1 **A model of the ecological system.**

reusable waste, W_r; and (3) nonrecyclable or reusable waste, W_E. The total waste ($W_P + W_C + W_E$) is then shipped elsewhere, W_O, or stored within the ecological system, W_S. Recycled or reusable waste is RW. The impacts from the processes are described as I_P from the government and commercial processes, I_C from the consumption processes, and I_E from the extraction processes. The resulting environmental quality is designated as EQ.

Important measures would include efficiencies, impacts, waste stored, and so on. For example, extraction efficiency is

$$\eta_e = \frac{M}{V + W_E}.$$

The efficiency of the commercial and governmental processes is

$$\eta_c = \frac{P}{\left(M - RM\right) + E + W_P}.$$

These measures can be used along with waste measures, impacts, and EQ to measure trends in a region.

23.3 Corporate Social and Environmental Indicators

Corporate performance indicators for social performance and environmental performance are needed for good management practice. These are needed if a corporation strives to manage the triple bottom line. The challenge is how to measure and track such intangibles as social performance and sustainability in a manner that reduces shareholder risk while enabling business managers to analyze environmental investments and risks within a financial framework consistent with that used for other business decision-making processes. Aside from the clear value to shareholders, such also could engender broad incentives for companies to view their operations through a sustainability lens.

There is continuing pressure on companies in many sectors to increase disclosure of their environmental and social impacts, and provide tangible proof when they make claims about proactive changes. Companies are asked to produce environmental and sustainability reports that can serve the needs of both internal and external audiences. Corporations are increasingly issuing reports of their social and environmental performance.

Table 23.4 provides a list of commonly used social performance indicators. Reports of social performance normally include such topics as community relations, human rights, workplace issues, ethics, and corporate governance.

Commonly used environmental indicators are provided in Table 23.5. They include chemical releases, water uses, and gases emitted as general measures of material uses and wastes emitted. Environmental spending is a process input and is a weak indicator of actual performance (an output). Many of the environmental indicators are only appropriate for manufactur-

Table 23.4 Common Social Performance Indicators

* Employee rights to freedom of speech and assembly
* Community impacts and relations
* Workplace issues
* Supplier relationships and standards
* Openers, disclosures, and accountability

Table 23.5 Common Corporate Environmental Indicators (Rank Order)

* Chemical releases
* Regulatory compliance
* Environmental spending
* Water use
* Greenhouse gases emitted
* Energy use

SOURCE: World Resources Institute.

ing companies. New ways are needed to measure the environmental practices of firms in the service sector, which can influence the behavior of upstream suppliers and downstream consumers.

Dow Chemical published a corporate Public Report (see *www.dow.com/environment/99rep*). This report covers the triple bottom line with a report on economic, social, and environmental performance. The report states:

> As part of our worldwide commitment to Responsible Care, we are working toward a vision of zero — as in no accidents, no injuries, and no harm to the environment. To help us reach that vision, we have made strong progress against aggressive targets — our Environment, Health & Safety 2005 Goals. And as we have grown and expanded in countries around the world, we have seen firsthand the tremendous social challenges that our communities face — from a lack of quality education and housing, to a need for fresh food and clean drinking water.

Dow pledges to incorporate principles of sustainable development and eco-efficiency into their business strategies.

One organization, the Global Reporting Initiative, strives to develop globally applicable guidelines for companies to report on their economic, social, and environmental performance (see *www.globalreporting.org*).

23.4 The Three-Factor Scorecard

The three-factor scorecard (TFSC) uses three factors to measure company performance: financial, environmental, and social (see Section 2.3). This scorecard is an adaption of the Balanced Scorecard.

The Balanced Scorecard (BSC) is a framework for integrating business measures derived from a company's strategy (Kaplan and Norton, 1996). The measures on a BSC are used by a company to articulate the strategy of the business, to communicate the strategy of the business, and to help align individual, organizational, and cross-departmental initiatives to achieve a common goal. A company uses the scorecard as a communication, information, and learning system.

The TFSC is shown in Figure 23.2. The measures used on the TFSC should provide a clear representation of the organization's long-term strategy for competitive success. For a company that adheres to the triple bottom line strategy, the measures used must implement this strategy clearly.

Each of the four perspectives has no more than six objectives and their associated measures. A company can utilize this approach with the use of key performance indicators to develop a measurement system that covers three broad categories of performance: economic, environment, and social.

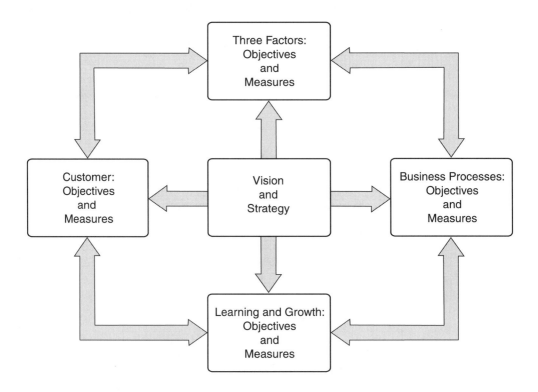

Figure 23.2 **The three-factor scorecard. The three factors are financial, social, and environmental.**

Table 23.6 **Sample Outline of Objectives and Measures for a Company Adopting a Sustainable Strategy**

Perspectives	Sample objective and measure
Vision and strategy	• To be a leader in the development of a company that supports a sustainable world • Use the triple bottom line companywide (three factors)
Customer	• Increase customer satisfaction with goals for excellent service and environmentally and socially responsible actions • Measure customer satisfaction
Learning and growth	• Increase employee learning and understanding of the triple bottom line philosophy • Measure employee understanding and alignment with objectives
Business processes	• Institute a triple bottom line decision system • Measure process efficiencies and environmental and social effectiveness
Three factors	
Financial	• Increase the return on capital • Measure return on capital
Social	• Increase the openness toward measures of the three factors • Measure transparency of three factors by surveys of all stakeholders
Environmental	• Develop a set of environmental measures of waste and pollution • Measure waste and pollution emissions

The advantage of a scorecard approach is that it trains business attention on a range of different performance measures. The challenge is in the choice and number of yardsticks or indicators to include and the difficulties in discerning overall performance from a broad range of indicators.

A sample TFSC for a company adopting a sustainable world, triple bottom line strategy might use a basic outline of objectives as shown in Table 23.6. Then under each perspective a sample objective and measure is provided.

A well-developed TFSC provides a company's strategy and vision and serves to create shared understanding. The TFSC can also focus the change process on the right indicators. Finally, the TFSC enables learning at all levels of the organization.

23.5 Measuring a Community's Quality of Life

By Michael R. Hagerty
Associate Professor
Graduate School of Management
University of California, Davis

Quality of life (QOL) can be measured — and improved — through community action. Many cities, states, and entire countries now monitor QOL and publish regular reports, with more beginning each year. Miringoff and Miringoff (1999) list 11 states and 28 cities that publish regular QOL reports. The state of Florida summarizes the purpose of their program as

"The Commission wants to answer the question, 'How are we doing?' . . . Where we are doing well, we can feel good. Where the results are not good, we can start by asking, 'Why?' and 'What can we do about it?'"

23.5.1 COMPONENTS OF QUALITY OF LIFE

Research on QOL began 40 years ago, when the National Opinion Research Center at the University of Chicago surveyed citizens on what things contribute most to their satisfaction with life. Since then, hundreds of surveys on citizens around the world have found quite similar components of QOL. In order of importance these are relationships with family, material well-being, health, a meaningful job, feeling part of one's local community, and personal safety.

Individual communities will differ in how they define each of these components, but all share certain characteristics. First, all define QOL as something more than just economic growth. Indeed, most QOL movements arise because people want to *balance* economic growth with the other areas that make life fulfilling. The U.S. Congress has taken up this call; *Congressional Reports* quotes the term "Quality of Life" more than 10 times per week in debate on the floor.

Second, QOL reports should measure outputs, rather than inputs. For example, the number of police on the street is an input, which may or may not improve personal safety. In contrast, the number of violent crimes reported in a community is an output, and is more relevant in a QOL report. Congress is now requiring that all federal agencies adopt this criterion. The Government Performance and Results Act requires all executive branch programs to be measured on results (outcomes) not intentions. Agencies must adopt a strategic planning model that focuses thinking externally and involves stakeholders and Congress. Budgets must be designed to reflect strategic missions. Real measurement of results must be used to justify appropriations and authorizations.

Third, communities should encourage *broad participation* in developing QOL goals, including diverse citizen groups, business groups, and local governments. Developing priorities in areas of QOL requires the input of the citizens it affects, and crafting solutions requires the skills of business and government. Many communities report that just the *process* of getting these groups together to plan community progress encourages communication and participation — a basic component of QOL.

Fourth, communities should establish *measurable goals* in each area of QOL, then track progress on each. Note that many of these recommendations are old friends to business executives — they were pioneered in for-profit companies to improve the quality of their own output, under names such as total quality management and management by objectives.

23.5.2 OREGON BENCHMARKS

One of the earliest and best-constructed QOL initiatives was in Oregon. The governor and legislature of Oregon in 1989 created a nonpartisan citizens panel, chaired by the governor, to craft a strategic plan for the state and measurable indicators of progress. After extensive citizen involvement and public review, the panel — the Oregon Progress Board — released its first strategic plan and 158 benchmarks in 1991. Within six months, the legislature had unanimously adopted these measures and directed the Progress Board to update them every 2 years. By 1993 the number of benchmarks had expanded to 272, including 43 classified as critical. By 2010, Oregon should have achieved three strategic goals: the best educated and prepared workforce, maintaining the state's natural environment and uncongested way of life, and a diverse, internationally oriented economy paying high wages.

Interim goals are established for intervening years. Targets are set in areas ranging from reducing violent crime to raising adult literacy levels to increasing the miles of Oregon rivers and streams that meet state and federal water quality standards. Among the benchmarks designated as "urgent" are those related to teenage pregnancy, air quality, and educational skills. The Progress Board has also designated "core" benchmarks to provide an enduring measure of Oregon's vitality. They include the percentage of Oregon adults with good health practices and the proportion of goods produced in Oregon that are sold abroad. One unexpected result of benchmarks is the degree to which it has fostered cooperation among government agencies and between the public and private sectors. In 1994 Oregon won the "Innovations in American Government Program" from Harvard University for their program. A sample of the benchmarks is given in Table 23.7.

Further Reading

Government Performance Information Consultants (*members.home.net/gpic/*). Contains information related to performance measurement and program evaluation. Has a particularly good hyperlinked list of "think tanks."

Miringoff, Marc L., and Marque Luisa Miringhoff. *The Social Health of the Nation: How America Is Really Doing,*. New York: Oxford University Press, 1999.

Oregon Progress Board (*www.econ.state.or.us/opb/benchmark_tables/index.htm*).

Table 23.7 Sample Oregon Benchmarks (Environment Section)

	1980	1990	1991	1992	1993	1994	1995	1996	1997	1998	2000	2010	Grade
Air													
79. Percentage of Oregonians living where the air meets government ambient air quality standards	30%	54%	51%	58%	100%	100%	100%	100%	100%		100%	100%	A
80. Carbon dioxide emissions as a percentage of 1990 emissions		100%	112%	125%	124%	132%	117%	119%			100%	100%	F
Water													
81. Percentage of Oregon wetlands in 1990 still preserved as wetlands		100%	100%	100%	100%	100%	100%		100%	100%	100%	100%	A
82. Stream water quality index													
a. Percentage of monitored stream sites with significantly increasing trends in water quality		8%					21%	32%	52%		25%	25%	A
b. Percentage of monitored stream sites with significantly decreasing trends in water quality		20%					8%	2%	0%		5%	0%	A
83. Percentage of assessed groundwater that meets drinking water standards	87%	95%		95%		94%		94%		95%	94%	94%	A
84. Percentage of key rivers meeting instream water rights													A
a. 9 or more months of year	53%	39%	50%	56%	72%	61%	94%	94%			60%	65%	A
b. 12 months a year	47%	44%	39%	22%	22%	28%	35%	70%			35%	40%	A

Land

												Grade
85. Percentage of Oregon agricultural land in 1970 still preserved for agricultural use	98%		98%		97%	97%	97%	97%		97%	97%	A
86. Percentage of Oregon forest land in 1970 still preserved for forest use	92%	90%	92%	92%	92%	91%	91%	92%		92%	92%	A
87. Pounds of Oregon municipal solid waste landfilled or incinerated per capita			1,519	1,501	1,516	1,511	1,570	1,640		1,506	1,495	F
88. Percentage of identified hazardous waste sites that are cleaned up or being cleaned up	67%	68%	71%	70%	67%	66%	69%	69%	68%	67%	56%	A
a. Tank sites	66%	67%	71%	69%	66%	65%	69%	69%	68%	67%	55%	A
b. Other hazardous substances	97%	75%	79%	76%	73%	70%	69%	71%	74%	70%	69%	A

Plants and Wildlife

												Grade
89. Percentage of wild salmon and steelhead populations in key sub-basins that are at target levels	48%	39%	30%	20%	11%	2%	2%	2%		13%	35%	F
90. Percentage of native fish and wildlife species that are healthy		76%	76%	76%	76%	75%	75%	72%	72%	77%	80%	F
91. Percentage of native plant species that are healthy		83%	86%	88%	86%	88%	85%	85%	85%	90%	95%	C–

Outdoor Recreation

												Grade
92. Acres of state-owned parks per 1,000 Oregonians	35	31	31	30	30	29	29	29		35	35	F

International Paper follows the principles of the Sustainable Forestry Initiative program for woodlands management. These principles include regenerating every acre of company land harvested within two years or allowing five years of natural reforestation. Rangers Bill Wallace and Bill Gregg take a walk in the woodlands near Georgetown, South Carolina. PHOTOGRAPH BY JACK KENNER. COURTESY OF INTERNATIONAL PAPER.

CHAPTER 24

Social Entrepreneurship and Investing

24.1 Social Entrepreneurship

An **entrepreneur** is a person who organizes, operates, and assumes the risk for a new business venture. The entrepreneur identifies an opportunity and conceives of a means of seizing the opportunity. The entrepreneurial process is shown in Table 24.1. The lead entrepreneur identifies the opportunity and conceives several strategies for seizing the opportunity. After selecting a strategy, he or she develops an initial business model and develops a team that can help transform that model into a business plan. Using the business plan, the team shows the plan to sources of financial, intellectual, and physical assets necessary to bring the new venture to fruition.

The next step is to implement the plan and adjust to its new realities as necessary. Finally, the entrepreneur and the team develop a plan for harvesting the wealth or value created by the venture. Leadership succession is also an issue that needs to be resolved as the firm grows.

Social entrepreneurship combines the idea of a social mission with the business-like qualities of sound business practices and methods. By contrast, economic entrepreneurs are largely motivated by the idea of wealth creation and the creation of a profit-seeking venture. The contrast between economic entrepreneurship and social entrepreneurship is shown in Table 24.2. In general, the social entrepreneur is primarily focused on delivering social benefits while the economic entrepreneur is focused on growth of revenues and profits.

Table 24.1 **The Entrepreneurial Process**

Step	Activity
1	Identify a problem or opportunity.
2	Conceive several strategies for solving the problem or seizing the opportunity.
3	Select a strategy.
4	Develop a business model.
5	Build an initial team.
6	Develop a business plan.
7	Access and obtain the necessary resources.
8	Implement the business plan.
9	Adjust the plan as required.
10	Develop a harvest plan or succession plan.

Social entrepreneurship can include social purpose business ventures, such as for-profit community development banks, and hybrid organizations mixing not-for-profit and for-profit elements, such as homeless shelters that start businesses to train and employ their residents. Social entrepreneurs look for the most effective methods of serving their social missions.

We have always had social entrepreneurs. They helped to build universities, welfare agencies, and myriad social welfare organizations. For social entrepreneurs, the social mission is clear and central. This mission affects how social entrepreneurs perceive and assess opportunities. Mission-related results become the central criterion, not wealth creation. Wealth is just a

Table 24.2 **Contrasts between Social and Economic Entrepreneurship**

Factor	Economic entrepreneurship	Social entrepreneurship
Opportunity or problem	Commercial need or opportunity	Social need
Desired outcomes:		
Financial	Profit	Modest surplus for investment
Investors	Return on investment	Social and environmental benefits
Source of investment	Capital markets	Donors or charitable investment
Harvest or succession plan	Sales of securities or merger	Succession of leadership
Governance	Board of directors	Board of directors
Private or public	Private or public	Public at least to donors and stakeholders
Legal form	Corporation	Corporation, association, or cooperative

means to an end for social entrepreneurs. With business entrepreneurs, wealth creation is a way of measuring value creation. This is because business entrepreneurs are subject to market discipline, which determines in large part whether they are creating value. If they do not utilize resources for more economically productive uses, they tend to be driven out of business. Markets are not perfect, but over the long haul, they work well as a test of private value creation, specifically the creation of value for customers who are willing and able to pay. An entrepreneur's ability to attract resources in a competitive marketplace is a good indication that the venture represents a more productive use of these resources than the alternatives it is competing against. If an entrepreneur cannot convince a sufficient number of customers to pay an adequate price to generate a profit, this is a strong indication that insufficient value is being created to justify this use of resources. A redeployment of the resources happens naturally because firms that fail to create value cannot purchase sufficient resources or raise capital. Firms that create the most economic value have the cash to attract the resources needed to grow.

Markets do not work as well for social entrepreneurs. In particular, markets do not do a good job of valuing social or environmental improvements, public goods and impacts, and benefits for people who cannot afford to pay. These elements are often essential to social entrepreneurship. As a result, it is much harder to determine whether a social entrepreneur is creating sufficient social value to justify the resources used in creating that value. However, the survival or growth of a social enterprise is not proof of its efficiency or effectiveness in improving social conditions. A social enterprise competes for donations, volunteers, and other support. Its results are difficult to measure. Examples of measures are costs/client, meals served per day and habitats saved.

Examples of social entrepreneurship are not limited to industrialized nations. Entrepreneurs in less developed nations are creating profitable new businesses. These efforts need to be appropriate for local markets. For example, Hindustan Lever, Ltd., of India developed an innovative heat shield technology that allows perishable goods to be transported across India in standard, nonrefrigerated trucks, reducing costs and energy use while reducing the risk of food spoilage and contamination.

An example of a social entrepreneur in Brazil is Rodrigo Baggio of Rio de Janiero (Mitchell, 2000). He built a system of 117 computer schools operating in the slums of 13 Brazilian states. This venture strives to teach youth about information technology and find them jobs.

Social entrepreneurs create social value through innovation and leveraging financial resources, regardless of source, for social, economic, and community development. Great organizations like Salvation Army and Goodwill have a long history of social entrepreneurship. The Social Entrepreneurs Alliance for Change (see *www.sea-change.org*) is an organization supporting social entrepreneurs.

24.2 Social Innovation

Social innovation is the process of introducing change through new ways of accomplishing social goals. A new paradigm for social innovation is the emergence of partnerships between private enterprise and public interest organizations that produces profitable and sustainable change for both sides (Kanter, 1999).

The need for new arrangements for public–private partnerships is the limitation of former methods focused on charity only. Charity consisted of donating money and little else. Social innovation is concerned with business and nonprofit partnerships that view community needs as opportunities to develop ideas and business methods to solve community problems. It is focused on attacking public sector problems to stretch the capabilities of companies.

When companies approach social needs in this way, they have a stake in the problems, and they treat the effort the way they would treat any other effort central to the company's operations. They use their best people and their core skills.

Perhaps the best example is the hotel chain Marriott and their training program, Pathways to Independence. The program for Welfare to Work participants offers training of the skills necessary to work in hotels and guarantees a job offer on completion of the program. This program helps meet a social welfare need and provides Marriott with trained workers.

Kanter (1999) identified six characteristics of successful private–public partnerships as listed in Table 24.3. The Marriott's program exemplifies these characteristics. Marriott had a clear business agenda to secure trained, stable workers. Both Marriott and state welfare agencies had the resources to commit to the program and Marriott worked with organizations in each locality such as Goodwill or a community college. Marriott and their partners put financial investment and other resources into the program. The partners were rooted in the community and had extensive links to other organizations. Finally, Marriott had a long-term commitment since this program was central to their success.

Marriott's Pathways program has produced big benefits. About 70% of Pathways' graduates are still employed by Marriott after a year, compared with only 45% of the welfare hires who did not participate. Pathways is considered to be such a source of competitive advantage for Marriott that the company keeps the program details proprietary.

Table 24.3 **Six Characteristics of Successful Partnerships**

- Clear business agenda that strongly relates to social needs
- Strong partners committed to change
- Investment by both parties
- Rootedness in the user communities
- Links to other organizations
- Long-term commitment to sustain and replicate the solution

Significant business contributions to the social sector use the core competencies of a business — the things it does best — for social innovation. Furthermore, the activities of the partnership are focused on results, while seeking measurable outcomes and demonstrated changes. The effort can then be sustained and replicated in other places. The community gets new approaches that build capabilities and point the way to permanent improvements. The business gets bottom-line benefits: new products, new solutions to critical problems, and new market opportunities.

Perhaps one of the finest examples of a social entrepreneur who sought effective business – social agency partnerships was John Sawhill (1936–2000). Sawhill served in several U.S. Department of Energy positions and as president of New York University from 1975 to 1979. He was named CEO of the Nature Conservancy in 1990 and served until his death in 2000. Under his guidance, the organization helped preserve more than 11 million acres of open land, set up 1,300 private nature reserves, and raised annual revenue to $780 million in 1999. He also increased membership in the group to more than 1.1 million members, making it the nation's largest conservation organization and 14th largest nonprofit institution. His view was that the Nature Conservancy should use the best business practices and partner with companies.

Sawhill set a big goal: to influence land use in much larger areas surrounding core habitats. This meant promoting compatible economic development, a difficult task. He put out the word that the conservancy sought partnerships with business. "We seek to work with a broad variety of people and organizations: individuals, businesses, government, other nonprofits, universities, you name it," Sawhill said (see *www.tnc.org*).

"Compatible development" to Sawhill meant buying lands valuable for their biological diversity, "not to manage them as nature preserves but rather to resell them with permanent restrictions prohibiting environmentally incompatible uses" (Lloyd, 2000). This is a very creative, businesslike strategy.

24.3 Social Investing

The idea of integrating one's concerns about social issues with one's investing decisions, whether out of religious beliefs or political convictions, has always been popular with a core group of investors. Today, socially conscious investing is moving into the mainstream as a result of two changes.

First is the emergence of a discerning consumer — one who makes informed and integrated decisions and more clearly understands the link between the power of the private sector and its influence on society. The second is a series of changes in the way Americans prepare for their financial futures, which has put more power into the hands of individual investors. As a result, a tremendous opportunity exists for financially competitive mutual funds that can help shareholders meet their investment goals while recognizing their social concerns.

In the past many investors looked only at the financial performance of an investment. Now, many use the triple bottom line — social, economic, and environmental factors — to evaluate their investment in companies or mutual funds.

Many investors have become interested in the relevance of environmental and social issues. Many investment firms believe that companies that care for the environment and can handle social challenges will outperform those that do not. Companies with strong environmental and social performance as well as economic performance will have reduced risks and provide more stable returns.

Total assets under management in socially screened portfolios have risen from $529 billion to more than $1.5 trillion in the period from 1997 to 2000. Socially responsible funds are growing at twice the rate of other mutual funds. In 1995, 55 funds were labeled socially responsible. By the end of 1999, that number had risen to 175. Social investing is clearly big business.

The idea of social investing can be defined as investing business with positive social and environmental performance coupled with strong financial performance. In the 1980s many believed the term meant simply avoiding investments in companies that were doing business in South Africa. Before long, a small number of investment managers began to exclude investments in firms associated with tobacco, alcohol, or firearms. The definition has become even broader in recent years. One has only to do a quick search on the Internet to find investment managers who offer funds that will satisfy many criteria.

Socially responsible investing has historically been based on negative screening, that is, on ruling out investments in certain types of businesses. It is difficult to set up appropriate screens for companies. To no small degree, the definition of socially responsible investing changes from person to person. Some individuals focus on the products the company makes, others focus on how employees are treated, still others focus on additional factors such as a company's impact on the environment.

Many mutual funds used alcohol, tobacco, gambling, military products, and nuclear power as their screens to eliminate companies. But General Electric, one of the world's most successful companies, would be eliminated in that situation because it has an extremely small division that makes products for the nuclear power industry. A far larger part of GE's business is its medical products division that produces such life-saving products as MRIs and CAT scanning machines.

The pendulum is now swinging from negative to positive screening. Investment managers are beginning to look at a far broader range of corporate behavior and then make investment decisions based on how responsible a company is as part of the global community. Analysts can use a three-factor scorecard as described in Section 23.4. Indicators could include toxics releases, water use, material intensity, energy intensity, and other measures.

Many believe that sound environmental management leads to reduced risk to the firm, and that this risk reduction may be valued by financial markets. Investments in environmental management lead to better short-term environmental performance as well as the prospect of further improvements in the future. These improvements confer a reduction in the firm's risk, which is the key factor that investors consider when deciding on the return that they will require for making a particular investment. Lower risks mean lower required returns and, therefore, lower costs for financing the activities of the firm.

More than $150 billion is invested in socially responsible U.S. mutual funds. There are about 60 mutual funds that use positive screens for selecting companies for their portfolio. Should a fund include WalMart? Some reject it because it sells firearms while others do not like the impacts of WalMart on small local stores.

The Calvert Benchmark of socially responsible U.S. companies is widely followed (see *www.calvert.com*). The index consists of 468 companies that are weighted on a market capitalization basis. To be included companies must meet Calvert's standards in the environment, workplace issues, product safety, community relations, weapons contracting, international operations and human rights, and respect for the rights of indigenous peoples.

The 10 largest companies in the index are American International Group, Cisco Systems, IBM, Intel, Lucent Technology, Merck, Microsoft, Oracle, Pfizer, and SBC Communications.

Dow Jones, which publishes *The Wall Street Journal* launched the Dow Jones Sustainability Group Index in Europe in 1999. It includes 229 companies in 68 industries that represent the top 10% that practice global sustainability using the three-factor measures of social, economic, and environmental performance (see *www.sustainability-index.com*). Companies selected by the Dow Jones Index include BMW, Fuji Photo, Unilever, and Honeywell.

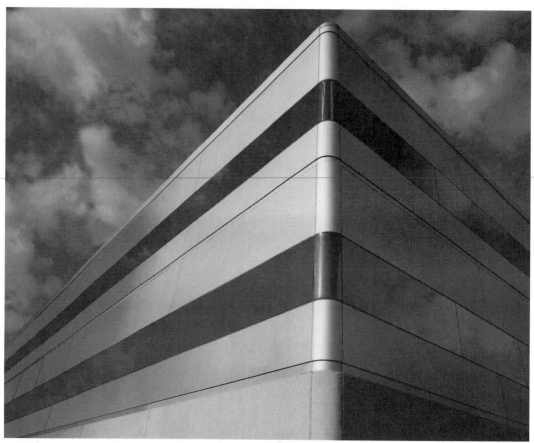

The twentieth century brought architecture to a peak of modern sleekness, with a corresponding rise in the use of artificial systems to maintain the light and air qualities of the internal environment. PHOTOGRAPH COURTESY OF CORBIS.

CHAPTER 25

Buildings, the Internet, and the Future of a Sustainable World

25.1 Buildings and Construction

In every nation of the world, the designers of the built environment — the planners, architects, engineers, constructors, government regulators, facility managers, and others who design and build facilities — shape the way the world population uses the physical infrastructure to live, travel, and work. As the economies of the world continue to expand, designers are faced with the challenge of leading future development in a way that protects and conserves environmental quality and the natural resource base. Future development undertaken in an environmentally and economically sustainable manner requires new ways of thinking in planning, designing, building, operating, and maintaining the built environment. The building and construction industry along with government is beginning to make fundamental changes in the way infrastructure is designed, built, utilized, maintained, and renewed, in order to achieve sustained global economic growth while enhancing environmental quality.

The demand for construction and building worldwide exceeds $3 trillion annually. However, change and improvement come slowly. This industry needs to remove or reduce barriers to improving building and construction practices in order to support sustainable development goals. Facilities are often designed using least-cost technologies that do not respond to opportunities to improve productivity and to enhance environmental quality. Also, there is an inability to achieve consensus on government design and construction policies that advance sustainable development and there are the frustrations of knowing better technologies are available but not having the capability to find and retrieve them. Furthermore, there is a lack of adequate international attention to building and construction research and practice. Buildings and facilities often must be designed and constructed using inappropriate building codes and fragmented regulatory systems, which does not allow for adopting new and better materials and practices. The construction process all too often is marked by adversarial relationships and distrust that make consensus building difficult, if not impossible, to achieve. Existing procedures for project selection, characterized by inflexible and inappropriate rules and regulations and codes and standards, also inhibit the ability to use innovative technologies and systems to further sustainable principles.

It is important to identify and utilize innovative design and construction techniques that utilize materials and systems that support the design of facilities that meet sustainable development standards. New approaches may include sound land use planning, materials substitution, and closed-loop construction (recycled or remanufactured modules). Buildings can be designed so that the heating and cooling requirements can be reduced by 10% to 15%. By adopting moderate shell efficiency measures, such as thicker insulation and better windows, new buildings could require an estimated 50% less heat and 25% less air-conditioning than today's average building. With tough government standards, buildings can be built to require an estimated 70% less heat and 40% less air conditioning.

Improving the efficiency of lighting in new commercial buildings is another technical option that can yield substantial reductions. The tough scenario measures — a combination of high-efficiency fluorescent bulbs

Table 25.1 **Practices and Technologies for Energy-Efficient Buildings**

- Manufactured wall systems with integrated superinsulation and superwindows optimized for orientation, external temperature, and internal needs
- Photovoltaic roof shingles with reflective roofing.
- Low-cost, high-performance solar water heaters and other advanced solar heating and cooling technologies
- Strategic positioning of trees to reduce cooling costs; fuel cells providing low-carbon energy, and energy storage
- Advanced high-efficiency lighting systems actively operating with an array of daylighting and site/task strategies to optimize luminosity and reduce energy consumption
- Smart technology to closely match energy and water supply for multifunctional and integrated appliances and buildings control systems, and automatic load modulation of heating and cooling systems in response to varying weather, environment, and occupant demands
- Improved sensors and controls, zoning and variable loading of the heating and cooling
- Healthful house construction that is radon resistant, nonallergenic, and makes use of recycled materials

and ballasts, improved reflectors, and better use of daylight — could lower lighting energy needs by 60% in these buildings.

By 2020, new practices and technologies may greatly reduce the energy use of buildings. Table 25.1 summarizes several methods for creating energy-conserving buildings.

25.2 In the Tradition of Architecture

By Charles DeLisio, AIA
Architect
Makato Architecture & Design

> . . . *natural beauty has a necessary place in the spiritual development of any individual or any society. I believe that whenever we substitute something man-made and artificial for a natural feature of the earth, we have retarded some part of man's spiritual growth.*
> — Rachel Carson

The consideration of sustainable design issues — environmental quality, energy and resource use, waste and pollution — leads to a broader discussion of current architectural work and theory. Despite a growing interest in renewable energy sources and a common understanding that environmental protection is imperative, most recent American architecture is climatically and regionally indifferent, including much of the work published and exhibited nationally. The influences of climate, geography, and culture that historically had helped to create a rich and lasting architecture are today dominated by technological development and its influences. With great expenditure of resources and at great cost to the environment, we are creating a predominantly mediocre built environment unworthy of the tradition of architecture.

Early in this century developments in technology began to change the path of architecture in the United States. The development of mechanical and electrical systems, combined with artificially inexpensive energy, removed the limitations of form, site, and materials that were previously necessary if a building was to be habitable. High ceilings were no longer necessary in hot regions to promote thermal stratification. Large glazing areas were no longer required to admit natural light, and operable windows and vents were no longer required to bring in fresh air. Architects no longer had to be concerned with providing natural light and air as these could be provided artificially. Some within the design community argued that the artificial environment was better for people than the natural environment — engineers were proud of their ability to overcome climate. As the natural influences on architecture were gradually discarded, artificial energy consumption increased dramatically causing a corresponding increase in the pollution of the natural environment.

This development and application of technology also changed the profession of architecture as specialists began to design and install these systems.

Architects no longer had direct responsibility for environmental control; engineering consultants became responsible for providing a suitable indoor thermal and visual environment. Industrial production in North America and western Europe added to the forces of change, as building product and material suppliers achieved influential roles in the use of materials and systems, and in architectural discussion.

These events created a fundamental shift in architecture leading to severe and unintended consequences: advancement of a universal design approach based on standard products and systems, degradation of the interior and exterior environment, disassociation from the natural world, and the reduction of architectural discourse to simple discussions of style. Generations of architectural students have not learned of important past works and sources, and the focus of many architects has become solely aesthetic design, visual form, and two-dimensional image. Many architects now accept working with a limited design palate.

The Pittsburgh Municipal Courts Building and CNG Tower, both in Pennsylvania, exemplify the common approach to architecture that is a result of these changes. These buildings are strongly visual, heavily invested with rich materials and equipment, and technically sophisticated. Other examples include the Apple Computer Company headquarters in Cupertino, California, the University Center at Carnegie Mellon University in Pittsburgh, and the renovated World Bank Headquarters in Washington, D.C. — all are materially rich and technologically well endowed but indifferent to region, climate, or tradition. While these are a diverse set of buildings, they make clear that even a tremendous use of resources is not sufficient for creating architecture that will engage on an emotional or spiritual level.

Contrast these buildings with The Salk Institute in California or The Museum of Roman Art in Spain, two well-known projects that offer a rich, direct, and sensitive environment and establish connections to site and place. The proposed Highcroft Center for the Performing Arts, to be built north of Pittsburgh, is also based on a more sensitive approach to design. This is work that is influenced by history, climate, and region; architecture that transcends technology, standardized products, or a universal design approach.

This common approach to architecture can also lead to a frivolous waste of resources. Within Fifth Avenue Place, in Pittsburgh, the designers attempted to create the feeling of a glass-covered atrium by means of a sophisticated artificial lighting system installed above the false "skylight" at the arcade level. Despite these complex and expensive efforts the resulting interior environment is far from convincing.

Another avoidable resource waste is the temporary, and disposable, storm protection required for many buildings in the southeastern United States. This is a result of indifference to climate and place as regionally appropriate building techniques, like wide roof overhangs and storm shutters, are often ignored by architects and builders.

When compared to "common practice," the costs of sustainable design are sometimes greater, many times neutral, and often very hard to evaluate. Resource economics is an emerging discipline, but advancing sustainable design based on economics alone may prove self-defeating; a new era of inexpensive energy may again lead to resource waste and indifference to region and climate, and the resulting loss of richness that occurs with this disconnection.

Architects influence or make decisions every day, at every scale, without having to meet a strict standard of "cost effectiveness." The suspended glass wall at the Carnegie Museum of Art courtyard is not the most cost-effective solution; the new Pittsburgh International Airport would have been more cost effective with a simpler roof form; the Union Trust Building central stained glass dome would have been cheaper if it were only a flat plaster ceiling; the most cost-effective building skin is not the ornamental granite and stainless steel cladding on the CNG Tower. These decisions, to build above absolute minimum standards, often result from ideas and discussions originated by architects. Architects have influence, and much of what is built depends on the ideas which architects choose to "bring to the table." Sustainable issues should not be held to a higher standard of cost effectiveness than any other architectural idea.

Architecture has always been about sensitive and creative building including sensitivity to the natural world. Louis Kahn's best work is about natural light and natural materials. Le Corbusier refined the *brise soleil,* was always aware of place, and warned against attempts to dominate climate through technology. There are traces of this sensitivity in the past work of architects Bruce Goff and Luis Barragan, and, today, in the work of Ricardo Legorreta, Norman Foster and Rafael Moneo. There are traces of this sensitivity throughout architectural writing, including Frank Lloyd Wright's essays, Ian McHarg's book *Design with Nature,* and the current writing of architectural critics Kenneth Frampton (*Towards a Critical Regionalism: Six Points for an Architecture of Resistance*) and William J. R. Curtis (*Contemporary Transformations of Modern Architecture*).

Regional, climatic, and traditional influences on architecture are recognized and remain strong throughout the world, but have become rare in American architecture. Recent, and highly visible, opportunities to build on these influences include The Carnegie Science Center in Pittsburgh and the proposed Museum of the American Indian in Washington, D.C., but these buildings are based on common and safe practices, despite the programmatic and symbolic opportunities. They seem only a set of simply composed spaces furnished with expensive materials, systems, and finishes, with little recognition of site, climate, or tradition.

Sustainable design will bring a refreshing change in current building standards, not necessarily in quality, but most certainly in levels of expectation and levels of comfort. The use of salvaged materials in new buildings, the avoidance of popular styles, and the creation of a naturally varying interior environment will require different thinking on the part of both clients and architects. Perhaps the lessons from Denise Scott Brown and Robert Venturi, or the messiness of *deconstruction,* will help in shifting these expectations. The expensive precision and *newness* of most recent work is not always necessary or desirable.

Architecture is about sensitive and creative building, including sensitivity to the natural environment and to creating a rich way of living in the world. Our impressions and perceptions of the natural world enrich life. A thunderstorm begins as the sight of dark clouds and lightning. The addition of other sensations, the sudden drop in temperature, the roar of thunder, and the feel of a strong, cool breeze, creates a rich environmental impression. Architects have created many rich places: The Carnegie Museum of Art in Pittsburgh; Thorn Crown Chapel; Fallingwater; The San Juan Capistrano Library; The Museum of Archaeology and History in Montreal; and the newly renovated National Airport in Washington, D.C. These are memorable places, work that is direct and lasting, architecture that is not of a style or of a kind. This is work that is aware of climate and place; work that is sensitive, engaging, and influential.

The current discussion of sustainable design is in the tradition of architecture. This tradition may now be enriched by other cultures or philosophies — the Amish rejection of technology, the writings of Rachel Carson, Islamic opposition to Western culture, Thomas Jefferson's resistance to industrial society, or Thoreau's desire to live ". . . deliberately." A sustainable architecture should be less about wasteful consumption and style, and more about issues that affect our lives, other lives, and future lives. There should be less domination and destruction of nature and more awareness of differing cultures. We will gain richer and more diverse buildings and communities, and reestablish architecture as a source of emotional and spiritual power.

25.3 Telecommunications and the Internet

Technologies to transport ideas and information across long distances have spread knowledge. Railways and steamships first allowed the speedy widespread dissemination of news and ideas. The development of organizations that used these technologies led to the postal service and the publishing industry. By 1850 postal mail emerged in Europe and the United States. By 1860 telegraphy provided another means of quickly communicating. In the early 20th century, the telephone emerged as a means of communication. Telecommunication became a fast, efficient means of spreading knowledge. Newspapers and books also became important in the 19th century and the publishing industry flourished.

Mobile (wireless) phones had emerged by 1980 and over 500 million portable phones were in use by 2000. Mobile phones bring the ability to communicate from the car, street or field. Communication systems change commerce and culture. Mobile phones are attractive in less developed countries where land lines are difficult to build.

The radio was commercialized in the early 1900s and provided news and information to most households by 1935. Technology has placed in the hands of all who would shape national policy a new tool, potent in the shorter run and of uncertain long-run impact: television. Vivid visual experience is now instantaneously transmitted from almost any point of the world and inexpensively received worldwide with a billion households potentially attentive.

The current wave of change stems from the marriage of the computer to these communications links. Telephone and television networks have long conveyed sound and pictures as analog waves. Now, however, many types of information — text, sound picture, or video — can be transmitted digitally, as compressed bits in the binary language of computers. Digitization is sweeping through the telephone, photography, remote sensing, broadcasting, film, and music industries.

The number of personal computers in use worldwide exceeds 400 million. During the period from 1995 to 2000 the Internet grew, with over 200 million personal computers connected to the Internet. The **Internet** is a network of individual networks that cooperate to direct traffic so that information can pass among them. The Internet has now brought the most potent tool of communication ever available to the individual. Using e-mail or a Web site, he or she can now publish, to one recipient, to hundreds, or to tens of thousands.

Table 25.2 summarizes the time for new technologies to reach penetration to 50% of the households in the United States. Notice that the Internet is expected to reach one-half of U.S. households by 2003. The Internet is driving efficiency gains in a variety of ways. They include allowing companies to reduce building construction, cut paper use, and perform other functions with less physical throughput. Some business-to-consumer e-commerce allows companies to substitute more efficient warehouses for retail stores. And business-to-business e-commerce increasingly is allowing companies to reduce inventories, overproduction, unnecessary capital purchases, mistaken orders, and the like — all with resulting energy efficiencies and reductions.

Table 25.2 **Penetration Time for New Technology**

New Technology	Penetration years to reach 50% of households
Telephone	68
Electricity	50
Radio	35
Television	25
Personal Computer	20
Internet	8

25.4 **Telecommuting and Sustainability**

Telecommuting is work carried out at home or at an office close to home (remote from central offices or production facilities) where the worker has no personal contact with coworkers but is able to communicate with them and perform work-related activities using computer and communication technologies.

About one-half of U.S. companies have telecommuting programs in effect. Information workers are more likely to have tasks that could be performed away from the office. Currently, information workers are generally estimated as being 50% or more of the labor force.

A variety of computer and communication technologies support telecommuting, such as telephones, computer modems, fax machines, electronic mail, and computer information networks. The standard equipment for most telecommuters is a PC and/or laptop computer, a modem, and a fax — and these are often available in one integrated hardware package. It may be necessary to have an additional telephone line installed, and it is now possible to link the worker's own telephone into the company telephone network during working times. Some people also find it useful to have a small photocopier, a pager, or a mobile phone. More sophisticated communication services, such as call waiting, call forwarding, three-way calling, speed calling, and caller ID, also make it easier for telecommuters to be away from the office.

Consider the tremendous savings now that millions of us are able to work from home — or at least, dial into the office more than drive there. Many businesses are now using the Internet to save office space, materials, and transportation costs by letting employees work from home. The number of home offices is now growing by about three million a year. Telecommuting reduces overhead for companies, and can also help cut traffic congestion, saving energy and cutting pollution in numerous ways.

Using e-mail and other electronic documents also saves energy, by saving paper. Making paper is energy intensive and can be extremely polluting. Newspapers, the country's biggest single paper consumer, are also going on-line. Arguably, of all technologies, telecommunications, along with renewable energy, has the most potential to deliver sustainability, and the vision of integrated optical communication networks, powered by renewables, is compelling. This is a sector at the heart of globalization and dematerialization. Teleservices will restructure entire economies, while the new "on-line, on-demand" regime heralds a leaner, greener age of consumerism. Whether the benefits of advanced telecommunications will simply get drowned in "information overkill" remains to be seen. For now, the industry's role as a sustainability leader seems positive.

25.5 Driving toward Sustainability

By Joel Makower
Editor
The Green Business Letter

Are automobiles compatible with "sustainability"?

At first glance, automobiles and the environment seem at direct odds with each other. Cars and other motor vehicles are linked to nearly every environmental problem faced by the planet and its inhabitants: air pollution, acid rain, global warming, water pollution, solid waste, overdevelopment, and the myriad impacts associated with building roads deeper and deeper into previously undeveloped land — deforestation, soil erosion, species loss, wetlands destruction, and on and on. That doesn't necessarily include the resources and energy used to build and operate vehicles, and the pollution created during their manufacture and disposal. All have a significant impact on the environment.

And yet most modern societies are built around the notion of mobility — getting to jobs and schools, moving goods and materials from place to place, shopping for and transporting purchases home, and in general living lives that enjoy the freedom and independence of movement. So, any notion of wiping automobiles from our social structure is ludicrous at best. The fact is, the opposite is true: Car ownership is expected to continue its upward climb in coming years.

Consider the numbers: In 1999, the global passenger car fleet stood at 520 million vehicles, according to data compiled by the Worldwatch Institute. And while that may seem significant — the number of cars has doubled since 1975 — it means only about 8% of the earth's population owns a vehicle. About 9 in 10 people don't — yet.

So, what happens when the newly open markets of India, China, Africa, Eastern Europe, and Latin America create perhaps a billion more middle-class denizens who desire increased mobility? Can the planet and its people support this magnitude of vehicle growth?

It's a tough question. A basic level of mobility is fundamental to a sustainable society. Mobility allows individuals to hold jobs that are beyond walking or bicycling distance from their homes. Mobility engenders the ability of individuals to conduct trade and commerce over a broader region, enabling, for example, rural farmers to sell their goods directly to big-city distributors at market rates, rather than relying on unscrupulous middlemen. Mobility facilitates people attending schools and universities. And mobility helps bring people to health-care providers, rather than the other way around, increasing individuals' chances of receiving proper care. Mobility, in short, is a key ingredient for lifting entire communities out of a desperate, hand-to-mouth existence into a more self-sufficient, economically stable lifestyle.

How do you bring mobility to billions without necessarily raising pollution and resource use proportionately? The global automobile manufacturers — at least, those that plan to be around in 25 years — are pondering that question, and their early responses are encouraging.

Though it isn't likely to show up in advertising slogans for some time, Ford, General Motors, Honda, Volvo, and other global automakers are beginning to view themselves not merely as vehicle manufacturers but as "mobility providers," or even "sustainable mobility providers." Some companies actually utter these phrases publicly; others merely use them internally. Either way, these are not simply semantic euphemisms; these words speak volumes about the new world of sustainable business.

"Sustainable mobility," for example, embodies the notion of "servicizing" a product. **Servicizing** refers to an increasingly utilized business model centered around selling the services a product delivers, rather than the physical product itself. One example comes from the world of automobile manufacturing and involves Ford and the automotive paint division of DuPont. In the 1990s, Ford recognized that a significant amount of the paint it was buying did not end up on vehicles. Because of inefficient manufacturing processes, paint escaped into air emissions or into water-based sludge. Ford paid twice for the paint — once to buy it and once to properly dispose of the wasted paint — neither of which brought value to Ford or its customers. So DuPont and Ford renegotiated their relationship to "servicize" the painting process. Now, rather than selling Ford a product — paint — DuPont now sells Ford a service — painted cars. This gives DuPont an incentive to paint the most amount of cars with the least amount of paint, greatly reducing Ford's waste, costs, and liability. A business model that shares the resulting savings rewards both companies for their increased efficiency.

Automotive painting is by no means the only example of servicizing. For example, a carpet company is leasing carpeting services — acoustics, aesthetics, padding underfoot, and all the rest — rather than selling the physical materials. As with other leases, the manufacturer owns the asset and takes responsibility for it at the end of the lease. In the case of the carpet company, they are learning how to turn old carpet, which used to be a liability, back into new carpet, an asset.

In servicizing, the notion of straightforward buying and selling softens and diversifies into a spectrum of property rights arrangements, including leasing, pooling, sharing, and take-back. Value is increasingly created and measured by the function provided, and for the manufacturer, the product increasingly becomes a means of delivering this function, rather than an end in itself.

That same notion can apply not just to painting cars, but to owning and driving them, too. Consider, for example, the Intelligent Community Vehicle System, or ICVS, created in the late 1990s by Honda. ICVS is a system of shared electric and hybrid-electric vehicles — from bicycles to small passenger cars — suitable for use in a campus-like or small-community environment. Participants receive an electronic "smart card" that enables them to use the system. So, for example, someone needing to use an electric-assisted bicycle would insert the smart card into a vending machine at a centrally located kiosk and receive a small battery that is easily inserted into the front of the bicycle near the handle bars. After reaching her desired destination, the user would return the battery to a nearby vending machine, with the appropriate charge levied to her smart card. Meanwhile, the system keeps track of the bicycle inventory and where they are located.

Using the small passenger cars in the ICVS system involves an even more sophisticated level of technology. An individual desiring to use a car inserts his smart card into a kiosk, at which time a vehicle automatically drives up and parks. The user's smart card serves as a means for both unlocking the doors and starting the engine. Then, when the user reaches his destination kiosk, he locks the car, inserts his smart card into the kiosk — and the car drives away and parks itself!

In effect, ICVS has "servicized" transportation by removing ownership of the physical good (the car) and providing only the desired service (the ride from place to place).

Honda isn't alone in thinking about such notions. William Clay Ford, chairman of Ford Motor Company (and great-grandson of Henry Ford, the company's founder), acknowledged in a 2000 speech that "We understand the triple bottom lines of sustainability must all be addressed for us to be successful. For example, we can't expand in potentially huge markets such as India and China — and provide a better life for millions of the world's poorest people — unless we can do it in a sustainable way. It's the only way we will be able to grow our business and reward our shareholders in developing and existing markets. . . . Frankly, we don't have a detailed map of how to get there yet, or even an exact picture of what sustainable mobility will look like. In addition to the automobile, it may include mass transit, or Internet access, or something we haven't even thought of yet."

Given the downright antagonistic relationship between environmentalists and the automobile industry, it seems a breath of fresh air to hear Mr. Ford portray such an environmentally friendly future for his company. But make no mistake, Ford, Honda, and the others have a long way to go before the notion of "sustainable mobility" becomes a reality. They must invent, refine, and mass market new technologies and create new global infrastructures that facilitate the use of shared vehicles, increased access to mass transit, or whatever other business models they pursue. They must work creatively with local and national governments to ensure their efforts will be supported by city planners, developers, and many others.

And, of course, there is the human factor, which can unwittingly thwart even the most well-intended corporate intentions. Getting people to accept the notion of sharing, not owning, vehicles could be challenging if the U.S./Japanese/European model of individual car ownership becomes the desired goal of individuals in a newly emerging Third World marketplace. The fact is, while the notion of sharing vehicles may be compelling and make perfect sense, it has never been tried on a large scale. Can automobile companies' considerable marketing clout help shape individuals' mobility aspirations? It remains to be seen.

As Honda would probably acknowledge, the notion of an Intelligent Community Vehicle System is meaningless unless you have an intelligent community.

25.6 Government and Sustainability's Triple Bottom Line: Promoting the "Three E's"

By Thomas Jacobson
Associate Professor
Department of Environmental Studies & Planning
Sonoma State University

The concepts "sustainability" and "sustainable development" have proven difficult to define. Despite these difficulties, a persistent descriptor of sustainability and sustainable development has been that they encompass "the Three E's": environmental quality, economic health, and social equity. These three components of sustainability might be described in another way, as "the triple bottom line." (See Chapter 23 regarding the triple bottom line.)

The principle of the Three E's has been embraced by many of those governmental entities that have ventured into the field of sustainability. Described below are some of the motivations for this movement, as well as examples from the federal, regional, and local levels. Methods of ensuring that sustainability concerns are addressed locally, and of measuring local progress with regard to those concerns, are also outlined.

25.6.1 MOTIVATIONS FOR EMBRACING THE THREE E'S MODEL OF SUSTAINABILITY

While there are many definitions for "sustainability" and "sustainable development," and these definitions have provided a source of substantial controversy, one of the most widely used and accepted definitions is the one articulated by the United Nations' World Commission on Environment and Development (the "Brundtland Commission") in 1987: "Sustainable development is development that meets the needs of the present without compromising the ability of future generations to meet their own needs." The focus on meeting human needs reflected here, though not shared by everyone in the sustainability movement, has characterized much of what has followed. It draws clear connections between the environment and economic well-being.

A crucial milestone in bringing the concept of sustainability into widespread usage was the 1992 United Nations Earth Summit in Rio de Janeiro, which adopted "sustainable development" as the principle that should guide the evolution of the world's communities and economies. The context for this adoption is significant. The Earth Summit was expressly concerned with how to address environmental concerns in conjunction with questions about worldwide economic development. This meeting in Rio recognized the link between the two concerns, and the importance of their being addressed together and in a coordinated fashion.

In this setting, social equity has been viewed as a necessary complement to the other two "E's." Fundamental to the descriptions of sustainable development described above are the idea of both intergenerational equity (the Brundtland Commission's "development that meets the needs of the present without compromising the ability of future generations to meet their own needs"), and concerns, expressed at the Earth Summit, about meeting the present and near-term needs of developing nations.

One of the goals of the Earth Summit was that the principles of sustainable development articulated in Rio take root in the policies of individual nations.

25.6.2 THE PRESIDENT'S COUNCIL ON SUSTAINABLE DEVELOPMENT

In 1993, the year following the U.N. Earth Summit in Rio, Congress took steps in that direction by passing the Earth Leadership Act. This act established the President's Council on Sustainable Development (PCSD; see *www.whitehouse.gov/PCSD/*).

Under the Clinton/Gore administration, the principle of sustainability as represented by the Three E's has been typical of much of the administration's environmental message: Environmental protection is not at odds with economic growth. In fact, the administration has maintained, environmental protection is fundamental to economic well-being.

In 1996, the PCSD issued a report, "Sustainable America — A New Consensus," stating that a sustainable America can only be achieved by creating sustainable communities characterized by the Three E's. The PCSD also called for the formation of multi-stakeholder regional efforts to promote sustainable development. One of these, the Bay Area Alliance for Sustainable Development, set in the nine-county region surrounding San Francisco Bay, was a response.

25.6.3 A REGIONAL MODEL FOR SUSTAINABILITY: THE BAY AREA ALLIANCE FOR SUSTAINABLE DEVELOPMENT

The Bay Area Alliance for Sustainable Development (BAASD), established in 1997, has brought a wide range of organizations and individuals together with the goal of developing and implementing an action plan based on the Three E's.

BAASD is notable in that, while it includes representatives of federal, state, and local governments, it also includes a wide range of nongovernmental organizations. Among its basic goals, however, is that individual governmental agencies will subscribe to the vision (as well as to specific proposals) that will result from the work of this broadly based collaboration.

BAASD is firmly grounded on the Three E's model of sustainability. In June 1997, BAASD adopted this vision statement: "We envision a Bay Area where the natural environment is vibrant, healthy and safe, where the economy is robust and globally competitive, and where all citizens have equitable opportunities to share in the benefits of a quality environment and a prosperous economy."

Toward this end, BAASD has produced a *Draft Compact for a Sustainable Bay Area* (April 20, 2000), which recommends strategies and actions aimed at progress toward each of the Three E's. The draft compact describes the Three E's as "equally important and interdependent."

25.6.4 INCORPORATING THE THREE E'S AT THE LOCAL LEVEL

One of the biggest challenges facing the movement toward sustainability is how to take global principles and make them concrete locally. This is a challenge that some city governments have chosen to undertake. The following describes a variety of approaches available to cities interested in promoting sustainability goals.[1] Some are variations on traditional planning and regulatory programs. Others reflect a range of incentives to encourage specific private sector, sustainability enhancing behaviors. Additional approaches rely on strategic partnerships with business and environmental groups. Still others focus on fostering (or at least not stifling) private sector innovation that promotes sustainability. Together, they represent opportunities to address a vision of sustainability unique to each city. In some cases, this vision may be based on a comprehensive Three E's model, while in other instances the vision may not be as broad.

Comprehensive Plans

In many states, cities are required to adopt **comprehensive plans**, otherwise called *general plans* or *master plans*. These documents address a wide array of concerns: land use, housing, transportation, resource protection, and so on. Typically, comprehensive plans include goal statements, policies for achieving those goals, and implementation programs tied to those policies.

Although the degree varies, most states grant a substantial amount of policy discretion to cities in adopting their plans. Some states, such as California, require that certain topics be addressed, but without requiring that specific policies be adopted or promoted. Other states, such as Oregon, mandate far more in the way of state-level policies that are to be adopted at the local level. Still, in every state, there is an opportunity to infuse the comprehensive plan with sustainability principles. For example, comprehensive plans can address promoting transportation alternatives to the automobile, solid waste reduction, revitalizing blighted neighborhoods at the city core, energy conservation, promoting biodiversity, water conservation, and a host of other topics central to sustainability.

The efficacy of using the comprehensive plan to promote sustainability is increased in those states that have adopted the "consistency requirement." In

---

[1] In some states, many of the same opportunities are available to county governments.

those states (e.g., Florida and California), most land use and public works decisions must be consistent with the comprehensive plan. This increases the power of the comprehensive plan in a dramatic way, making it the single most powerful planning tool in a community.

Zoning, Subdivision, and Other Regulations

As the "workhorses" for implementing local land use and environmental policy, zoning, subdivision, and other regulations have the potential for promoting sustainability. A few examples are subdivision regulations requiring drought-resistant landscaping in new developments or higher levels of insulation in new construction; zoning ordinances mandating higher density, mixed-use development close to transit; and regulations limiting development in areas of sensitive or unique habitat.

Promoting Sustainability through Municipal Operations

Government, including government at its most local levels, is big business. Cities are major purchasers — of paper products, vehicle fleets, office machinery, and so on. Each purchase is an opportunity to engage in sustainable practices and support sustainable industries. Cities are major builders — of office space, libraries, parks, roads, water and sewer systems, and so on. Each construction project is an opportunity to utilize "green building" and other sustainable technologies and to encourage businesses that are adopting sustainable practices. Cities are also educators and employers, with the ability to promote the development of human capital.

Offering Incentives to the Private Sector

Some cities have begun to address providing incentives to privately sponsored development projects that promote sustainability goals. For example, development fees might be reduced for a project that incorporates an energy-saving design or that will otherwise reduce resource use or waste.

The city of San Diego instituted a growth control program in the 1980s that, while not expressly aimed at furthering sustainability, adopted a version of this approach. It reduced infrastructure fees charged to new development in those areas where the city most wished to see new development occur — "infill" projects in already developed areas of the city. Meanwhile, development proposed for the city's edges was required to pay the full cost of the infrastructure needed to serve it. Interestingly, the program was a victim of its own success. So much development was attracted to the city's inner areas that the city could not afford to continue providing the subsidy required, and revamped its approach to managing growth.

Government/Environmental/Business Collaboration

An interesting development has been an increase in the level of cooperation between local governments, environmental groups, and business interests in developing and implementing policies aimed at addressing the Three E's.

For example, in California, the Sierra Business Council, the Bay Area Council, and the Silicon Valley Manufacturing Group, all business-oriented organizations, have allied themselves with some environmental groups, and together worked with cities to put together strategies linking, variously, compact development, improved public transportation systems, and protected open space and natural systems.

Asserting Leadership by Not Saying No

Increasingly, critics are pointing out the degree to which cities have acted as obstacles to sustainable development. One example is the tendency for cities to reject compact, transit-oriented development because it represents greater density and, thus, runs into local opposition. Another is a resistance to accepting innovations in design and technology (e.g., with regard to wastewater treatment), despite their potential for promoting environmental goals. While government responsiveness to community concerns about perceived threats to quality of life, or to potential health risks, is clearly valid and important, there is a growing call for local governments to be more open to well-thought-out proposals for innovation.

25.6.5 MEASURING PROGRESS WITH THE THREE E'S

Cities are beginning to recognize that it is not enough to make lofty proclamations regarding sustainability, or to merely adopt policies that speak to those goals. Sustainability "indicators" have been adopted in some jurisdictions, while there is also some discussion of adopting methods for assessing the impacts of specific municipal decisions on sustainability.

Indicators

A fundamental component of local sustainability programs is a schedule of "indicators" of progress. These can take a variety of forms, depending largely on the immediate concerns of the community. For example, the city of Santa Monica, California, has included acres of parkland, per capita water usage, number of street trees, and percentage of its bus fleet running on natural gas, as indicators of the city's progress toward its sustainability goals. The Bay Area Alliance for Sustainable Development has proposed a wide array of indicators across the Three E's, with potential implications for local governments. Among them are persons below the poverty line; workers with jobs earning less than a living wage; housing units needed in job surplus areas; vehicle miles traveled per capita; energy use per capita; solid waste per capita; days in violation of air quality standards; and so on.

Sustainability Impact Analysis

Another approach under discussion is to establish mechanisms to evaluate the impacts of municipal decisions on sustainability. In a way, this is a version of the "Seventh Generation" question often attributed to the Iroquois people of eastern North America. These Native Americans are said to have asked, before making any important decision affecting their communities, what effect the decision would have on the next seven generations.

This principle could be implemented by cities by developing a set of questions to be asked as a means of determining the impacts of municipal decisions on the Three E's. For instance, a requirement could be to analyze all investments in the community's transportation system for whether they would have an equitable impact across income groups. This could avoid, for instance, funding transit that primarily serves higher income commuters if it would mean reducing service to lower income neighborhoods. While analysis of this type need not mandate a particular outcome, it could help to ensure that decisions are more fully informed, similar to the environmental impact analysis required by the National Environmental Policy Act and its state-level counterparts.

25.6.6 CONCLUSIONS

The Three E's, or triple bottom line, principle characterizes much of what is being undertaken by those government entities engaged in the sustainabililty movement. The options that are available to all levels of government in promoting the Three E's are many and growing, and will likely require a mix of traditional and innovative approaches to governance.

25.7 The Problem of the Last Straw: The Case of Global Warming[2]

By INDUR M. GOKLANY
Independent Consultant
Vienna, Virginia

A major argument implicitly or explicitly advanced for making immediate and significant reductions in greenhouse gas (GHG) emissions is that human-induced climate change on top of other environmental problems may, like the last straw, overwhelm human and natural systems, particularly with respect to natural ecosystems, forests and biodiversity. Another argument is that because of the inertia of the climate system, the long residence time of carbon dioxide in the atmosphere, and the relatively slow turnover in the energy system — the major contributor to GHG emissions — reductions must be made now, before evidence of human-induced climate change and its impacts become more compelling. This section provides a different approach to addressing the problem of the last straw and to the issues posed by the inertia of the climate and energy systems.

---

[2]This section is based on the author's article titled "Potential Consequences of Increasing Atmospheric CO_2 Concentration Compared to Other Environmental Problems," in *Technology*, Vol. 7, Suppl. 1, pp. 189–213.

25.7.1 THE RELATIVE IMPORTANCE OF CLIMATE CHANGE COMPARED TO OTHER ENVIRONMENTAL CHANGES

Over the last century or more, according to the Intergovernmental Panel on Climate Change (IPCC), an international panel established to examine the science and impacts of climate change, the globe has warmed 0.3–0.7°C, perhaps due to man's influence (IPCC, 1996a). During this period, matters have worsened with respect to some climate-sensitive environmental indicators or sectors of the economy, but not because of anthropogenic warming itself. Specifically:

- Sea level has risen a modest 8 in. in the past century, but it is not clear what portion of that rise, if any, is due to global warming. Regardless, its impacts on coasts and coastal resources are secondary compared to those due to other human activities such as development; subsidence due to extraction of oil, gas, and water resources; overfishing; agricultural runoff; and damming of rivers upstream of estuaries (IPCC, 1996a, 1996b).
- Forested area declined by 190 million hectares (Mha) in developing nations between 1980 and 1995 largely because increases in food demand outstripped increases in agricultural yield while forest cover in developed nations expanded by 20 Mha mainly because of technology-based high-yield agriculture. Conversion of forests and other habitat to agricultural uses is the greatest current and future threat to global biodiversity, and to carbon stores and sinks.

For other critical climate-sensitive sectors and indicators, matters have actually improved:

- Global agricultural productivity has never been higher. An acre of cropland sustains about twice as many people today than it did in 1900. People have never been fed better or more cheaply. Between 1961 and 1995, food supplies per person increased 19% although the population increased 84%; and between 1969–1971 and 1990–1992, people in developing countries suffering from chronic hunger declined from 35% to 21%.
- Deaths due to "climate-sensitive" infectious and parasitic diseases are now the exception rather than the rule in the richer countries, and are declining in most developing countries due to better nutrition and public health measures. Accordingly, between 1950–1955 to 1990–1995, overall death rates in developing countries dropped from 19.8 to 9.3 per 1,000 population, and global life expectancy at birth increased from 46.4 to 64.7 years.
- While increased population and wealth has put more property at risk, which has helped increase U.S. property losses due to floods and hurricanes during this century, there are no clear trends in losses in terms of wealth. More importantly, based upon 9-year moving averages,

deaths due to hurricanes, tornados, floods, and lightning have decreased from 46% to 97%, compared to their earlier peaks during the 20th century, while death rates declined between 60% and 99%.

Thus, despite any warming, by virtually any climate-sensitive measure of human well-being, the average person's welfare improved during the 20th century. While some credit for increasing agricultural and forest productivity is probably due to higher CO_2 concentrations and higher wintertime temperatures, most of these improvements in climate-sensitive indicators are due to technological progress, driven by market- and science-based economic growth, technology, and trade. Such progress has also reduced the vulnerability of the human enterprise to climate change. As a result, technological progress has so far had a greater impact on the climate-sensitive sectors than has climate change itself.

But what about the future?

Table 25.3 summarizes the impacts of global warming in the foreseeable future. It indicates the potential (though not necessarily, probable) impacts due to warming. Table 25.3 is based on, and assumes the validity of the analyses underlying, the IPCC's 1995 Impact Assessment (IPCC 1996b). It shows that:

- In the absence of warming, global production would have to increase 83% from 1990 to 2060 to meet the additional food demand from a larger and wealthier population (Reilly *et al.,* 1996). Such an increase is more likely if sufficient economic resources are available to develop, acquire, and operate the technologies and infrastructure necessary to increase food production (Goklany, 1998). Due to global warming, agricultural production may decline in developing — but increase in the developed — countries, resulting in a net change in global production of +1% to −2% in 2060 (Reilly *et al.,* 1996). Thus, a future food crisis is more likely if economic growth and technological change falter or if trade, which enables food surpluses to move voluntarily to deficit areas, is inhibited, than if warming occurs (Goklany, 1998).

- If all else remains the same, by 2050, global forest area may *increase* 1% to 9% due to global warming alone. But if land use change due to greater agricultural and other human demands is also considered, forest cover may *decline* by 25%, putting enormous pressure on the world's biodiversity (Solomon *et al.,* 1996).

- Sea level may rise 3 to 19 inches, with a "best estimate" of 10 inches by 2060, and about twice that by 2100 (Warrick *et al.,* 1996). The global cost estimate for protecting against a 20-in. rise in 2100 is about $1 billion per year or less than 0.005% of global economic product.

- By 2060, incidences of malaria, a metaphor for climate-sensitive infectious and parasitic diseases, may increase by about 5% to 8% of the base rate in the absence of warming. That increase may double by 2100 (McMichael *et al.,* 1996; Goklany, 2000).

Table 25.3 **Projected Climate Change Impacts Compared to Other Environmental Problems**

		Impact/effect	
Climate-sensitive sector/indicator	Year	Baseline, includes impacts of environmental problems other than climate change	Impacts of climate change, on top of the baseline
Agricultural Production	2060 for baseline >2100 for climate change	Must increase 83%, relative to 1990	Net global production would change −2.4% to +1.1%; but could substantially redistribute production from developing to developed countries
Global forest area	2050	Decrease 25–30 (+)%, relative to 1990	Reduced loss of global forest area
Malaria incidence	2060	500 million	25 to 40 million additional cases
	2100	500 million	50 to 80 million additional cases
Sea level rise	2060	Varies	<25 cm (or 10 inches)
	2100	Varies	<50 cm (or 20 inches)
Extreme weather events	2060 or 2100	Not applicable	Unknown whether magnitudes or frequencies of occurrence will increase or decrease

SOURCES: IPCC, 1996a, 1996b; Reilly *et al.*, 1996; Solomon *et al.*, 1996; McMichael *et al.*, 1996; Warrick *et al.*, 1996; Goklany, 1998, 2000.

- The frequency and magnitude of extreme weather events, such as tornados, hurricanes and cyclones, may or may not increase (IPCC, 1996a).

Thus, Table 25.3 suggests that stabilizing GHG concentrations immediately, unlikely though it might be, would do little or nothing over the next several decades to solve those problems which are the major reasons for wanting to do something about anthropogenic warming, except for the case of sea level rise. Land and water conversion will continue almost unabated, with little or no reduction in the threats to forests, biodiversity, and carbon stores and sinks; the food security of a larger world population will not have been substantially improved, if at all; and incidence rates of infectious and parasitic diseases will be virtually unchanged.

Moreover, poorer nations are generally expected to be most vulnerable to warming, not because their climate change is expected to be inherently greatest, but because of a deeper disease: poverty. They lack economic resources to easily develop or afford the technologies needed to adapt to or cope with any adversity, or purchase food surpluses produced elsewhere to make up for projected shortfalls (IPCC, 1996b). However, even if climate change were halted, poorer nations will continue to be vulnerable to all kinds of adversity, whether due to natural or man-made causes.

Since the above conclusions are based on the IPCC's 1995 impacts assessment, they are relatively robust, unless those estimates were substantially underestimated. But that seems unlikely. First, the studies underlying the IPCC's impact assessment generally do not adequately account for technological change and human adaptability. They are often more critical to estimating the impacts of environmental change than are changes in meteorological and climatic variables; underestimating them would increase the negative, while reducing the positive, impacts of global warming. Second, recent analyses of agricultural and sea level impacts of climate change indicate that they could be less adverse than reported by the IPCC. Moreover, recent studies of mortality due to or associated with extreme heat and cold continue to suggest at least partial cancellation of effects globally. Also, for some sectors, the IPCC's underlying impact studies assumed a greater climatic change than its own best estimates, for example, agriculture. In addition, empirical data on CO_2 concentrations suggest that carbon fertilization is real and probably already taking place which, by itself, ought to moderate CO_2 growth rates, at least for a while; increase timber production, at least in managed forests; and help reduce loss of forest cover. Fifth, empirical data indicate that the atmospheric methane growth rate, which had been increasing since 1945, may have peaked in the early 1980s and might stabilize in the next decade. Between 1984 and 1996, the growth in atmospheric concentrations of methane was estimated to have slowed by about 75%. Sixth, recent calculations of radiative forcing by GHGs suggest that the forcing used in the IPCC's climate change projections may have been overestimated by a net 10% because an underestimate for CFCs was more than offset by an overestimate of the CO_2 forcing. Finally, the decades-long slowdown in the world's population growth rate ought to reduce estimates of future impacts of climate change. Thus, for instance, the IIASA's 1996 central estimate for the population in 2100 was 10.4 billion, against the 11.3 billion central estimate assumed in many of the IPCC's projections.

In summary, both historical trends and future projections of impacts indicate that anthropogenic climate change is not now — nor is it likely to be in the foreseeable future — as urgent as other global environmental or public health problems. However, it has been argued that climate change on top of the other environmental problems may be the straw that breaks the camel's back, particularly with respect to forests, ecosystems, and biodiversity, which suggests that immediate action ought to be taken to curtail GHG emissions.

25.7.2 ADDRESSING THE WHOLE PROBLEM RATHER THAN A PORTION

There are several approaches to dealing with the problem of the last straw, none of which have to be mutually exclusive. The first, more common approach — and one consistent with the above argument — is to concentrate only on reducing or eliminating the last straw. Another approach would be to lighten the entire burden before the last straw falls. Consider malaria, for instance. Under the first approach — focusing on the last straw — one would try to eliminate the 50–80 million climate change related cases in 2100 by eliminating climate change, while under the second approach one would attempt to reduce the total number of cases, whether it is 500 million this year or 550–580 million in 2100 (see Table 25.3). The latter approach has several advantages:

- Even a small reduction in the baseline (i.e., nonclimate change related) rate could provide greater aggregate public health benefits than a large reduction in the additional cases due to climate change. Assuming exponential growth in the relative number of additional malaria cases due to climate change, reducing the number of baseline malaria cases by an additional 0.2% per year between now and 2100 would more than compensate for any increases due to climate change.

- Resources employed to reduce the base rate would provide substantial benefits to humanity decades before any significant benefits are realized from limiting climate change. Considering that a million Africans die from malaria annually and that it costs $8 to save a life-year from malaria, humanity would be better served if $1 billion were spent now to reduce malaria rather than on limiting climate change to curb, at least in part, potential increases in malaria decades hence.

- Given the uncertainties noted previously regarding impacts assessments, the benefits of reducing the base rate are much more certain than those related to limiting climate change.

- The technologies developed and public health measures implemented to reduce the base rate would themselves serve to limit additional cases due to climate change when, and if, they occur.

- Reducing the base rate would serve as an insurance policy against adverse impacts of climate change whether that change is due to anthropogenic or natural causes, or if it comes more rapidly than the IPCC's "best estimates." In effect, by reducing the base rate today, one would also be helping solve the cumulative malaria problem of tomorrow, whatever its cause.

- Because of the inertia of the climate system, it is unrealistic to think that future climate change could be completely eliminated even if

GHG emissions were to be frozen immediately at today's level. Given the lack of progress in reducing GHG emissions in response to the Kyoto Protocol, such a freeze is most unlikely. Moreover, full adherence to the protocol would reduce projected temperature increase for 2100 by less than 10%.

The second approach to dealing with the problem of the last straw is more comprehensive. It reduces the cumulative rather than merely a portion of the impact, and a relatively small portion at that. Removing a few straws before the proverbial last straw drops effectively increases the camel's ability to bear the burden that it would otherwise have to bear. In essence, the second approach would reduce the vulnerability of human and natural systems regardless of the sources of stress, and strengthen the ability to cope not only with climate change but other, currently more urgent, environmental stressors.

The same logic holds for other climate-sensitive sectors where problems are expected to be dominated for the next several decades by nonclimate change related factors. As Table 25.3 indicates, these sectors include agriculture, food security, forests, ecosystems, and biodiversity.

25.7.3 DEALING WITH TODAY'S PROBLEMS WITHOUT IGNORING THE FUTURE

It might also be argued that regardless of how urgent climate change might be over the next several decades, unless GHG emission reductions commence now it may be too late, because of the inertia of the climate system, to do much about warming if one waits for its impacts to become serious. In other words, climate change may not be urgent today or tomorrow, but it could be crucial the day after. Table 25.3, however, suggests that even if it takes 50 years to replace the energy system from start to finish, we could wait a couple of decades before initiating control actions beyond what would be obtained automatically through secular improvements in technology. Notably, carbon intensities of currently developed countries (in terms of CO_2 emissions per GDP) have declined 1.3% per year since 1850.

Moreover, as Table 25.3 indicates, even if climate change is completely halted, the major, imminent threats to global forests, ecosystems, biodiversity, carbon sinks and stores, and global food security would persist, and the underlying problems of malaria or other infectious and parasitic diseases would be undiminished. In fact, the world would be beset by the very same catastrophes that control of climate change hopes to avoid. The basic issue, therefore, is one of how to solve the urgent problems of today and tomorrow, while also enhancing our ability to address the serious problems of the day after.

There are two complementary approaches to addressing this issue. First, we can focus on fixing current environmental problems that might be aggravated by climate change. With respect to the interrelated problems of agriculture, food security, deforestation, and biodiversity, this means attacking their major causes, namely, the conversion of land and water to satisfy the demands of a larger and wealthier population for food, fiber, and timber. Some analysts contend that it is, therefore, necessary to decrease demand by reducing populations and/or modifying dietary and consumption habits. However, this is easier said than done. In a democratic society, where individuals and families are free, within the constraints of the market, to choose the number of their offsprings, their diets, and their consumption patterns, it is unlikely that such recommendations will have a significant impact since they ignore human nature. An alternative approach — and one, arguably, more likely to succeed because it accepts the flaws of human nature — would be to produce in an environmentally sound manner more food, timber, and other products per unit of land or water diverted to human use. Such increases in productivity would limit conversion of land and water to human uses while helping meet human demands adequately.

Assuming that global population will be 9.6 billion in 2050, that food supplies per capita would increase at the historical 1969–1971 to 1989–1991 rate, and that new cropland will, on average, be just as productive as cropland in 1993, estimates are that if net agricultural productivity does not increase, in 2050 cropland would have to increase globally by at least 121% or 1,753 Mha (an area equal to the combined land mass of China and Brazil) beyond the 1993 level of 1,448 Mha. Much of this would necessarily have to come from forested areas. On the other hand, a productivity increase of 1% per year from 1993 to 2050 would reduce additional cropland requirements to 368 Mha, while a productivity increase of 1.5% per year would give back 78 Mha of cropland to forests and other uses.

Such increases in productivity are plausible given that there are numerous existing-but-unused opportunities to enhance productivity in an environmentally sound manner, and that technological change has yet to run its course. However, to capitalize on these opportunities and to increase productivity, it is essential to have economic growth in order to generate the investments needed for researching, developing, acquiring, and operating more productive technologies, as well as any additional infrastructure necessary for the efficient functioning of the food and agricultural sector. By 2050, an estimated $250 billion may have to be invested annually in developing countries' food and agricultural sectors.

To increase agricultural and forest productivity, some of these investments should be used to bolster R&D, for instance, on precision farming, integrated pest management, and methods to reduce post-harvest and end-use crop and timber losses. Greater emphasis should also be placed on increasing productivity in suboptimal conditions which might become more prevalent due to climate change, such as drought (due to higher temperatures and redistribution of precipitation), prolonged submergence, higher salinity (due to greater evaporation and saltwater intrusion in coastal agricultural areas), and higher carbon dioxide. And biotechnology can play a crucial role here.

Such measures, in addition to enhancing food security and limiting forest conversion, would also reduce CO_2 emissions, habitat loss, and fragmentation of the landscape which would otherwise add to the substantial existing barriers to "natural" adaptation (via migration and dispersion) of species if climate changes. Notably, Article II of the United Nations Framework Convention on Climate Change (FCCC) refers, among other things, to allowing ecosystems to adapt naturally to climate change. Finally, increased agricultural productivity would lower the demand for cropland, which would reduce land prices, thereby decreasing the costs of purchasing or reserving land for conservation, carbon sequestration, or both. Harkening back to the "last straw" metaphor, these measures are analogous to clipping each straw so that many more could be added in the future.

Another measure that would address both existing and future problems is freer and unsubsidized trade. Developing nations currently import 10% of the grain they consume. Their future food deficits are expected to grow because their increases in food demand are expected to outstrip their increases in productivity, and could be further aggravated under global warming. On the other hand, climate change could increase developed countries' surpluses. Any increase in developing countries' food deficits due to climate change can be addressed in exactly the same way as current imbalances in production (and comparative advantage) are addressed today, namely, through trade. In the parlance of the last straw metaphor, trade allows the entire burden to be shared among two or more camels. Trade allows surpluses to flow voluntarily to deficit areas. But to be able to afford such trade, developing countries will need to "grow" the nonfood sectors of their economies. This is yet another reason for increasing economic growth, particularly in developing countries.

The second complementary approach to addressing social and environmental problems — whether they arise today, tomorrow or the day after, or whether they are due to climate change or another agency — would be to reduce the vulnerability of society by a generic increase in its resilience to adversity.

If we look around the world today, we find that almost every indicator of human or environmental well-being improves with wealth. Poorer countries are hungrier and more malnourished; their air and water is more polluted; and they are more prone to death and disease from climate-sensitive infectious and parasitic diseases. Consequently, they have higher mortality rates and lower life expectancies. This is because they are more vulnerable to any adversity since they are short on the fiscal and human capital resources needed to create, acquire and use new and existing technologies to cope with that adversity. Similarly, poorer countries are expected to have the greatest vulnerability to climate change, not because their climate change is expected to be larger, but because they lack the resources to adapt adequately. Thus, economic growth, by enhancing technological change, would make society more resilient and less vulnerable to adversity in general, and to climate change in particular. Just as strengthening a camel's back increases its ability to cope with a heavier load regardless of its contents, so does economic growth strengthen a society's ability to cope with adversity regardless of its cause.

25.7.4 CONCLUSIONS: THE PROBLEM OF THE LAST STRAW

Although for the next several decades, climate change may not be as urgent as other environmental and public health problems, it could be the proverbial last straw, particularly for natural ecosystems and biodiversity. However, given the magnitude of the impacts of climate change compared to other environmental problems, it may be futile to intercept that last straw if, as seems more likely, the cumulative load proves to be unsustainable well before that last straw drops.

Beyond focusing on that straw alone, there are at least three approaches to dealing with the camel's burden: (1) reduce the cumulative burden by removing a few straws or shortening each straw; (2) strengthen its back so that it can bear a heavier burden; and (3) spread the burden among more camels.

The first approach calls for measures to solve current problems that may be exacerbated by climate change by targeting the root causes of those problems. Some of these measures are also capable of addressing multiple problems. For instance, increasing agricultural and timber productivity would also reduce hunger, habitat loss, and CO_2 emissions.

The second approach calls for strengthening the institutions that support the forces of economic growth and technological change. These forces help perpetuate, if not set in motion, a virtuous cycle that helps solve environmental problems and reduces society's vulnerability to adversity in general. Specifically, economic growth provides the means for creating and affording cleaner, more productive technologies. In turn, technological change reinforces economic growth. Moreover, economic growth creates conditions conducive to a long-term reduction in population growth rates which, in turn, will help limit GHG emissions.

The third approach consists of unsubsidized trade. Because it facilitates movement of food from surplus to deficit areas, it is critical to global food security. It also discourages exploitation of marginal resources, helps disseminate new technologies, and boosts economic growth which, in turn, reinforces the second approach.

Thus, these approaches, used in tandem, would solve current problems, provide immediate and substantial benefits to humanity, limit GHG emissions, and reduce vulnerability and increase adaptability to climate and other environmental changes. By increasing adaptability, they would also raise the level at which atmospheric GHG concentrations may become, to use the terminology of the FCCC, "dangerous," which would give humanity and the rest of nature more breathing room and reduce the eventual costs of controlling GHG emissions. In fact, there can be no optimal approach to addressing climate change that does not explicitly incorporate the approaches outlined above.

25.8 Sustainable Consumption[3]

BY NORMAN MYERS
Honorary Visiting Fellow
Green College
Oxford University

In order to achieve all-round sustainability of our economies and lifestyles, we need to revise our consumption patterns. Hence the significance of two conferences to be held in Tokyo in late May 2000. The first, entitled "The Transition to Sustainability," will place much emphasis on consumption and is organized by the Inter-Academy Panel, composed of 80 scientific academies around the world and headed by the U.S. National Academy of Sciences. The second, "Business and Environment," is primarily a gathering of business leaders and is equally important because the corporate community holds an outsize key to achieving sustainable consumption.

A first step toward sustainable consumption is to recognize that consumption patterns will inevitably change in the future, if only by force of environmental circumstance — notably global warming, among a host of environmental problems. As that future arrives, we must ensure that there is an increase in consumption by the three billion people with incomes of less than $3 per day. At the same time, the 800 million people in developing and transition countries who earn enough to move into the high-consuming classes should be able to enjoy the fruits of their newfound affluence. How to enable them to do so without undue disruption of environmental systems, especially those of global scope such as the climate?

One answer may lie with the business sector, often fingered as the source of consumption-derived environmental problems. Were human communities to deploy all of the ecotechnologies that are already available from innovative business (such as energy efficiency, pollution controls, waste management, recycling, cradle-to-cradle products, and zero-emissions industry), we could enjoy twice as much material welfare while consuming only half as many natural resources and causing only half as much pollution and waste (Hawken *et al.*, 1999). Ecotechnologies are now worth $600 billion per year, on a par with the global car industry. There are big profits ahead for truly enterprising businesses.

Three policy initiatives could promote the transition to sustainable consumption. We could abandon gross national product (GNP) as an indicator of economic well-being; it suggests to the consumer that our economies need take no account of sustainability. In the United States, per capita GNP rose by 49% during 1976–98, whereas per capita "genuine progress" (the economy's output with environmental and social costs subtracted and added

---

[3]This section is reprinted with permission from *Science*, Vol. 287, Mar. 31, 2000, p. 2419. Copyright 2000 American Association for the Advancement of Science.

weight given to education, health, etc.) declined by 30% (Cobb *et al.,* 1999). Several alternative indicators, such as Net National Product, are being developed by Canada, Britain, Sweden, the Netherlands, and Austria. Second, we could ensure that prices reflect all environmental and social costs. For example, U.S. society ultimately pays at least $6 to burn a gallon of gasoline (through pollution, road accidents, traffic congestion, etc.). Pricing gasoline realistically would curtail the excessive car culture and open up huge market demand for improved public transportation. Similar considerations apply to the true price of a hamburger, a shirt, and even a house.

Yet (and herein lies the third initiative) consumers are encouraged to practice environmental ignorance thanks to subsidies that, for example, support fossil fuels 10 to 15 times more than clean and renewable sources of energy such as solar energy and wind power. Were all these subsidies to be phased out and a marketplace with a level playing field established, the energy alternatives would soon become commercially competitive. There are hosts of other subsidies that promote the car culture, over-intensive agriculture, wasteful use of water, overlogging of forests, and overharvesting of marine fisheries. Worldwide they total almost $1.5 trillion, or twice as much as all military spending. They induce massive distortions in our economies and do massive harm to the environment. Although it will be hard to change consumption patterns, they may be more plastic than currently supposed. (For example, during a recent 20-year period, almost 40 million Americans gave up smoking.) Similarly, it may be possible to change many production patterns — the flip side of the consumption coin.

However hard it will be to live with the drastic changes required, it will not be so hard as to live in a world drastically impoverished by the environmental injuries of current consumption.

25.9 **Technology and Environmentalism**

By Braden Allenby
Environment, Health, and Safety Vice President
AT&T

If one were choreographing the dance between technology and environment, it would be a strange one indeed. Contrary to today's ahistorical postmodernism, the dance has existed since early humans began driving megafauna extinct in Australia; smelting metal in China, Greece, or Rome; or deforesting Europe and North Africa. Nor are end-of-pipe, command-and-control regulations a modern invention: As early as 1306, London adopted ordinances limiting the burning of coal for air quality reasons. Another example is the English Alkali Act of 1863, passed to control emissions of gaseous hydrochloric acid from the LeBlanc method for producing sodium carbonate. This act imposed a form of "best available control technology," or BACT, in the form of acid absorption towers designed by William Gossage.

Such examples, and their modern analogs, illustrate the continuing dialogue between environmentalism and technology. Environmentalism is a powerful movement for addressing simple and easily observed problems: clean air and water, waste-site cleanup, preservation of valued landscapes, protection of endangered species, and even bans on materials whose problematic impacts can be easily demonstrated (e.g., lead in gasoline). The successes achieved in these cases are real and important.

Surveys continually show strong political support for environmentalism, but this primarily extends to easily seen problems — one of the reasons much activism focuses on manufacturing, where environmental issues are both easily detected and relatively easily fixed. When issues involve complex systems with time cycles measured in decades or centuries rather than months, and where scientific uncertainty is relatively high, public support falls off rapidly. This explains in part why Americans strongly support environmentalism generally but are deeply split over the Kyoto process. This is an obvious complexity when attempting to craft long-term policies to address these fundamental environmental perturbations. A more basic problem arises from the evolution of environmentalism, which has generally positioned environmental issues as overhead — something to be taken care of only after primary missions are accomplished. In firms, this is represented by end-of-pipe technologies such as scrubbers or water-treatment plants, which are relatively independent of product or process design. The mental model behind this approach encourages simplistic, often ideological, approaches to complex problems and is problematic when applied to complex, real-world systems.

Consider, for example, the costs of a wrong decision in an end-of-pipe scenario: In general, it will result in a little wasted capital or an inadequate level of protection that, once recognized, can be fixed by a quick and simple switchout of the control technology. The penalties for being wrong are relatively small and easily fixed. End-of-pipe technology, in other words, is a simple system. But this changes when core technologies, linked with other systems and embedded in a complex cultural and economic matrix, are mandated. The complexity of these systems is orders of magnitude greater than with control technologies and the costs of wrong decisions, to the economy and to the environment, can be far greater. Thus, for example, if rather than a scrubber one mandates a change in chemistry of gasoline, one impacts not just a factory, but a complex technological system, and the costs of being wrong, as in the case of the required addition of MTBE to gasoline, can be billions of dollars and billions of gallons of unnecessarily contaminated groundwater.

This leads to what should be a simple rule for all: prior to encouraging a fundamental technological change, it should be standard practice to at least try to identify the resulting real-world impacts. This should be the case whether it is a firm, a government, or an NGO urging the change: it should apply to genetically modified organisms, to policies encouraging biomass plantations (and thus possible increased distortion of the nitrogen cycle), and to NGO demands to ban key industrial materials (What will replace them? And how can one know what is better unless one has answered this question?). In doing so, the complexity of the real world, not the simplistic language of ideology, should be our guide.

BIBLIOGRAPHY

Adriaanse, Albert, *et al.* (1997). *Resource Flows: The Material Basis of Industrial Economies.* Washington, DC: World Resources Institute.

Arrow, K., *et al.* (1995). *Science* **268,** 520; Pimental, D., *et al.* (1997). *BioScience* **47**(11), 747; Costanza, R. R., *et al.* (1997). *Nature* **387,** 253; Pearce, D. (1998). *Environment* **40**(2), 23.

Ayres, Robert U. (1998). *Turning Point: An End to the Growth Paradigm.* London: Earthscan Publications Ltd.

Bailey, C. (1988). "The Social Consequences of Tropical Shrimp Mariculture Development." *Ocean and Shoreline Management,* **11,** pp. 31–44.

Bailey, Ronald. (2000). *Earth Report 2000.* New York: McGraw Hill.

Baldwin, S., H. Geller, G. Dutt, and N. H. Ravindranath. (1985). "Improved Wood-burning Cookstoves — Signs of Success." *Ambio* **14,** 280–287.

Baldwin, S. F., (1987) *Biomass Stoves, Engineering Design, Development, and Dissemination.* Arlington, VA: Volunteers in Technical Assistance.

Barnes, D. F., K. Openshaw, K. Smith, and R. van der Plas. (1994). *What Makes People Cook with Improved Biomass Stoves?* World Bank Technical Paper No. 242, Energy Series.

Baron, David P. (2000). *Business and the Environment.* Upper Saddle River, NJ: Prentice Hall.

Bauer, Peter T. (1981). *Equality and the Third World and Economic Delusion.* Cambridge, MA: Harvard University Press, p. 206.

Bell, Daniel. (1999). *The Coming of Post-Industrial Society.* New York: Basic Books.

Berger, J. J. (1998). *Charging Ahead: The Business of Renewable Energy and What It Means for America.* Berkeley, CA: University of California Press, pp. 114–117.

Bertolucci, Michael. (1998, Sep. 9). Presentation at Hewlett-Packard Laboratories.

Billings, W. D. (1952). "The Environmental Complex in Relation to Plant Growth and Distribution." *Quarterly Review of Biology* **27,** 251–265.

Bittman, M. (1996). "Today's Fish: Straight from the Farm." *New York Times*, September 18.

Bloom, David E., and David Canning. (2000). "The Health and Wealth of Nations." *Science,* February 18, pp. 1207–1209.

Boserup, E. (1970). *Women's Role in Economic Development.* London: Allen and Unwin.

Botkin, Daniel, and Edward Keller. (1998). *Environmental Science.* New York: Wiley.

British Petroleum. (1999). *British Petroleum Statistical Review of World Energy.* London: British Petroleum Corporate Communications Services.

Bronk, Richard. (1998). *Progress and the Invisible Hand.* Boston: Little Brown.

Brower, M., and W. Leon. (1999). *The Consumer's Guide to Effective Environmental Choices.* New York: Three Rivers Press.

Brown, L. R. (1995). *Who Will Feed China?* New York: W. W. Norton.

Brown, Lester. (2000). *State of the World 2000.* New York: Norton.

Buchanan, Scott. (1982). Reported in *So Reason Can Rule.* New York: Farrar, Strauss, Giroux, p. 12.

Bunch, D., M. Bradley, T. Golob, and R. Kitamura. (1993). "Demand for Clean-Fuel Vehicles in California: A Discrete-Choice Stated Preference Pilot Project," *Transportation Research A,* **27A**(3), 237–253.

Campbell, Colin, and Jean Laherrere. (1998). "The End of Cheap Oil." *Scientific American,* March, pp. 78–83.

Cassman, K.G. (1999). "Ecological Intensification of Cereal Production Systems: Yield Potential, Soil Quality, and Precision Agriculture," *Proceedings of the National Academy of Science* **96,** 5952–5959.

Casten, Thomas. (1998). *Turning Off the Heat.* New York: Prometheus.

Chen, B. H., C. J. Hong, M. R. Pandey, and K. R. Smith. (1990). *World Health Statistics Quarterly,* **43,** 127–138.

Ciriacy-Wantrup. (1968). *Resource Conservation: Economics and Policies,* 3rd ed. Berkeley, CA: University of California Press.

Clark, Colin. (1985). [Book review] *Population and Development Review,* **11,** 120.

Clayton, Anthony, and Nicholas Radcliffe. (1996). *Sustainability, A Systems Approach.* Boulder, CO: Westview Press.

Cobb, C., G. S. Goodman, and M. Wackernagel. (1999). *Why Bigger Isn't Better: The Genuine Progress Indicator.* San Francisco: Redefining Progress.

Cohen, Joel E. (1995). *How Many People Can the Earth Support?* New York: W. W. Norton.

Comiskey, B., J. D. Albert, H. Yoshizawa, and J. Jacobson. (1998). "An Electrophoretic Ink for All-Printed Reflective Electronic Displays." *Nature* **394,** 253–255.

Conway, Gordon. (1999). *The Doubly Green Revolution.* Ithaca, NY: Cornell University Press.

Conway-Schempf, N., and L. Lave. (1996). "Pollution Prevention through Green Design," *Pollution Prevention Review* **Winter,** 11–20.

Coppock, R. (1998). "Implementing the Kyoto Protocol," *Issues in Science and Technology* **14**(3), 66–74.

Costanza, R., ed. (1991). *Ecological Economics: The Science and Management of Sustainability.* New York: Columbia University Press.

Costanza, R., and L. Cornwell. (1992). "The 4P Approach to Dealing with Scientific Uncertainty." *Environment* **34,** 12–20.

Costanza, R., and H. E. Daly. (1987). "Toward an Ecological Economics." *Ecological Modeling* **38**, 1–7.

Costanza, R., and H. E. Daly. (1992). "Natural Capital and Sustainable Development." *Conservation Biology* **6**, 37–46.

Costanza, R., H. E. Daly, and J. A. Bartholomew. (1991). "Goals, Agenda and Policy Recommendations for Ecological Economics," in *Ecological Economics: The Science and Management of Sustainability* (R. Costanza, ed.). New York: Columbia University Press.

Costanza, R., L. Wainger, C. Folke, and K.-G. Maler. (1993). "Modeling Complex Ecological Economic Systems: Toward an Evolutionary, Dynamic Understanding of People and Nature." *BioScience* **43**, 545–555.

Daily, G. C., ed. (1997). *Nature's Devices: Societal Dependence on Natural Systems.* Washington, DC: Island Press.

Daily, Gretchen C., Ann H. Ehrlich, and Paul R. Ehrlich. (1994, July). "Optimum Human Population Size." *Population and the Environment* **15**(6), 469–475.

Dalla Costa, John. (1998). *Ethical Imperative.* Reading, MA: Addison-Wesley.

Daly, Herman. (1996). *Beyond Growth: The Economics of Sustainable Development.* Boston: Beacon Press.

Daly, Herman E., and John Cobb. (1989). P. 109 in *For the Common Good.* Boston: Beacon Press.

Dambach, B. F., and B. R. Allenby. (1995). "Implementing Design for Environment at AT&T." *Total Quality Environmental Management* **4**, 51–62.

Davis, M. G. (1986). "Climatic Instability, Time Lags, and Community Disequilibrium," in *Community Ecology* (J. Diamond and T. J. Case, eds.). New York: Harper & Row.

DeSimone, Livio, and Frank Popoff. (1997). *Eco-Efficiency.* Cambridge, MA: The MIT Press.

Dillon, P., and K. Fischer. (1992). *Environmental Management in Corporations: Methods and Motivations.* Medford, MA: Tufts University Press.

Directed Technologies, Inc., Air Products and Chemicals, BOC Gases, The Electrolyser Corp., and Praxair, Inc. (1997, July). *Hydrogen Infrastructure Report.* Prepared for Ford Motor Company under USDOE Contract No. DE-AC02-94CE50389.

Dougherty, D., and C. Hardy. (1996). Sustained product innovation in large, mature organizations: Overcoming innovation-to-organization problems, *Acad. Mgt. J.,* **39**(5), 1120-1153.

Economist (The), October 30, 1990.

Economist (The). "A Century of Progress." April 15, 2000, p. 86.

Elkington, J. (1994). "Towards the Sustainable Corporation: Win-Win-Win Business Strategies for Sustainable Development," *California Management Review* **36**(2), 67, 91.

Elkington, John. (1998). *Cannibals with Forks.* Gabriola Island, BC, Canada: New Society Publishers.

Ellul, Jacques. (1973). *The Technological Society.* New York: Random House.

Esty, D. C. (1994). *Greening the GATT: Trade, Environment, and the Future.* Institute for International Economics, p. 10, footnote 2.

Ettlie, J., W. P. Bridges, and R. D. O'Keefe. (1984). "Organization Strategy and Structural Differences for Radical vs. Incremental Innovation." *Management Science* **30**(6), 682–695.

Ezzati, M., D. M. Kammen, and B. M. Mbinda. (2000). "Comparison of Emissions and Residential Exposure from Traditional and Improved Cookstoves in Kenya." *Environmental Science & Technology,* in press.

Fiksel, J. (1996). *Design for Environment: Creating Eco-Efficient Products and Processes.* New York: McGraw-Hill.

Financial Times, "Busting Through with Ballard," Winter 1998, pp. 40–45.

Florida, R. (1996). "Lean and Green: The Move to Environmentally Conscious Manufacturing." *California. Management Review* **39,** 80–105.

Florida, R., and M. Atlas. (1997). *Report of Field Research on Environmentally-Conscious Manufacturing In the United States.* Pittsburgh, PA: Carnegie Mellon University.

Food and Agriculture Organization. (1991). *Food Balance Sheets.* Rome: Food and Agriculture Organization of the United Nations.

Food and Agriculture Organization (1993). *Aquaculture Production 1985–1991.* Rome: Food and Agriculture Organization of the United Nations.

Food and Agriculture Organization (1997). *The State of World Fisheries and Aquaculture.* Rome: Food and Agriculture Organization of the United Nations.

Fussler, Claude. (1996). *Driving Eco-Innovation.* London: Pitman Publishing.

Georg, S., I. Ropke, and U. Jorgensen. (1992). "Clean Technology — Innovation and Environmental Regulation." *Environmental Resource Econ.* **2,** 533–550.

Giampietro, M., and D. Pimentel. (1993). In *The Tightening Conflict: Population, Energy Use, and the Ecology of Agriculture* (L. Grant, ed.). Teaneck, NJ: Negative Population Growth, Inc.

Gilliland, M. W. (1977). "Energy Analysis: A Tool for Evaluating the Impact of End Use Management Strategies on Economic Growth." Paper presented at International Conference on Energy Use Management, Tucson, AZ.

Gleick, P. H. (1993). *Water in Crisis.* New York: Oxford University Press.

Global Leaders for Tomorrow. (2000). "Pilot Environmental Sustainability Index." Yale Center for Environmental Law, January 31.

Goklany, I. M. (1998). "The Importance of Climate Change Compared to Other Global Changes," pp. 1024–1041 in *Proceedings of the Second International Specialty Conference: Global Climate Change — Science, Policy, and Mitigation/Adaptation Strategies,* October 13–16, 1998, Crystal City, VA. Sewickley, PA: Air & Waste Management Association.

Goklany, I. M. (2000). "Potential Consequences of Increasing Atmospheric CO_2 Concentration Compared to Other Environmental Problems." *Technology,* in press.

Goldemberg, J., T. B. Johansson, A. K. N. Reddy, and R. H. Williams. (1985). "Basic Needs and Much More with One Kilowatt per Capita." *Ambio* **14,** 190–200.

Goodstein, Eban. (1999). *Economics and the Environment,* 2nd ed. New York: John Wiley and Sons.

Graedel, T. E., and B. R. Allenby. (1995). *Industrial Ecology.* Upper Saddle River, NJ: Prentice Hall.

Grainger, R. J. R., and S. M. Garcia. (1996). FAO Fisheries Technical Paper 359. Rome: Food and Agriculture Organization of the United Nations.

Gramlich, E. M. (1992). *A Guide to Benefit-Cost Analysis.* Upper Saddle River, NJ: Prentice Hall.

Green, S., M. Gavin, and L. Aiman-Smith. (1985). "Assessing a Multidimensional Measure of Radical Technological Innovation." *IEEE Transactions on Eng. Management* **42**(3), 203–214.

Greising, D. (1994, Aug. 8). "Quality: How to Make It Pay." *Business Week,* pp. 54–59.

Griffin, A., and J. R. Hauser. (1983). "The Voice of the Customer." *Marketing Science* **12**(1), 1–27.

Gujja, B., and A. Finger-Stich. (1996). "Shrimp Aquaculture's Impact in Asia." *Environment*, September, pp. 12–39.

Haerdter, R., W. Y. Chow, and O. S. Hock. (1997). "Intensive Plantation Cropping, a Source of Sustainable Food and Energy Production in the Tropical Rain Forest Areas in Southeast Asia." *Forest Ecology and Management* **91**(1), 93–102.

Hall, Charles A., Cutler J. Cleveland, and Robert Kaufmann. (1986). *Energy and Resource Quality: The Ecology of the Economic Process.* New York: John Wiley.

Halvorson, T., and P. Farris. (1997). "Onsite Hydrogen Generator for Vehicle Refueling Application," *Proceedings of the '97 World Car Conference,* January 19–22, 1997, Riverside, CA, pp. 331–338.

Hardin, G. (1968). "The Tragedy of the Commons," *Science* **162**, 1243–1248.

Hardin, Garrett. (1985). *Filters Against Folly.* New York: Viking Press, Chap. 3.

Hardin, Garrett. (1990). "An Ecological View of Ethics," in *The Church and Contemporary Cosmology* (James B. Miller and Kenneth E. McCall, eds.). Pittsburgh, PA: Carnegie Mellon University Press, p. 345.

Hardin, Garrett. (1991). "Paramount Positions in Ecological Economics," in *Ecological Economics* (Robert Costanza, ed.). New York: Columbia University Press, pp. 47–57.

Hardin, Garrett. (1999). *The Ostrich Factor: Our Population Myopia.* New York: Oxford University Press, Chap. 4.

Harris, J. M. (1996). "World Agricultural Futures: Regional Sustainability and Ecological Limits." *Ecological Economics* **17**(2), 95–115.

Harrop, G. (1995). *The Future of the Electric Vehicle: A Viable Market?* London: Pearson Professional, pp. 109–110.

Hart, Stuart. (1997). "Beyond Greening." *Harvard Business Review,* January, pp. 66–76.

Hart, Stuart, and Mark Milstein. (1999). "Global Sustainability and the Creative Destruction of Industries." *Sloan Management Review,* pp. 23–33.

Hauser, J. R., and D. Clausing. (1988). "The House of Quality." *Harvard Business Review* **66**(3), 63–73.

Hawken, P., A. B. Lovins, and L. H. Lovins. (1999). *Natural Capitalism: The Next Industrial Revolution.* Boston: Little, Brown.

Hayek, F. A. (1945). "The Use of Knowledge in Society." *American Economic Review* **35**(4), 519–530.

Hazeltine, Barrett, and Christopher Buel. (1998). *Appropriate Technology.* San Diego: Academic Press.

Hendrickson, C. T., and F. C. McMichael. (1992). "Product Design for the Environment." *Environmental Science and Technology* **26,** 844.

Herring, H. B. (1994). "900,000 Striped Bass, and Not a Fishing Pole in Sight." *New York Times*, November 6.

Heywood, V. H., ed. (1995). *Global Biodiversity Assessment.* Cambridge: Cambridge University Press; Stanners, D., and P. Bourdeau. (1995). *Europe's Environment: The Dobris Assessment.* Copenhagen: European Environment Agency; Middleton, N. J., and D. S. G. Thomas. (1997). *World Atlas of Desertification.* U.N. Environment Program. New York: Edward Arnold; Alexandratos, N., ed. (1995). *World Agriculture: Towards 2010, An FAO Study.* Chichester, UK: Wiley, and Rome:

Food and Agriculture Organization of the United Nations; *A 2020 Vision for Food, Agriculture, and the Environment.* (1995). Washington, DC: International Food Policy Research Institute.

Hoffman, Andrew J. (1997). *From Heresy to Dogma: An Institutional History of Environmentalism.* San Francisco: New Lexington Press.

Hoffman, Banesh. (1972). *Albert Einstein, Creator and Rebel.* New York: Columbia University Press, p. 12.

Horrigan, J. B., F. H. Irwin, and E. Cook. (1998). *Taking a Byte out of Carbon: Electronics Innovation for Climate Protection.* Washington, DC: World Research Institute.

Horvath, A., C. Hendrickson, L. Lave, F. McMichael, and T.-S. Wu. (1995). "Toxics Emissions Indices for Green Design and Inventory." *Environmental Science and Technology* **29,** 86–90.

Howard, Alice, and Joan Magretta. (1995). "Surviving Success." *Harvard Business Review,* September–October, pp. 110–118.

Howarth, Richard B. (1996). "Status Effects and Environmental Externalities," *Ecological Economics* **16**(1), 25–34.

Hubbert, M. King. (1971, Sep.). "The Energy Resources of the Earth." *Scientific American* **225**(3), 60–70.

Huber, Peter. (1999). *Hard Green.* New York: Basic Books.

Hunt, C., and E. Auster. (1990). "Proactive Environmental Management: Avoiding the Toxic Trap." *Sloan Management Review* **Winter,** 7–18.

IEEE. (1999, Feb.). "IEEE Recommended Practice for Utility Interface of Residential and Intermediate Photovoltaic (PV) Systems," IEEE Standards Coordinating Committee 21, Photovoltaics, Draft 10.

Industry, Fresh Water and Sustainable Development. (1997). Geneva: World Business Council for Sustainable Development.

IPCC. (1996a). *Climate Change 1995: The Science of Climate Change.* Oxford: Cambridge University Press.

IPCC. (1996b). *Climate Change 1995: Impacts, Adaptations and Mitigation of Climate Change.* Oxford: Cambridge University Press.

Kahn, Herman (with Julian Simon). ed. (1984). *The Resourceful Earth.*

Kahn, Herman, William Brown, and L. Martel. (1976). *The Next 200 Years — A Scenario for America and the World.* New York: William Morrow.

Kammen, D. M. (1995a), "Cookstoves for the Developing World," *Scientific American* **273**, 72–75.

Kammen, D. M. (1995b), "From Energy Efficiency to Social Utility: Improved Cookstoves and the *Small Is Beautiful* Model of Development," in *Energy as an Instrument for Socio-Economic Development* (J. Goldemberg and T. B. Johansson, eds.). New York: United Nations Development Programme, pp. 50–62.

Kanter, Rosabeth. (1999). "From Spare Change to Real Change." *Harvard Business Review,* May–June, pp. 127–132.

Kaplan, G. (2000). "Industrial Electronics." *IEEE Spectrum,* January, pp. 104–109.

Kaplan, Robert, and David Norton. (1996). *The Balanced Scorecard.* Cambridge, MA: Harvard Business School Press.

Kneese, Allen V. (1988). "The Economics of Natural Resources." *Population and Development Review* **14**(suppl.): 289, 302.

Knight, F. H. (1921). *Risk, Uncertainty and Profit.* New York: Harper and Row.

Korten, David C. (1999). *The Post-Corporate World.* San Francisco: Berrett-Kochler, 1999.

Krugmann, H. (1987). *Review of Issues and Research Relating to Improved Cookstoves.* IDRC-MR 152e. Ottawa, Canada: International Development Research Centre.

Krupnick, Alan, and Dallas Burtraw. (1996). "The Social Costs of Electricity: Do the Numbers Add Up?" *Resource and Energy Economics* **18,** 423–466.

Kurani, K., T. Turrentine, and D. Sperling. (1994). "Demand for Electric Vehicles in Hybrid Household: An Exploratory Analysis." *Transport Policy* **1**(4), 244–256.

Laitner, Skip, Stephen Bernow, and John DeCicco (1998). "Employment and Other Macroeconomic Benefits of an Innovation-Led Climate Strategy for the United States." *Energy Policy* **26**(5), 425–432.

Lal, R., and B. A. Stewart. (1990). *Soil Degradation.* New York: Springer-Verlag.

Lave, L., E. Cobas-Flores, C. Hendrickson, and F. C. McMichael. (1995). "Using Input–Output Analysis to Estimate Economy-Wide Discharges." *Environmental Science and Technology* **29,** 420A–426A.

Lawrence, A., and D. Morell. (1995). "Leading-Edge Environmental Management: Motivation, Opportunity, Resources, and Process," in *Research in Corporate Social Performance and Policy, Supplement 1* (J. Post, D. Collins, and M. Starik, eds.). Greenwich, CT: JAI Press, pp. 99–126.

Layard, R., ed. (1977). *Cost-Benefit Analysis.* New York: Penguin Books.

Leonard-Barton, D. (1990). "Core Capabilities and Core Rigidities: A Paradox in Managing New Product Development." *Strategic Management Journal* **13,** 111–125.

Lipman, T. E., and M. A. Delucchi. (1996). "Hydrogen-Fueled Vehicles." *International Journal of Vehicle Design* **17**(5/6), Special Issue.

Lipp, S., G. Pitts, and F. Cassidy, eds. "A Life Cycle Environmental Assessment of a Computer Workstation," in *Environmental Consciousness: A Strategic Competitiveness Issue for the Electronics and Computer Industry.* Austin, TX: Microelectronics and Computer Technology Corporation.

Lloyd, Jennifer. (2000). "Conservationist John C. Sawhill." *Investors Daily,* May 24, p. 7.

Los Alamos National Laboratory, *Fuel Cells — Green Power,* p. 24-25. Available at *www.education.lanl.gov/resources/fuelcells.*

Lovins, Amory, L. H. Lovins, and Paul Hawken. (1999). "A Road Map for Natural Capitalism." *Harvard Business Review,* June, pp. 145–158.

Lubchenco, J. (1998). "Entering the Century of the Environment: A New Social Contract for Science," *Science* **279,** 491–497.

Lynn, G. S., J. G. Morone, and A. S. Paulson. (1996). "Marketing and Discontinuous Innovation: The Probe and Learn Process." *California Management Review* **38**(3), 8–37.

MacKenzie, J. (1994). *The Keys to the Car: Electric and Hydrogen Vehicles in the 21st Century,* New York: World Resources Institute.

Magretta, John. (1997) "Growth through Global Sustainability." *Harvard Business Review,* January, pp. 79–88.

Makofske, W. J., and E. F. Karlin, eds. (1995). *Technology and Global Environmental Issues.* New York: HarperCollins.

Makower, J. (1993). *The e Factor: The Bottom-Line Approach to Environmentally Responsible Business.* New York: Times Books.

Mankiw, N. Gregory. (1997). *Principles of Economics.* Orlando, FL: The Dryden Press.

Margulis, L., and D. Sagan. (1999). "Second Nature: Welcome to the Machine." *UMASS* **4**(1), 24–29.

Markels, M. (1995). "Fishing for Markets: Regulation and Ocean Farming." *Regulation*, **3.**

Markvart, T., ed. (1994). *Solar Electricity,* Chichester, UK: John Wiley & Sons.

Masters, Gilbert. (1997). *Introduction to Environmental Engineering and Science.* Upper Saddle River, NJ: Prentice Hall.

May, R. M. (1978). "The Evolution of Ecological Systems." *Scientific American* **239**(3), 160–175.

McDermott, C. M., and R. Handfield. (1997). *The Parallel Approach to New Product Development and Discontinuous Innovation.* Working paper. Troy, NY: The Lally School of Management and Technology, Rensselaer Polytechnic Institute.

McMichael, A. J., *et al.* (1996). "Human Population Health," pp. 561–584 in *Climate Change 1995: Impacts, Adaptations and Mitigation of Climate Change.* Oxford: Cambridge University Press.

Meeks, F. (1990). "Would You Like Some Salmon with Your Big Mac?" *Forbes,* December 24.

Messenger, R., and J. Ventre. (1999). *Photovoltaic Systems Engineering.* Boca Raton, FL: CRC Press.

Miringhoff, Marc L., and Marque Luisa Miringhoff. (1999). *The Social Health of the Nation: How America Is Really Doing.* New York: Oxford University Press.

Mishan, E. J. (1968). *The Costs of Economic Growth.* London: Staples Press.

Mitchell, Emily. (2000). "Getting Better at Doing Good." *Time,* February 21, pp. B9–B12.

Moore, Taylor. (1999). "Wind Power: Gaining Momentum." *EPRI Journal* **Winter,** 8.

Moravec, Hans. (1999). "Rise of the Robots." *Scientific American,* December, pp. 124–135.

Morone, J. G. (1993). *Winning in High-Tech Markets.* Boston: Harvard Business School Press.

Morrison, Phillip, and Kosta Tsipis. (1998). *Reason Enough to Hope.* Cambridge, MA: The MIT Press.

Munk, N. (1995). "Real Fish Don't Eat Pellets." *Forbes,* January 30.

Murray, H. (1995). "Just Compensation." *Far Eastern Economic Review,* March 9, pp. 14–15.

Musgrave, R. A. (1969). "Cost-Benefit Analysis and the Theory of Public Finance." *J. Econ. Lit.* **7,** 797–806.

Mydans, S. (1996). "Thai Shrimp Farmers Facing Ecologists' Fury." *New York Times,* April 28.

Myers, N. (1994). "Tropical Deforestation: Rates and Patterns," in *The Causes of Tropical Deforestation* (K. Brown and D. W. Pearce, eds.). Vancouver, BC: University of British Columbia Press, pp. 27–41.

Myers, Norman, and Julian L. Simon. (1994). *Scarcity or Abundance?* New York: Norton, pp. 133–134.

Myers, R. A., N. J. Barrowman, J. A. Hutchings, and A. A. Rosenberg. (1995). "Population Dynamics of Exploited Fish Stocks at Low Population Levels." *Science,* **269,** pp. 1106–1108.

Nadler, D., and M. L. Tushman. (1990). "Beyond the Charismatic Leader: Leadership and Organizational Change." *California Management Review* **32**(2), 77–97.

National Academy of Sciences. (1994). *Population Summit of The World's Scientific Academies.* Washington, DC: National Academy of Sciences Press.

National Fire Protection Association. (1998). *NFPA 70 National Electrical Code, 1999 Edition.* Quincy, MA: NFPA.

Naylor, R. L., *et al.* (1998). *Science* **282,** 883.

Nesheim, M. C. (1993). "Human Nutrition Needs and Parasitic Infections," in *Parasitology: Human Nutrition and Parasitic Infection* (D. W. T. Crompton, ed.). Cambridge: Cambridge University Press, pp. S7–S18.

Nkonoki, S. R., and B. Sorensen. (1984). "A Rural Energy Study in Tanzania: The Case of Bundilya Village." *Natural Resources Forum,* **8,** 51–62.

Nohria, N., and R. Gulati. (1996). "Is Slack Good or Bad for Innovation?" *Acad. Management Journal* **39**(5), 1245–1264.

Noori, H., and R. Radford. (1990). *Management of Manufacturing Technologies,* New York: McGraw Hill.

Nordhaus, William D., and Edward C. Kokkelenberg. (1999). *Nature's Numbers.* Washington, DC: National Academy Press.

Office of Technology Assessment. (1981). *An Assessment of Technology for Local Development.* Washington, DC: U.S. Government Printing Office.

Office of Technology Assessment. (1992). *Green Products by Design.* Washington, DC: U. S. Government Printing Office.

Office of the President of the United States. (1997). *Report to the President on Federal Energy Research and Development for the Challenges of the Twenty-First Century.* Washington, DC: Government Printing Office.

Ogden, J. (1999a). "Developing a Refueling Infrastructure for Hydrogen Vehicles: A Southern California Case Study." *International Journal of Hydrogen Energy* **24,** 709–730.

Ogden, J. (1999b). "Prospects for Building a Hydrogen Energy Infrastructure," *Annual Reviews of Energy and the Environment,* **24,** 227–279.

Ogden, J., T. Kreutz, and M. Steinbugler. (1998, Oct.). "Fuels for Fuel Cell Vehicles: Vehicle Design and Infrastructure Issues," Society of Automotive Engineers Technical Paper No. 982500.

Ogden, J., M. Steinbugler, and T. Kreutz. (1999). "A Comparison of Hydrogen, Methanol and Gasoline as Fuels for Fuel Cell Vehicles." *Journal of Power Sources* **79,** 142–168.

Openshaw, K. (1979). "A Comparison of Metal and Clay Charcoal Cooking Stoves." Paper presented at the Conference on Energy and Environment in East Africa, Nairobi, Kenya.

Openshaw, K. (1982). *The Development of Improved Cooking Stoves for Urban and Rural Households in Kenya.* Stockholm, Sweden: Beijer Institute/Royal Swedish Academy of Sciences.

Oregon Progress Board. *www.econ.state.or.us/opb/benchmark.tables/index.htm.*

Organization for Economic Cooperation and Development. (1997) *Economic Globalization and the Environment* and *Sustainable Development: OECD Policy Approaches for the 21st Century.*

Ottman, Jacquelyn. (1993). *Green Marketing.* Lincolnwood, IL: NTC Business Books.

Pacey, Arnold. (1992). *The Maze of Ingenuity.* Cambridge, MA: The MIT Press.

Paoletti, M. G., and D. Pimentel. (1996). "Genetic Engineering in Agriculture and the Environment." *BioScience* **46**(9), 665–673.

Patel, M. R. (1999). *Wind and Solar Power Systems.* Boca Raton, FL: CRC Press, Chap. 10.

Pederson, S., C. Wilson, and B. Stotesbery, eds. (1996). *Electronics Industry Roadmap.* Austin, TX: Microelectronics and Computer Technology Corporation.

Peirce, William. (1996). *Economics of the Energy Industries.* Westport, CT: Praeger.

Perrings, Charles. (1987). *Economy and Environment.* Cambridge: Cambridge University Press, p. 48.

Peterson, G., C. R. Allen, and C. S. Holling. (1998). "Ecological Resilience, Biodiversity, and Scale." *Ecosystems* **1**, 6–18.

Pezzey, John. (1992). "Sustainability: An Interdisciplinary Guide," *Environmental Values,* 1–4, 321–362.

Pimentel, D., and M. Pimentel. (1996). *Food, Energy and Society.* Niwet, CO: Colorado Press.

Pimentel, D., *et al.* (1994). "Natural Resources and Optimum Human Population." *Population and the Environment* **15**(5), 347–369.

Pimentel, D., C. Harvey, P. Resosudarmo, K. Sinclair, D. Kurz, M. McNair, S. Crist, L. Sphritz, L. Fitton, R. Saffouri, and R. Blair. (1995). "Environmental and Economic Costs of Soil Erosion and Conservation Benefits." *Science* **267**, 1117–1123.

Pimentel, D., C. Wilson, C. McCullum, R. Huang, P. Dwen, J. Flack, Q. Tran, T. Saltman, and B. Cliff. (1996). "Economic and Environmental Benefits of Biodiversity." *BioScience.* **47**(11), 747–757.

Pimentel, D., J. Houser, E. Preiss, O. White, H. Fang, L. Mesnick, T. Barsky, S. Tariche, J. Schreck, and S. Alpert. (1997). "Water Resources: Agriculture, the Environment, and Society." *BioScience* **47**(2), 97–106.

Pimentel, D., O. Bailey, P. Kim, E. Mullaney, J. Calabrese, L. Walman, F. Nelson, and X. Yao. (1999). "Will Limits of the Earth's Resources Control Human Numbers? *Environment, Development, and Sustainability* **1**(1), 19–39.

Pinstrup-Andersen, P., *et al.* (1997). *The World Food Situation: Recent Developments, Emerging Issues and Long-Term Prospects.* Food Policy Statement 26. Washington, DC: International Food Policy Research Institute.

Population Reference Bureau. (1998). *World Population Data Sheet.* Washington, DC: PRB.

Porter, M. E., and C. van der Linde. (1995). "Green and Competitive: Ending the Stalemate." *Harvard Business Review* **73**, 120–134.

Porter, Michael E. (1995). "Green and Competitive." *Harvard Business Review,* September, pp. 120–133.

Portney, Paul, and John Weyant, eds. (1999). *Discounting and Intergenerational Equity.* Washington, DC: Resources for the Future.

Postel, S. (1992). *Last Oasis: Facing Water Scarcity.* New York: W. W. Norton.

Postel, S. (1999a). *Pillar of Sand: Can the Irrigation Miracle Last?* New York: W. W. Norton.

Postel, Sandra. (1999b). "When the World's Wells Run Dry." *World Watch,* September, pp. 30–38.

Postel, S. L., G. C. Daily, and P. R. Ehrlich. (1996). "Human Appropriation of Renewable Fresh Water." *Science* **271** 785–788.

Postrel, Virginia. (1998). *The Future and Its Enemies.* New York: The Free Press.

Potts, Malcolm. (2000). "The Unmet Need for Family Planning." *Scientific American,* January, pp. 89–93.

Prahalad, C. K., and G. Hamel. (1999, May–June). "The Core Competence of the Corporation." *Harvard Business Review,* 79–91.

Quandt, C. O. (1995). "Manufacturing the Electric Vehicle: A Window of Technological Opportunity for Southern California." *Environment and Planning A* **27**(6), 835–862.

Read, P. (1994). *Responding to Global Warming.* Zed Books Limited.

Reddy, A. K. N., R. H. Williams, and T. B. Johansson. (1996). *Energy after Rio: Prospects and Challenges.* New York: United Nations Publications.

Reilly, J., *et al.* (1996). "Agriculture in a Changing Climate: Impacts and Adaptations," pp. 427–467 in *Climate Change 1995: Impacts, Adaptations and Mitigation of Climate Change.* Oxford: Cambridge University Press.

Reilly, William K. (1999). "Private Enterprises and Public Obligations." *California Management Review,* Summer, pp. 17–26.

Reinhardt, Forest L. (2000). *Down To Earth.* Boston: Harvard Business School Press.

Ristenen, Robert. (1999). *Energy and the Environment.* New York: Wiley.

Roberts, L., ed. (1998). *World Resources 1998–1999.* New York: Oxford University Press.

Rohatgi, A., S. Narasimha, *et al.* (1996). In *Proceedings of the 25th IEEE PV Spec. Conference,* p. 741.

Sagoff, Mark. (1995). "Carrying Capacity and Ecological Economics," *BioScience* **45**(9), 610–619.

Sailor, W.C., *et al.* (2000). "A Nuclear Solution to Climate Change?" *Science,* May 19, p. 1177.

Scheffler, Samuel, ed. (1988). *Consequentialism and Its Critics.* New York: Oxford University Press.

Schelling, T. C. (1997, Nov.–Dec.). "The Cost of Combating Global Warming." *Foreign Affairs,* pp. 8–14.

Schmidheiny, Stephen, and Frederico Zorraquin. (1998). *Financing Change.* Cambridge, MA: The MIT Press.

Schumacher, E. F. (1973). *Small Is Beautiful.* New York: Harper and Row.

Schwartz, Peter, *et al.* (1999). *The Long Boom.* Reading, MA: Perseus Books.

Science. (1995). "Cities as Disease Vectors," p. 270.

Sen, Amartya. (1999a). *Development as Freedom.* New York: Knopf.

Sen, Amartya. (1999b). "Economics—The Ethical Dimension." *Scientific American,* January, p. 19.

Serageldin, I. (1999). "Biotechnology and Food Security in the 21st Century." *Science* **285,** 387–389.

Service, R. (1998). "Miniaturization Puts Chemical Plants Where You Want Them." *Science* **282,** 400.

Shaw, George Bernard. *Man and Superman.* 1903. Act IV.

Sheats, J. R., H. A. Antoniadis, M. Hueschen, W. Leonard, J. Miller, R. Moon, D. Roitman, and A. Stocking (1996). "Organic Electroluminescent Devices." *Science* **273,** 884–888.

Sheridan, D. (1983, Feb.). "The Colorado — An Engineering Wonder without Enough Water." *Smithsonian,* pp. 45–54.

Shetty, P. S., and N. Shetty. (1993). "Parasitic Infection and Chronic Energy Deficiency in Adults." *Parasitology* **107**(suppl), S159–S167.

Shnayerson, M. (1996). *The Car that Could: The Inside Story of GM's Revolutionary Electric Vehicle.* New York: Random House.

Simon, Julian E., Calvin Beisner, and John Phelps, eds. (1995). *The State of Humanity.* Cambridge, MA: Blackwell Publishers Ltd.

Simon, Julian. (1980). *The Ultimate Resource.* Princeton, NJ: Princeton University Press.

Slywotzky, Adrian, and David Morrison. (1997). *The Profit Zone.* New York: Times Business.

Smil, Vaclav. (1996). *Environmental Problems in China: Estimates of Economic Costs.* East-West Center Special Report 5. Honolulu, HA: East-West Center.

Smil, Vaclav. (1999). *Energies*. Cambridge, MA: The MIT Press.

Smil, Vaclav. (2000). *Feeding the World*. New York: The Free Press.

Smith, A. (1776). *An Inquiry into the Nature and Causes of the Wealth of Nations*. Indianapolis, IN: Liberty Classics.

Smith, K. R. (1991). *The Hearth as System Central*. Draft ESMAP Report.

Smith, K. R. (1994). "The Health Impact of Cookstove Smoke in Africa," in *African Development Perspectives Yearbook 3* (Research Group on African Development Perspectives Bremen, eds.). Munster, Germany: LitVerlag, pp. 417–434.

Smith, K. R., G. Shuhua, H. Kun, and Q. Daxiong. (1993). "100 Million Biomass Stoves in China: How Was It Done?" *World Development* **18**, 941–961.

Snow, C. P. (1993). *The Two Cultures*. Cambridge: Cambridge University Press, pp. 1–51.

Solomon, F., *et al*. (1996). "Wood Production under Changing Climate and Land Use," pp. 492–496 in *Climate Change 1995: Impacts, Adaptations and Mitigation of Climate Change*. Oxford: Cambridge University Press.

Solow, Robert. (1974). "The Economics of Resources or the Resources of Economics." *Journal of Economic Literature* **6**, 11.

Southgate, D. (1992, November). "Shrimp Mariculture Development in Ecuador: Some Resource Policy Issues." Working Paper No. 5, Environment and Natural Resources Policy and Training Project, University of Wisconsin, Madison.

Sperling, D., and Susan Shakeen. (1995). "Transportation and Energy." *American Council for an Energy Efficient Economy*. Washington, DC: Government Printing Office.

Steinbugler, M. (1996). "How Far, How Fast, How Much Fuel: Evaluating Fuel Cell Vehicle Configurations." Paper presented at the Commercializing Fuel Cell Vehicles Conference, Intertech Conferences, September 17–19, 1996, Hyatt Regency O'Hare, Chicago.

Stewart, Thomas. (1997). *Intellectual Capital*. New York: Doubleday.

Strong, Maurice (Chairman). (1992). *Agenda 21*. Rio de Janeiro: United Nations Commission on Environment & Development.

Swift, B. (1998). "A Low-Cost Way to Control Climate Change." *Issues in Science and Technology* **14**(3), 75, 77.

Swift, B. (1998, p. 77). The U. S. Acid Rain Program is a notable model for broader multilateral credit-trading scheme for another reason. The use of high-quality monitoring, the implementation of a public Allowance Tracking System, and the imposition of steep penalties have led to 100 percent compliance among U.S. utilities. Similar tough enforcement regimes will be needed to avoid rampant cheating on any multilateral global-warming agreement.

Swift, Jonathan. (1726). *Gulliver's Travels*. Travel III, Chap. II.

Taylor, Alex. (1999). "Oil Forever." *Fortune*, November, pp. 193–194.

Tenner, Edward. (1996). *Why Things Bite Back?* New York: Vantage.

Thomson, Sir William (Lord Kelvin). (1891). *Popular Lectures and Addresses*. London: Macmillan, vol. 1, p. 80.

Thurow, Lester. (1998). *Building Wealth*. New York: HarperCollins.

Tilman, D. (1999). "Global Environmental Impacts of Agricultural Expansion: The Need for Sustainable and Efficient Practices," *Proceedings of the National Academy of Science* **96**, 5995–6000.

Todd, R. (1994). "Zero-Loss Environmental Accounting Systems," in *The Greening of Industrial Ecosystems* (B. Allenby and D. Richards, eds.). Washington, DC: National Academy Press, pp. 191–200.

Tushman, M. L., and P. C. Anderson. (1997). *Managing Strategic Innovation and Change.* Oxford, UK: Oxford University Press.

Underwriters Laboratories. (1997). *Standard for Power Conditioning Units for Use in Residential Photovoltaic Power Systems.* UL Subject 1741.

U.S. Bureau of the Census. (1998). *Statistical Abstract of the United States 1993.* 200th ed. Washington, DC: U.S. Government Printing Office.

U.S. Department of Energy. (1990, Aug.). *Assessment of Costs and Benefits of Flexible and Alternative Fuel Use in the US Transportation Sector.* DOE/PE-0095P. Washington, DC: U.S. DOE Policy, Planning and Analysis.

U.S. Environmental Protection Agency (1993). *Life Cycle Assessment: Inventory Guidelines and Principles.* EPA/600/R-92/245. Washington, DC: EPA.

U.S. Environmental Protection Agency. (1996). *1994 Toxics Release Inventory Public Data Release.* Washington, DC: Office of Pollution Prevention and Toxics, EPA.

U.S. News and World Report. (1997, Dec. 22). "Hot Air Treaty."

U.S. Senate, 95th Congress, 1st Session. Public Utility Regulatory Policy Act of 1977, Washington, D.C., Government Printing Office, 1977.

Vitousek, P. M., H. A. Mooney, J. Lubchenco, and J. M. Melillo. (1997). "Human Domination of Earth's Ecosystems." *Science* **277,** 494–499.

Walker, B. H., *et al.* (1999). Pp. 329–375 in *The Terrestrial Biosphere and Global Change* (B. Walker *et al.,* eds.). Cambridge: Cambridge University Press.

Wall Street Journal (The). (1997, Oct. 12). "Exxon Urges Developing Nations to Shun Environmental Curbs Hindering Growth."

Wall Street Journal (The). (1998, Apr. 16). "Global-Warming Debate Gets No Consensus in Industry."

Walubengo, D. (1995). "Commercialization of Improved Stoves: The Case of the Kenya Ceramic Jiko (KCJ)," in *Stove Images: A Documentation of Improved and Traditional Stoves in Africa, Asia, and Latin America* (B. Westhoff and D. Germann, eds.). Brussels, Belgium: Commission of the European Communities.

Warrick, R. A., *et al.* (1996). "Changes in Sea Level," pp. 359–405 in *Climate Change 1995: The Science of Climate Change.* Oxford: Cambridge University Press.

Watson, R. T., *et al.* (1998). *Protecting Our Planet — Securing Our Future.* U.N. Environment Program. Washington, DC: National Aeronautics and Space Administration, and World Bank.

Weber, M. (1996). "The Fish Harvesters: Farm-Raising Salmon and Shrimp Makes Millionaires, and Also Creates Dead Seas." *E Magazine,* November/December.

Whittaker, R. H. (1975). *Communities and Ecosystems,* 2nd ed. New York: Macmillan.

Wiens, J. A., J. F. Addicott, T. J. Case, and J. Diamond. (1986). "Overview: The Importance of Spatial and Temporal Scale in Ecological Investigations," in *Community Ecology* (J. Diamond and T. J. Case, eds.). New York: Harper & Row.

Williams, R. H. (1996, Jan.). *Fuel Decarbonization for Fuel Cell Applications and Sequestering of the Separated CO_2.* Report No. 296. Princeton, NJ: Princeton University Center for Energy and Environmental Studies.

World Bank (1992). *World Development Report 1992.* New York: Oxford University Press.

World Bank. (1993). *World Development Report: Investing in Health.* New York: Oxford University Press.

World Health Organization. (1992). *Annual Statistics.* Geneva: WHO.

World Health Organization. (1996). *Micronutrient Malnutrition: Half the World's Population Affected.* Geneva: WHO, No. 78, pp. 1–4.

World Health Organization (1999). *World Health Report.* Geneva: WHO.

World Resources Institute. (1992). *World Resources 1992.* New York: Oxford University Press.

World Resources Institute. (1993). *World Resources 1993–94.* New York: Oxford University Press.

World Resources Institute. (1994). *World Resources 1994–95.* New York: Oxford University Press.

World Resources Institute. (1999). *World Resources 1998–99: A Guide to the Global Environment.* New York: Oxford University Press.

Youngquist, W. (1997). *Geodestinies: The Inevitable Control of Earth Resources Over Nations and Individuals.* Portland, OR: National Book Company.

Yunus, Muhammad. (1998). "Alleviating Poverty Through Technology." *Science,* October 16, pp. 409–410.

INDEX